Springer
Proceedings in Physics   88

Springer-Verlag Berlin Heidelberg GmbH

Physics and Astronomy

ONLINE LIBRARY

http://www.springer.de/phys/

Springer Proceedings in Physics

L. Tacconi   D. Lutz (Eds.)

Starburst Galaxies: Near and Far

Proceedings of a Workshop
Held at Ringberg Castle, Germany,
10–15 September 2000

With 141 Figures and 12 Tables

 Springer

Dr. Linda Tacconi
Dr. Dieter Lutz

Max-Planck-Institut für extraterrestrische Physik
Postfach 1312
85741 Garching
Germany

Cover picture: M 82 (NGC 3034) FOCAS (B, V, Hα)
The image was obtained by the Subaru Telescope, National Astronomical Observatory of Japan
© 2000 National Astronomical Observatory of Japan, all rights reserved

Library of Congress Cataloging-in-Publication Data

Starburst galaxies, near and far: proceedings of a workshop, held at Ringberg Castle, Germany, 10–15 September 2000 /
Dieter Lutz, Linda Tacconi, eds. p. cm. –
(Springer proceedings in physics, ISSN 0930-8989; v. 88) Includes bibliographical references.
ISBN 978-3-642-62562-6   ISBN 978-3-642-56538-0 (eBook)
DOI 10.1007/978-3-642-56538-0
1. Starbursts–Congresses. 2. Stars–Formation–Congresses. I. Tacconi, Linda, 1959– II. Lutz, Dieter, 1960– III. Series.
QB806.5 .S68 2001   523.8'8–dc21   2001020857

ISSN 0930-8989

ISBN 978-3-642-62562-6

http://www.springer.de

© Springer-Verlag Berlin Heidelberg 2001
Originally published by Springer-Verlag Berlin Heidelberg New York 2001

Typesetting: Camera ready copy from the authors/editors
Cover concept: eStudio Calamar Steinen
Cover production: *design & production* GmbH, Heidelberg

SPIN: 10790403   54/3141/di - 5 4 3 2 1 0

Preface

Studies of star formation at high redshift have produced an astonishing number of results over recent years that were only possible with the latest generation of large ground-based and space telescopes. In trying to decide how to highlight all of these in a conference, the organizers decided that it would be most informative to present the wealth of high redshift results in the context of the much firmer foundation of star formation in the local universe. The Scientific Organizing Committee decided to concentrate on the properties of the most extreme star formation environments, those of starburst galaxies, and bring together scientists studying the starburst phenomenon at all redshifts and size scales.

With these goals in mind, we held the workshop "Starburst Galaxies: Near and Far" during 10–15 September 2000. The venue for the workshop was Ringberg Castle, located near Tegernsee in the alpine foothills of Bavaria, Germany. Nearly 70 astronomers from universities, observatories, and research institutes from all over the world were in attendance. The Max-Planck-Institut für extraterrestrische Physik (MPE), Garching, near Munich, Germany sponsored the conference.

The Scientific Organizing Committee wanted to take advantage of the intimate atmosphere of the Schloß Ringberg as much as possible, to foster spirited discussion amongst the diverse group of participants. We allocated large amounts of time for discussion periods, poster viewing and coffee breaks, and tried to keep the program as interactive as possible. Each scientific session consisted of several talks concluded by a discussion period that was led by a designated discussion leader. The discussion leader introduced the main question or topic to be discussed and then made sure that the discussion stayed lively but contained. We then decided to summarize three of the main topics of the meeting, namely the universality of the initial mass function (IMF), the formation of ellipticals and bulges, and ways of exploring the cosmic star formation history, in the form of three 'mini-debates'. In this forum two or three review speakers introduced each highlight topic by giving to-the-point presentations of the complementary and/or contrasting views. These introductions were followed by extended open discussion periods led by a debate 'moderator', and involved the entire audience. Finally, the debate speakers had time for a short rebuttal and summary.

There was also plenty of time for socializing with friends and colleagues throughout the workshop. The conference opened with a buffet supper at the castle, where participants and their guests met to catch up on old times and already begin discussion. Arguably the social (or otherwise) highlight of the week was the excursion to the picturesque village of Pertisau on the Achensee in Austria, and the tempestuous hike into the surrounding mountains.

These proceedings hopefully provide the reader with a synopsis of the scientific discussions that took place during the conference week. They start with the introductory talk by George Rieke, who poses most of the important questions that were addressed during the week. This is followed by papers presenting results and reviews on the determination of the initial mass function in the local universe. There are then a large number of papers dealing with how star formation occurs in different environments, and then those describing the global properties of starburst galaxies, both from an observational and a theoretical perspective. The book closes with papers that place what we know about star formation at high redshift into the context of cosmic star formation history.

The co-chairs give their warmest thanks to both the Scientific and Local Organizing Committees, without whose help this conference would not have been possible. We are particularly indebted to Ric Davies and Lowell Tacconi-Garman for help with the proceedings. We wish to thank all of the participants for an extremely productive and enjoyable conference. The workshop was made possible through the financial support of the Max-Planck-Institut für extraterrestrische Physik. We are very grateful to B. Scheiner and A. Preda for their help with the finances. Special thanks go to A. Hörmann, B. Remberger, and the rest of the staff at Schloß Ringberg for once again happily accommodating our special requests, flawlessly handling every detail of our stay at Ringberg, and for making us feel like kings and queens of the castle.

Garching,   *Linda Tacconi*
May 2001   *Dieter Lutz*

Contents

Part IV   Star Formation and the ISM

Part VI   Cosmic Star Formation History

List of Participants

Kurt Adelberger
Harvard-Smithsonian
Center for Astrophysics
60 Garden St.
Cambridge, MA 02138, USA
kadelber@cfa.harvard.edu

Itziar Aretxaga
INAOE
Aptdo. Postal 51 y 216
72000 Puebla, Pue., Mexico
itziar@inaoep.mx

Andrew Baker
MPE
Postfach 1312
85741 Garching, Germany
ajb@mpe.mpg.de

Marco Barden
MPE
Postfach 1312
85741 Garching, Germany
marco@mpe.mpg.de

Frank Bertoldi
MPI für Radioastronomie
Auf dem Hügel 69
53115 Bonn, Germany
Bertoldi@MPIfR-Bonn.MPG.de

Andrew W. Blain
Institute of Astronomy
Madingley Road
Cambridge CB3 0HA, UK
awb@ast.cam.ac.uk

Robert Blum
Cerro Tololo
Interamerican Observatory
Casilla 603
La Serena, Chile
rblum@noao.edu

Gustavo Bruzual
Centro de Investigaciones
de Astronomia
Apartado Postal 261
Mérida 5101-A, Venezuela
bruzual@cida.ve

Andreas Burkert
Max-Planck-Institut für Astronomie
Königstuhl 17
69117 Heidelberg, Germany
burkert@mpia-hd.mpg.de

Chris L. Carilli
NRAO
P.O. Box 0
Socorro, NM 87801, USA
ccarilli@nrao.edu

Catherine Cesarsky
ESO
Karl-Schwarzschild-Str. 2
85748 Garching, Germany
ccesark@eso.org

Stephane Charlot
MPA
Karl-Schwarzschild-Str. 1
85748 Garching, Germany
charlot@mpa-garching.mpg.de

Luis Colina
Instituto de Fisica de Cantabria
(CSIC-UC)
Facultad de Ciencias
Avda de Los Castros S/N
39005 Santander, Spain
colina@ifca.unican.es

Ric Davies
MPE
Postfach 1312
85741 Garching, Germany
davies@mpe.mpg.de

Michael A. Dopita
Research School
of Astronomy and Astrophysics
The Australian National University
Cotter Road
Weston Creek ACT 2611, Australia
Michael.Dopita@anu.edu.au

Frank Eisenhauer
MPE
Postfach 1312
85741 Garching, Germany
eisenhau@mpe.mpg.de

Donald F. Figer
Space Telescope Science Institute
3700 San Martin Drive
Baltimore, MD 21218, USA
figer@stsci.edu

Natascha M. Förster Schreiber
CEA/DSM/DAPNIA/SAp
C.E. Saclay
Orme-des-Merisiers, Bât. 709
91191 Gif-sur-Yvette CEDEX, France
forster@discovery.saclay.cea.fr

Alberto Franceschini
Universitá di Padova
Vicolo Osservatorio 5
35122 Padova, Italy
franceschini@pd.astro.it

Uta Fritze-von Alvensleben
Universitätssternwarte Göttingen
Geismarlandstr. 11
37083 Göttingen, Germany
ufritze@uni-sw.gwdg.de

Giuseppe Gavazzi
Universitá di Milano - Bicocca
Dipartimento di Fisica
Via Celoria 16
20133 Milano, Italy
gavazzi@uni.mi.astro.it

Reinhard Genzel
MPE
Postfach 1312
85741 Garching, Germany
genzel@mpe.mpg.de

Andrea M. Gilbert
Astronomy Department
601 Campbell Hall
University of California
Berkeley, CA 94704, USA
agilbert@astro.berkeley.edu

Gerry Gilmore
Institute of Astronomy
Madingley Road
Cambridge CB3 0HA, UK
gil@ast.cam.ac.uk

Rosa M. González Delgado
Instituto de Astrofísica de Andalucía
Camino Bajo de Huetor, Apdo. 3004
18080 Granada, Spain
rosa@iaa.es

James R. Graham
Department of Astronomy
University of California
Berkeley, CA 94720, USA
jgraham@astro.berkeley.edu

Andy Harris
University of Maryland
Department of Astronomy
College Park, MD 20742, USA
harris@astro.umd.edu

Günther Hasinger
Astrophysikalisches Institut Potsdam
An der Sternwarte 16
14482 Potsdam, Germany
ghasinger@aip.de

David Hughes
INAOE
Apartado Postal 51 y 216
72000 Puebla, Pue., Mexico
dhughes@inaoep.mx

Shardha Jogee
Astronomy Department, MS 105-24
Caltech
Pasadena, CA 91125, USA
sj@astro.caltech.edu

Guinevere Kauffmann
MPA
Karl-Schwarzschild-Str. 1
85748 Garching, Germany
gamk@mpa-garching.mpg.de

Henry A. Kobulnicky
University of Wisconsin
Department of Astronomy
475 N. Charter St.
Madison, WI 53706, USA
chip@astro.wisc.edu

Ariane Lançon
Observatoire de Strasbourg
11, rue de l'Université
67000 Strasbourg, France
lancon@astro.u-strasbg.fr

Matthew D. Lehnert
MPE
Postfach 1312
85741 Garching, Germany
mlehnert@mpe.mpg.de

Dieter Lutz
MPE
Postfach 1312
85741 Garching, Germany
lutz@mpe.mpg.de

Jorge Melnick
European Southern Observatory
Alonso de Cordova 3107
Casilla 19001
Santiago 19, Chile
jmelnick@eso.org

Sabine Mengel
MPE
Postfach 1312
85741 Garching, Germany
mengel@mpe.mpg.de

Gerhardt R. Meurer
Dept. of Physics and Astronomy
The Johns Hopkins University
3400 North Charles Street
Baltimore, MD 21218, USA
meurer@pha.jhu.edu

Chris Mihos
Dept. of Astronomy
Case Western Reserve University
10900 Euclid Av.
Cleveland, OH 44106, USA
hos@burro.astr.cwru.edu

Alan Moorwood
ESO
Karl-Schwarzschild-Str. 2
85748 Garching, Germany
amoor@eso.org

Edward C. Moran
University of California, Berkeley
601 Campbell Hall
Berkeley, CA 94729-3411, USA
edhead@jester.berkeley.edu

Mark Morris
Division of Astronomy
University of California, Los Angeles
Los Angeles, CA 90095-1562, USA
morris@astro.ucla.edu

Bianca M. Poggianti
Osservatorio Astronomico di Padova
Vicolo dell'Osservatorio 5
35122 Padova, Italy
biancap@pd.astro.it

Alvio Renzini
ESO
Karl-Schwarzschild-Str. 2
85748 Garching, Germany
arenzini@eso.org

George H. Rieke
Steward Observatory
University of Arizona
Tucson, AZ 85750, USA
grieke@as.arizona.edu

Dimitra Rigopoulou
MPE
Postfach 1312
85741 Garching, Germany
dar@mpe.mpg.de

David B. Sanders
Institute for Astronomy
University of Hawaii
2680 Woodlawn Drive
Honolulu, HI 96822, USA
sanders@ifa.hawaii.edu

Daniel Schaerer
Laboratoire d'Astrophysique
Observatoire Midi-Pyrénées
14, av. E. Belin
31400 Toulouse, France
schaerer@ast.obs-mip.fr

Nick Scoville
Astronomy 105-24
Caltech
Pasadena, CA 91125, USA
nzs@astro.caltech.edu

Philip M. Solomon
Physics and Astronomy
SUNY, Stony Brook
Stony Brook, NY 11794, USA
psolomon@astro.sunysb.edu

Henrik Spoon
ESO
Karl-Schwarzschild-Str. 2
85748 Garching, Germany
hspoon@eso.org

Amiel Sternberg
Tel Aviv University
Ramat Aviv 69978, Israel
amiel@wise.tau.ac.il

Linda Tacconi
MPE
Postfach 1312
85741 Garching, Germany
linda@mpe.mpg.de

Jonathan C. Tan
Dept. of Astronomy
UC Berkeley
Berkeley, CA 94720, USA
jt@astron.berkeley.edu

Rodger I. Thompson
Steward Observatory
University of Arizona
Tucson, AZ 85721, USA
rthompson@as.arizona.edu

Christy Tremonti
Dept. of Physics and Astronomy
The Johns Hopkins University
3400 North Charles Street
Baltimore, MD 21218, USA
cat@pha.jhu.edu

William D. Vacca
Institute for Astronomy
2680 Woodlawn Drive
Honolulu, HI 96822, USA
vacca@minerva.ifa.hawaii.edu

Wil van Breugel
University of California
LLNL, L-413
Livermore, CA 94550, USA
wil@igpp.ucllnl.org

Paul P. van der Werf
Leiden Observatory
P.O. Box 9513
2300 RA Leiden, The Netherlands
pvdwerf@strw.leidenuniv.nl

Sylvain Veilleux
Dept. of Astronomy
University of Maryland
College Park, MD 20742, USA
veilleux@astro.umd.edu

Bradley C. Whitmore
Space Telescope Science Institute
3700 San Martin Drive
Baltimore, MD 21218, USA
whitmore@stsci.edu

Lin Yan
The Carnegie Observatories
813 Santa Barbara Street
Pasadena, CA 91101, USA
lyan@ipac.caltech.edu

1 Alvio Renzini
2 Catherine Cesarsky
3 Rosa González Delgado
4 Linda Tacconi
5 Lin Yan
6 Itziar Aretxaga
7 Paul van der Werf
8 James Graham
9 Phil Solomon
10 Alberto Franceschini
11 Brad Whitmore
12 Sabine Mengel
13 Uta Fritze-von Alvensleben
14 Frank Eisenhauer
15 Dieter Lutz

16 Rodger Thompson
17 Nick Scoville
18 Andrew Harris
19 Frank Bertoldi
20 Shardha Jogee
21 Dave Sanders
22 Mark Morris
23 Bill Vacca
24 Lee Armus
25 Don Figer
26 Andrew Blain
27 Jonathan Tan
28 Stephane Charlot
29 Marco Barden
30 Luis Colina

31 Matt Lehnert
32 Helmut Dannerbauer
33 Gerhardt Meurer
34 Amiel Sternberg
35 Chris Mihos
36 Ed Moran
37 Jorge Melnick
38 Ariane Lançon
39 MikeDopita
40 Natascha Förster-Schreiber
41 Daniel Schaerer
42 Ric Davies
43 David Hughes
44 Andrew Baker
45 Bianca Poggianti

46 Kurt Adelberger
47 Andrea Gilbert
48 Sylvain Veilleux
49 Chris Carilli
50 Wilvan Breugel
52 Christy Tremonti
53 Bob Blum
54 Chip Kobulnicky
55 George Rieke
56 Alan Moorwood
57 Henrik Spoon
58 Reinhard Genzel

Scientific Organizing Committee
Stephane Charlot
Alberto Franceschini
Reinhard Genzel
James Graham
Tim Heckman
Rob Ivison
Rob Kennicutt
Matthew Lehnert
Dieter Lutz (co-chair)
Max Pettini
Daniel Schaerer
Amiel Sternberg
Linda Tacconi (co-chair)

Local Organizing Committee
Andrew Baker
Marco Barden
Helmut Dannerbauer
Richard I. Davies
Reinhard Genzel
Susanne Harai-Ströbl
Matthew Lehnert
Dieter Lutz (co-chair)
Sabine Mengel
Henrik Spoon
Linda Tacconi (co-chair)
Lowell Tacconi-Garman

Part I

Introduction

Starbursts Near and Far: An Overview

George H. Rieke

Steward Observatory, University of Arizona, Tucson, AZ 85750, USA

Abstract. A survey of the conference participants produced a large list of "key" questions regarding starbursts, with little agreement on which ones are most important. In fact, all of them are of great interest. However, the list shows how viewpoints are shaped by familiar techniques and approaches rather than a broad view of the field. The goal of the conference should therefore be to reconcile views that differ because of the narrowness of our approaches rather than from true scientific uncertainties and limitations. Toward this end, I discuss a number of issues: 1.) the differing optical/UV and infrared views of starbursts; 2.) differing estimates of the effective temperatures of the hot stars; 3.) the mystery of why massive starbursts do not show high effective temperatures; 4.) determination of the initial mass function in starbursts; 5.) the nature of LINERs; and 6.) whether the huge masses deduced for the ISM in starburst regions are correct.

1 Introduction

One comes to a conference like this one to discuss the major questions in starburst studies. But what are they? Of course, we all know, but to be sure I conducted an informal and unscientific email survey of the conference participants just prior to coming. The results are below, somewhat edited to consolidate questions and shorten the list.

- What feedback mechanisms regulate the evolution of starbursts, and do they impose a maximum luminosity?
- How does the radio-FIR correlation work?
- What is the connection between optical/UV and IR/submm-detected starbursts?
- What supports molecular clouds against collapse? Conversely, what starts them toward collapse?
- How does the starburst environment affect star formation, such as in the initial mass function?
- What connection, if any, is there between star formation and active galactic nuclei?
- Why are the highest mass O stars not detected in IR-embedded starbursts?
- What is the long term evolution of starbursts? Periodic? Once only? Is there continuous gas flow or intermittent or once only to support them?
- How do starbursts at z = 1–2 compare with the local population?
- Are starburst outflows responsible for the chemical enrichment and collisional heating of the inter-cluster medium and intergalactic medium?

• What fraction of present-day stars form in starbursts as opposed to "steady state"? When? How do we define a starburst?

• What is the star formation history of the Universe?

• What is the relative importance of star formation and black hole/AGN energy generation in the early Universe?

• What is the cosmological evolution of luminous infrared galaxies and ultra-luminous infrared galaxies?

• How and when do galaxy spheroids form? Do we see them in Ly-break and submm detections?

• Do high-z starbursts (rather than quasi-stellar objects) photoionize the Ly-alpha forest?

• Chicken or egg: what forms first, black holes or stars?

The survey brings to mind the parable of the blind men and the elephant. Although we are each exploring important aspects of elephantine anatomy, our approaches can be sufficiently different that they hinder our progress on the physiology of the beast. Our challenge is to reconcile views that differ because of the narrowness of our approaches rather than from true scientific uncertainties and limitations. Toward this goal, I will touch on a personal selection of issues.

2   Unifying our Optical and Infrared Views

Our studies of star formation tend to fall into two extremes, with emphasis on optical/UV techniques for low density regions and infrared ones for high density. Table 1 (after [16]) emphasizes the differences in typical properties of these two types of region.

Table 1. Optical/UV vs. Infrared View of Starbursts.

Wavelength range	Optical/UV	Infrared
Region within galaxy	Spiral disks	Circumnuclear starbursts
Radius	1–30 kpc	0.1–2 kpc
Star formation rate	< 20 $M_\odot$ yr^{-1}	Up to 1000 $M_\odot$ yr^{-1}
Luminosity	Up to 10^{11} $L_\odot$	Up to 10^{13} $L_\odot$
Gas mass	Up to 10^{11} $M_\odot$	Up to 10^{11} $M_\odot$
Gas density	Up to 100 $M_\odot$ pc^{-2}	Up to 10^5 $M_\odot$ pc^{-2}
Optical depth (0.5μm) 0–2		1–1000
Star formation density 0–0.1 $M_\odot$ kpc^{-2} yr^{-1}		1–1000 $M_\odot$ kpc^{-2} yr^{-1}

The dense regions are inaccessible to the UV/optical because of extinction and the low density regions are inaccessible to the IR because of technical limitations. New facilities in the IR/submm should eventually allow a more unified understanding of star formation. In the meantime, we have to work with two seriously incomplete views of star formation in galaxies, with the likelihood of systematic biases in each case.

An example where the two approaches come into play is the star forming rate at high redshift. Deep optical (rest UV) images such as the Hubble Deep Fields show a plethora of galaxies, many of them small and of modest luminosity individually. Submm surveys find a small number of galaxies, evidently large and highly luminous. The luminosity densities from the two samples, and hence the star formation densities in a fixed volume of space, appear to be comparable. However, it has proven frustratingly difficult to make identifications from one spectral region with objects in the other. Given the difficulties, it seems likely that the few solid associations may represent exceptional objects. The result is that the star formation rate at high redshift ($z > 2$) may be uncertain by a factor of three (e.g., [37]).

3 Equivalent Temperature of Ionizing Stars

The hardness of the ionizing field is a critical constraint on the age of the hot stellar population in a starburst. Strong lines suitable for determining the effective temperature are relatively rare even in the rich optical spectra. The most commonly used measures combine a forbidden and a H recombination line and as a result are subject to biases such as the effects of electron temperature and metallicity. Over the temperature range $\sim$35,000 to $\sim$40,000, the He^+/H ratio is an accurate measure, although the accessible HeI lines are faint and can be difficult to measure accurately [17].

Alternately, effective temperatures can be measured from mid infrared fine structure line ratios [28,40]. The mid infrared lines appear to give systematically lower effective temperatures than the usual optical methods [40,42]. The HeI line at 1.7μm is very weak, but where it can be measured it should provide the most reliable possible determination through a virtually extinction-independent measure of He^+/H [43]. It appears to agree more closely with the mid-infrared approach than the optical one.

Although this comparison suggests a discrepancy of "only" about 4000 K in stellar effective temperature, that difference corresponds to a significant difference in main sequence mass and hence starburst age.

4 Where are the Hot Stars in Starbursts?

Assuming that the IMF includes very massive stars, one would expect a very young starburst to have them in significant numbers. However, the mid infrared fine structure lines show very few luminous, high metallicity galaxies with very hot stars, although some low metallicity galaxies show such a signature [28,40]. The dust absorption of the ultraviolet spectrum is expected to harden it [1]. Thus, if anything, the effective temperature of the stars may be over-estimated in these regions. The explanation for the behavior of the fine structure lines is unclear, particularly if the indications that they give a reliable temperature indication in the preceding section are confirmed.

5   Do Starbursts have Extreme Initial Mass Functions?

The high luminosity outputs of nuclear starbursts press their mass budgets to allow for the major constituents of the nuclear region: 1.) the old stars in the previously existing galaxy nucleus; 2.) the newly formed starburst population of stars; and 3.) the massive interstellar medium. The stringent mass constraints allow a probe of the global form of the initial mass function that is not possible in other regions. An IMF that forms too many low mass stars may not leave enough mass for the other components of the galaxy nucleus.

It was noticed early in the study of M82 that the proposed forms for the local IMF included too large a portion of low mass stars for starburst models required to fit within the dyamical mass [23,26]. Rieke et al. [26] considered a broad variety of possible forms for the IMF and selected "IMF8" as an optimum fit, since it could live within the mass constraints and fit all the observed parameters of the galaxy. Because the starburst in M82 is old enough that the most massive stars have left the main sequence, it would appear that there is little leverage to determine the IMF above about 30 $M_\odot$. However, stars above this mass produce copious amounts of oxygen, and IMF8 has a gentle cutoff toward the highest masses to avoid producing too much of this element to be consistent with the relatively normal abundances in the M82 interstellar medium (see also [25,45]). Such a cutoff has recently been suggested for the IMF on other grounds [11].

The type of IMF proposed for M82 has been termed "top heavy" and it has come under considerable criticism (e.g., [30]). The high mass slopes of the IMF in a variety of massive star forming regions have been found to follow roughly the Salpeter value ([11] and references therein). It is well determined in the solar neighborhood that the IMF is roughly flat in logarithmic mass units up to about 1 $M_\odot$, after which it rolls over to a Salpeter-like or steeper falloff. Thus, a new picture has emerged of a common form for the IMF that is flat to about 1 $M_\odot$ and has a Salpeter slope above this value.

This form of IMF is actually virtually identical to IMF8 so far as its performance in starbursts is concerned, except that it does not have the high mass rolloff to limit oxygen production. In particular, if the IMFs are all normalized to the same total mass, the portion of stars with 10 $M_\odot$ < M < 30 $M_\odot$ that dominates starburst properties is virtually the same for the two proposed starburst IMFs. However, the local form [31] falls far below them in this mass range. Thus, the controversy over a "top heavy" IMF has changed into a question of whether the local IMF is "bottom heavy" and if so, why.

Although the IMF is difficult to determine in any of these environments, continued effort to understand it is worthwhile. There are a number of reasons that it might differ between starbursts and less intensely star forming regions. The turnover in the local IMF near 1 $M_\odot$ is thought to be related to the Jeans mass in the collapsing protostars. Perhaps the Jeans Mass is higher in starbursts than in the solar neighborhood. This behavior would be a natural result of the higher temperature of the ISM in these regions, plus the steep dependence of the Jeans Mass on T (as T^2) [10]. In this case, the starburst would have a similar slope at high masses but it would turn over at a higher mass than IMFs in other star forming regions. Alternatively, there are arguments that the star formation in starbursts is triggered by cloud-cloud collisions, and some qualitative theoretical indications that these conditions favor formation of massive stars [33,46]. In this case, the slope of the IMF at high mass would be flatter in starbursts than in other, lower density star forming regions.

Another cause of differences could be the time duration of star formation. The observed age distributions in local star forming regions are consistent with nearly a delta function burst of star formation. In a starburst, an extended region is exposed to the effects of ongoing star formation for millions of years. Thus, even though the IMF in high density local star forming regions seems to be "normal", it is possible that the starburst IMF is modified because of the greater duration of the star formation.

Or perhaps it is the local IMF that is "weird". There are no fits to it that are as top heavy as the Salpeter power law, as Salpeter was aware: "It is not clear whether the steeper drop for masses larger than 10 $M_\odot$ is a real effect, since in this region masses and bolometric corrections are not known very accurately and the number of such stars near the galactic plane is small." [29]. It is therefore ironic that it has turned out to be such a good approximation to the IMF in more distant and denser star forming regions.

6   The LINER - Starburst Connection

Since Heckman identified Low Ionization Nuclear Emission Region (LINER) galaxies, their nature has been actively investigated. It has become clear that they represent a number of types of object. About 1/4 of them have broad Hα [15], unresolved ultraviolet sources [20], and X-ray properties suggestive of the presence of weak AGNs [18]. Others show spectral signatures of hot stars [5,21]. Still others can be fitted by the expected properties of an aging starburst, with a combination of photoionization by hot stars and shock excitation by supernovae to account for their emission line spectra [2,12].

The varieties in classification even extend to individual galaxies. As an example, NGC 4569 has recently been described as:

• An aging starburst, predominantly 8–11 Myr old, with substantial shock excitation, from fitting optical and infrared spectra [2]

• A starbust of moderate age, 5–6 Myr old, from fitting the emission line spectrum in the ultraviolet [13]

• A very young starburst, 3–5 Myr old (although the lack of detected Wolf-Rayet features may be a problem), based on fitting the optical/ultraviolet emission line spectrum [5]

• As containing a compact nuclear X-ray source, possibly indicating the presence of an AGN [7].

None of these papers presents a complete picture of the galaxy. The first one ignores the ultraviolet spectrum, the second and third ignore the infrared spectrum, the first three ignore the X-ray, and the fourth ignores the optical and infrared. The main lesson from this work is that the nature assigned to the activity depends critically on the spectral region under study. In addition, it seems likely that a multiple burst star formation model would fit all the constraints invoked in the first three works better than the models they present. Studies of other galaxies (e.g., Förster Schreiber, this conference) show that such multiple bursts may be common in starbursts.

Associations with starbursts are being found increasingly for LINERs, as indicated in the references above. The fact that $\sim 1/3$ of large spirals have LINER spectral characteristics [15], and that we have no other candidates for the post starburst phase, are likely to indicate that some portion of them are in fact a later step in the evolution of starbursts. LINER-like spectra can occur as the hot stars fade and supernova shocks dominate the emission line spectra [2,12], or due to formation of large numbers of planetary nebulae [39]. A full model of this process remains a challenge, however.

7   A Flawed Paradigm?

With the use of mm-wave interferometry. strong, very compact CO sources have been found in the cores of nearly all luminous infrared galaxies. Gas masses from the standard CO/H_2 ratio are huge and the corresponding extinctions of $A_V \sim 100$–1000 are enough to hide almost anything from optical and infrared investigation (see summary by [6]). However, there are an increasing number of reasons to question the large gas masses and the accompanying huge extinctions:

• The relatively high densities and temperatures of the molecular clouds in these regions may lead to bright CO and an overestimate of the H_2 mass [19]

• The dynamical masses from near infrared spectroscopy of the $2.3\mu m$ CO bandhead and Brγ are significantly less than the molecular hydrogen mass from the standard ratio, suggesting masses are overestimated by factors of 4–10 [34,3]

• Comparison of the dynamical mass from mm-wave CO line profiles and with the gas mass from far infrared measures and $C^{18}O$ suggests the standard values are high by a factor of ~ 3–6 [22,35,9]

• The CO/H_2 ratio appears to be a factor of 4–10 lower in our Galactic Center than given by the standard ratio [8]

• Dynamical stability of circumnuclear gas rings can require an order of magnitude less gas, from the Toomre Q condition [3]

• The extinction levels appear to be an order of magnitude less than derived using the standard ratio (Arp 220 [4]; NGC 1614 [3]).

If the molecular masses have in fact been overestimated, there are a number of interesting consequences. It would require that the conversion of gas to stars in starbursts be far more efficient than is typical in nearby star forming regions in the disk of the Milky Way, with typical efficiencies $\geq 50\%$. It also makes it harder to hide an AGN with a large column of extinction. For example, the lack of hard X-rays in these galaxies [24,27] is challenging to explain if they harbor powerful AGNs but the obscuration is five times less than previously assumed.

8 Conclusions

This personal selection of issues is not complete, but it already suggests that we have plenty of work ahead of us to refine our understanding of starbursts, near and far.

References

1. P.A. Aanestad: ApJ **338**, 162 (1989)
2. A. Alonso-Herrero, M.J. Rieke, G.H. Rieke, J.C. Shields: ApJ **530**, 688 (2000)
3. A. Alonso-Herrero, C.W. Engelbracht, M.J. Rieke, G.H. Rieke, A.C. Quillen: ApJ, in press (2000)
4. K.R. Anatharamaiah, F. Viallefond, N.R. Mohan, W.M. Goss, J.H. Zhao: ApJ **537**, 613 (2000)
5. A.J. Barth, J.C. Shields: PASP **112**, 753 (2000)
6. P.M. Bryant, N.Z. Scoville: AJ **117**, 2632 (1999)
7. E.J.M. Colbert, R.F. Mushotzky: ApJ **519**, 89 (1999)
8. G. Dahmen, S. Huttemeister, T.L. Wilson, R. Mauersberger: A&A **331**, 959 (1999)
9. D. Downes, P.M. Solomon: ApJ **507**, 6157 (1998)
10. B. Elmegreen: ApJ **486**, 944 (1997)
11. B.G. Elmegreen: ApJ **539**, 342 (2000)
12. C.W. Engelbracht, M.J. Rieke, G.H. Rieke, D.M. Kelly, J.M. Achtermann: ApJ **505**, 639 (1998)
13. J.R. Cabel, F.C. Bruhweiler: BAAS **195**, #117.01 (1999)
14. T.M. Heckman: A&A **87**, 152 (1980)
15. L.C. Ho, A.V. Filippenko, W.L. Sargent, C.Y. Peng: ApJS **112**, 391 (197)
16. R.C. Kennicutt: ARAA **36**, 189 (1998)
17. R.C. Kennicutt, F. Bresolin, H. French, P. Martin: ApJ **537**, 589 (2000)
18. A. Koratkar, S.E. Deustua, T. Heckman, A.V. Filippenko, L.C. Ho, M. Rao: ApJ **440**, 132 (1995)
19. P. Maloney, J.H. Black: ApJ **325**, 389 (1988)
20. D. Maoz, A.V. Filippenko, L.C. Ho, H.-W. Rix, J.N. Bahcall, D.P. Schneider, F.D. Macchetto: ApJ **440**, 91 (1995)
21. D. Maoz, A. Koratkar, J.C. Shields, L.C. Ho, A.V. Filippenko, A. Sternberg: AJ **116**, 55 (1998)
22. R. Mauersberger, C. Henkel, R. Wielebinski, T. Wiklind, H.-P. Reuter: A&A **305**, 421 (1996)
23. G.H. Rieke, M.J. Lebofsky, R.I. Thompson, F.J. Low, A.T. Tokunaga: ApJ **238**, 24 (1980)
24. G.H. Rieke: ApJ **331**, L5 (1988)
25. G.H. Rieke: in *Massive Stars in Starbursts*, eds. C. Leitherer, N. Walborn, T. Heckman, C. Norman (Cambridge, Cambridge Univ. Press, 1989), p. 205
26. G.H. Rieke, K. Loken, M.J. Rieke, P. Tamblyn: ApJ **412**, 99 (1993)
27. G. Risaliti, R. Gilli, R. Maiolino, M. Salvati: A&A **357**, 13 (2000)
28. P.F. Roche, D.K. Aitken, C.H. Smith, M. Ward: MNRAS **248**, 606 (1991)

29. E.E. Salpeter: ApJ **121**, 161 (1955)
30. S. Satyapal, et al.: ApJ **483**, 148 (1997)
31. J.M. Scalo: Fund. Cosmic Physics **11**, 1 (1986)
32. M. Schmidt: ApJ **129**, 243 (1959)
33. N.Z. Scoville, D.B. Sanders, D.P. Clemens: ApJ **310**, L77 (1986)
34. L.M. Shier, M.J. Rieke, G.H. Rieke: ApJ **470**, 222 (1996)
35. P.M. Solomon, D. Downes, S.J.E. Radford, J.W. Barrett: ApJ **478**, 144 (1997)
36. H. Sugai, Y. Taniguchi: AJ **103**, 1470 (1992)
37. J.C. Tan, J. Silk, C. Balland: ApJ **522**, 579 (1999)
38. Y. Taniguchi, Y. Ohyama: ApJ **508**, L13 (1998)
39. Y. Taniguchi, Y. Shioya, T. Murayama: AJ **120**, 1265 (2000)
40. M.D. Thornley, et al.: ApJ **539**, 641 (2000)
41. W.D. Vacca, P.S. Conti: ApJ **401**, 543 (1992)
42. L. Vanzi, G.H. Rieke: ApJ **479**, 694 (1997)
43. L. Vanzi, G.H. Rieke, C.L. Martin, J.C. Shields: ApJ **466**, 150 (1996)
44. J.M. Vilchez, B.E.J. Pagel: MNRAS **231**, 257 (1988)
45. B. Wang, J. Silk: ApJ **406**, 580 (1993)
46. A.P. Whitworth, A.S. Bhattal, S.J. Chapman, M.J. Disney, J.A. Turner: MNRAS **268**, 291 (1994)

Part II

The Initial Mass Function

The Initial Mass Function in the Galactic Center

Donald F. Figer

Space Telescope Science Institute, Baltimore, MD 21218;
Johns Hopkins University, Baltimore, MD 21218, USA

Abstract. The Galactic Center contains the most massive young clusters in the Galaxy and serves as the closest example of a massive starburst region. Our recent results suggest that the Galactic Center environment produces massive clusters with relatively flat initial mass functions, as might be expected on theoretical grounds. I will discuss these recent results, along with evidence for star formation in the immediate vicinity of the super massive black hole at the Galactic Center. The results of this work might be useful in extrapolating to other galactic centers with similar conditions, as well as other starburst regions.

1 Introduction

The Galactic Center region spans the central few hundred parsecs of the Galaxy and represents about 0.04% of the volume in the Galactic disk. Although small in size, the region contains 10% of the molecular material and young stars in the Galaxy, as evidenced by molecular line and radio maps [1]. The molecular clouds in the center tend to have high densities that are greater than $10^4\,\mathrm{cm}^{-3}$; more rarefied clouds are destroyed by the strong tidal field produced by the large stellar mass peaking in the central parsec. In addition, the infrared energy distributions produced by the clouds indicate very high temperatures, $T_{\mathrm{cloud}} = 70\,\mathrm{K}$; and molecular line observations indicate extraordinary turbulent velocities (FWHM) of about $25\,\mathrm{km\,s}^{-1}$, even on scales as small as 0.3 pc. In accordance with approximate pressure balance, the magnetic field strengths appear to also be high, on the order of a milliGauss. Reference [1] provides an excellent review of the GC environment.

Despite this tumultuous environment, the GC contains stars which produce $L = 10^9\,L_\odot$, and $L_{\mathrm{Ly-cont}} = 1$ to $3\times10^{52}\,\mathrm{s}^{-1}$, values dominated by older stars in the case of the former and younger stars in the case of the latter. Given this fertile breeding ground for young stars, we embarked on a program to identify all the sites of recent star formation in the GC and measure the initial mass function for particular sites, where possible [2–7]. This paper gives a summary of the results obtained so far, with a particular emphasis on the most recent result, a measurement of the initial mass function (IMF) in the Galactic Center.

2 Star Formation in the Galactic Center

The environment described in the introduction would appear to be somewhat hostile for star formation. Indeed, [8] notes that the Jeans mass in the GC is extraordinarily large ($\sim 10^5 \, \mathrm{M_\odot}$). Such high Jeans masses are at odds with the stellar observations, but the turbulent velocities are measured on size scales $\sim 0.3\,\mathrm{pc}$, a size that corresponds to $30\,\mathrm{M_\odot}$ (assuming $n \sim 10^4 \, cm^{-3}$). One way to avoid this inconsistency between the high Jeans mass and the existence of stars with more normal masses is to invoke highly compressive events, i.e. cloud-cloud collisions.

2.1   Major Sites of Star Formation

There are three young massive clusters in the GC: the Central cluster [9–15], the Arches cluster [16–18,6], and the Quintuplet cluster ([19–21,7]. They are nearly identical to each other in stellar content, except for age, the Arches being about half the age (2 Myr) of the other two.

The Central cluster refers to the collection of young ($\tau_{age} < 10\,\mathrm{Myr}$) stars in the central parsec. In the previous two decades, over 30 evolved massive stars having $M_{init} > 20\,M_\odot$ have been identified in this cluster, including 9 WR stars, 20 stars with Ofpe/WN9-like K-band spectra [22], several red supergiants [23], and many luminous mid-infrared sources [24]. The detailed spectroscopic analyses by [25,22] found that the brightest HeI emission-line stars in the GC have extremely strong stellar winds ($\dot{M} \sim 5$ to $80 \times 10^{-5}\,M_\odot\,yr^{-1}$) and relatively small outflow velocities ($v_\infty \sim 300$ to $1000\,km\,s^{-1}$). These values, taken together with the inferred stellar temperatures (17,000 K to 30,000 K) and luminosities (1 to $30 \times 10^5\,L_\odot$) suggest that the young cluster is about 4-6 Myr old and provides the majority of the ionizing flux ($10^{50.5}\,s^{-1}$) and heating ($10^{7.5}\,L_\odot$) which is evident in the central parsec [26].

The Arches contains about 160 O-stars ($M_{init} > 20\,M_\odot$), more than any other cluster in the Galaxy, and has a total mass of $10^4\,M_\odot$ [6]. References [17] and [16] first recognized the cluster for its dozen or so massive emission-line stars. Futher imaging by [18] revealed that the cluster contains a large number of O-stars and is responsible for heating and ionizing the surface of a nearby molecular cloud. A determination of the cluster age (2 Myr) and IMF was subsequently made by [6] using HST/NICMOS observations.

The Quintuplet cluster has about $2 \times 10^4\,M_\odot$ in stars [6], but its massive stars are more evolved than those in the Arches [7,27], consistent with 4 Myr age of the former. Indeed, the cluster contains more Wolf-Rayet stars than any cluster in the Galaxy, and over a dozen stars in earlier stages of evolution, i.e., LBV, Ofpe/WN9, and OB supergiants. Further, it contains two "Luminous Blue Variables" [28], including the "Pistol Star." The ionizing flux from the cluster is $10^{50.9}$ photons s^{-1}, or roughly what is required to ionize the nearby "Sickle" HII region (G0.18−0.04). The total luminosity from the massive cluster stars is $\approx 10^{7.5}\,L_\odot$, enough to account for the heating of the nearby molecular cloud, M0.20−0.033. The cluster is unique in the Galaxy in stellar content and age, except, perhaps, for the young cluster in the central parsec of the Galaxy.

2.2 Star Formation Near the Black Hole

Reference [29] presented $2.0-2.4$ μm spectroscopy of the central 0.1 square arc-seconds of the Galaxy, showing that the brightest stars in this region are hotter than typical red giants (also see [30] and [31]). Taken together with photometry and colors, they concluded that these objects are likely OB main sequence stars. It is unlikely that the stars are late-type giants stripped of their outer envelopes because such sources would be much fainter than those observed. In addition, [32], [33], and [34] note that these objects are not likely to be remnants of stellar mergers, unless the merger rate is elevated with respect to what is currently estimated. Given their location very near to the black hole, and their apparent relative youth, it is unlikely that they have drifted inward via dynamical friction. If we assume such a young age, and high stellar mass per star, then we are faced with the fact that large amounts of mass $(20\,M_\odot)$ have been compressed to very high densities $(> 4 \times 10^{11}\,\mathrm{cm}^{-3})$ very near to the black hole (several thousand AU).

3 The IMF in the Galactic Center

The IMF describes the relative number of stars produced in a star forming event as a function of initial stellar mass. It is often expressed as a single power law over mass ranges above $1\,M_\odot$, with a form $\mathrm{d}(\mathrm{Log}\,N)/\mathrm{d}(\mathrm{Log}\,M_{\mathrm{init}}) = \Gamma = \alpha + 1$; $\Gamma = -1.35$ for the Salpeter case [35]. The IMF for most young clusters can be described by a power law with Salpeter index, although significant variations are evident $(-0.7 > \Gamma > -2.1)$ [36]. Whether these variations are statistically significant, and whether environmental factors affect the IMF, remains to be proven.

Hubble Space Telescope (*HST*) Near-infrared Camera and Multi-object Spectrometer (NICMOS) observations of the Arches and Quintuplet clusters were obtained [6] and used to identify main sequence stars in the Galactic Center with initial masses well below $10\,M_\odot$, leading to the first determination of the initial mass function (IMF) for any population in the Galactic Center. The mass function derived for the Arches cluster is shown in Fig. 1. The central region $(r < 3'')$ is excluded from the analysis because our statistics are already less than 50% complete there for $M_{\mathrm{init}} < 35\,M_\odot$; however, note the dotted-line histogram which does include the inner region for the most massive stars. For the annulus with $3'' < r < 9''$, we find a slope which is significantly greater than -1.0, and so is one of the flattest mass functions ever observed for $M_{\mathrm{init}} > 10\,M_\odot$. If all of the data are forced to fit a single line, the slope is near -0.6, but we note that interesting structure may be present in the mass function in the form of near-zero-slope plateaus for $15\,M_\odot < M_{\mathrm{init}} < 50\,M_\odot$, and for $M_{\mathrm{init}} > 50$, i.e., the intrinsic mass function may be relatively flat at higher masses. In comparison, the average IMF slope for 30 clusters in the Milky Way and LMC is ≈ -1.4 for $\log(M_{\mathrm{init}}) > 0.7$, although a few clusters have $\Gamma \approx -0.7$ [36]. Some of these clusters also suggest a flattening of the IMF at higher masses, although the ac-

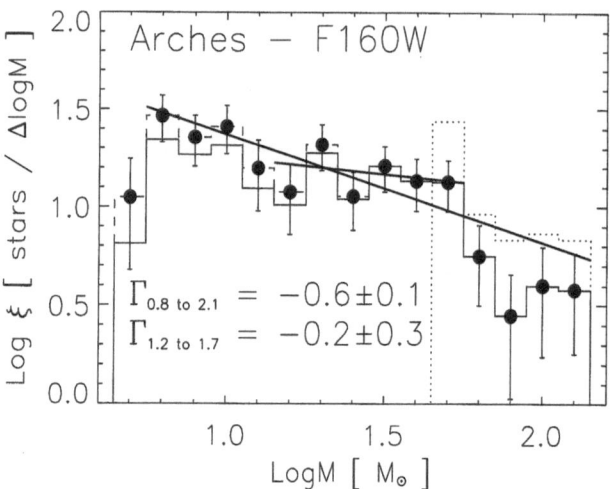

Fig. 1. Mass function for Arches cluster as measured in the F160W image, reproduced from [6]. Lines have been fit over two mass ranges. The slopes of both are relatively flat. The dotted histogram shows the number of massive stars over the whole cluster, including the inner region.

tual slopes in these comparison clusters are in general much more biased toward lower masses. Finally, we find that there are $\gtrsim 10$ stars with $M_{init} > 120\,M_\odot$.

There are five possible explanations for the flat slope and apparent plateaus: 1) inaccuracies in the M_{init} vs. magnitude relation, 2) inaccuracy in estimating cluster age, metallicity, and/or mass-loss rates, 3) effects of binaries, 4) effects of dynamical evolution, and 5) a real and interesting property of the IMF. If any of the first four possibilities is important enough to affect the slope of the observed mass function, then Figure 1 does not truly represent the IMF of the Arches cluster.

The first three possiblities are unlikely to have an effect and have been addressed by [6]. As for the fourth possibility, mass segregation takes place on a short time scale. In view of the theoretical results in [38] and [39], [6] divided the cluster into two annuli, an inner annulus ($2\overset{''}{.}5$ to $4\overset{''}{.}5$), and an outer annulus ($4\overset{''}{.}5$ to $7\overset{''}{.}5$), in order to assess a possible change in Γ as a function of radius (see Fig. 8 in [6]). Indeed, there is a significant change of slope from -0.2 to -0.8 over this annular range, consistent with expectations. Note that these figures show that the primary result of a flat IMF in the Arches cluster is not sensitive to the choice of annulus. The fifth possibility is most interesting for its implications concerning star formation in the GC. The unusually flat slope is consistent with the expectations that environmental conditions near the GC tend to favor the formation of high mass stars, relative to star formation taking place elsewhere in the Galaxy. However, before such a conclusion can be drawn and made quantitative, dynamical evolutionary effects discussed above need to be carefully modelled. As discussed in [8], the characteristics of the interstellar medium in the GC – particularly the large turbulent velocities, high cloud temperatures, strong magnetic fields, and large tidal forces – lead to a relatively large Jeans mass, compared to that elsewhere in the Galaxy.

4 Super Star Clusters in the GC

Reference [6] find a total cluster mass for the Arches cluster of $\approx 10^4\,M_\odot$ and about twice this value for the Quintuplet cluster. Both clusters are potential super star clusters, objects noted for their relative youth ($\tau_{age} < 20\,Myr$), compactness ($r_{half-light} < 1\,pc$), and large mass ($10^4\,M_\odot$ to $10^6\,M_\odot$) [40]. The total mass estimate for the Arches cluster is greater than that of NGC 3603, but it is still below the masses of Globular clusters. Using the new estimates, and Table 5 in [7], it seems that R136, the Quintuplet cluster and the Arches cluster are similar in mass, but the Arches cluster is about an order of magnitude more dense than R136, $\rho_{Arches} \gtrsim 3 \times 10^5\,M_\odot\,pc^{-3}$. In fact, the core of the Arches cluster appears to be more dense than most globular clusters. While these clusters appear to be as massive as small globulars, they will not survive long enough to become globulars. Indeed, there are no similar, but older, clusters that have been identified within the central 50 pc [2]. Fokker-Planck simulations by [38] suggest that these clusters will evaporate in $\lesssim 10$ Myr due mainly to strong tidal forces, a result confirmed by [41] and [39]. In fact, we may be seeing two snapshots in an otherwise identical evolutionary sequence, i.e., it is possible that the Quintuplet and Arches clusters are similar in their initial properties while the former is currently more extended due to its older age and the effects of dynamical evolution.

5 Conclusions

The first direct determination of the IMF has been made in the Galactic Center for the Arches cluster, and it appears to be significantly flatter than the Salpeter value. This result implies that the IMF law is not universal and may be affected by environmental conditions, although a similar result was found for NGC3603 [42]. We are continuing attempts to minimize possible errors in this determination. For instance, Keck spectroscopy has recently been obtained to verify the mass-magnitude relations. Finally, higher resolution imaging is needed to extend the IMF down to lower masses and to determine the lower-mass cutoffs.

References

1. M. Morris & E. Serabyn: ARAA **34**, 645 (1996)
2. D. F. Figer, PhD Thesis, UCLA (1995)
3. D. F. Figer, I. S. McLean, M. Morris: ApJ **447**, L29 (1995)
4. D. F. Figer, M. Morris, & I. S. McLean: in "4th ESO/CTIO Workshop on The Galactic Center," ed. R. Gredel, ASP, 263 (1996)
5. D. F. Figer, F. Najarro, M. Morris, et al.: ApJ **506**, 384 (1998)

6. D. F. Figer, S. S. Kim, M. Morris, et al.: ApJ **525**, 750 (1999)
7. D. F. Figer, I. S. McLean, M. Morris: ApJ **514**, 202 (1999)
8. M. Morris: ApJ **408**, 496 (1993)
9. E. E. Becklin, G. Negebauer: ApJ **151**, 145 (1968)
10. W. J. Forrest, M. A. Shure, J. L. Pipher, C. A. Woodward, in "The Galactic Center," ed. D. Backer, AIP Conf. Proc. 155, 153 (1987)
11. A. Krabbe, R. Genzel, S. Drapatz, V. Rotaciuc: ApJ **382**, L19 (1991)
12. D. A. Allen, in "The Nuclei of Normal Galaxies," eds. R. Genzel & A. I. Harris (Dordrecht: Kluwer), 293 (1994)
13. S. Libonate, J. L. Pipher, W. J. Forrest, M. L. N. Ashby: ApJ **439**, 202 (1995)
14. P. Tamblyn, G. H. Rieke, M. M. Hanson, et al.: ApJ **456**, 206 (1996)
15. R. Genzel: in "Star2000," ed. R. Spurzem, in press
16. T. Nagata, C. E. Woodward, M. Shure, N. Kobayashi, AJ **109**, 1676 (1995)
17. A. S. Cotera, E. F. Erickson, S. W. J. Colgan, et al.: ApJ **461**, 750 (1996)
18. E. Serabyn, D. Shupe, & D. Figer: Nature **394**, 448 (1998)
19. H. Okuda, H. Shibai, T. Nakagawa, et al.: ApJ **351**, 89 (1990)
20. T. Nagata, C. E. Woodward, M. Shure, J. L. Pipher, H. Okuda: ApJ **351**, 83 (1990)
21. I. S. Glass, A. Moneti, A. F. M. Moorwood: MNRAS **242**, 55P (1990)
22. F. Najarro, A. Krabbe, R. Genzel, et al.: A&A **325**, 700 (1997)
23. S. V. Ramirez, K. Sellgren, J. S. Carr, S. Balachandran, R. D. Blum, D. M. Terndrup, & A. Steed, BAAS, **195**, #62.05 (2000)
24. R. Genzel, N. Thatte, A. Krabbe, H. Kroker, L. Tacconi-Garman: ApJ **472**, 153 (1996)
25. F. Najarro, D. J. Hillier, R. P. Kudritzki, A. Krabbe, R. Genzel, D. Lutz, S. Drapatz, T. R. Geballe: A&A **285**, 573 (1994)
26. A. Krabbe, et al.: ApJ **447**, 95 (1995)
27. D. F. Figer, R. M. Rich, M. Morris, I. S. McLean, E. Serabyn, R. Puetter, A. Yahil: ApJ **525**, 759 (1999)
28. T. R. Geballe, F. Najarro, D. F. Figer: ApJ **530**, 97 (2000)
29. D. F. Figer, E. E. Becklin, I. S. McLean, A. M. Gilbert, J. R. Graham, J. E. Larkin, N. Levenson, H. I. Teplitz, M. K. Wilcox, M. Morris: ApJ **533**, L49 (2000)
30. R. Genzel, A. Eckart, T. Ott, F. Eisenhauer: MNRAS **291**, 219 (1997)
31. A. Eckart, T. Ott, R. Genzel: A&A **352**, 22 (1999)
32. M. Rieke: in preparation (2000)
33. V. C. Bailey, M. B. Davies: MNRAS **308**, 257 (1999)
34. T. Alexander: ApJ **527**, 835 (1999)
35. E. E. Salpeter: ApJ **121**, 161 (1955)
36. J. Scalo: in "The Stellar Initial Mass Function," eds. G. Gilmore and D. Howell, **142** Proc. of 38th *Herstmonceux Conference*, (San Francisco: ASP), 201 (1998)
37. G. Meynet, A. Maeder, G. Schaller, D. Schaerer, C. Charbonnel: A&AS **103**, 97 (1994)
38. S. S. Kim, H. M. Lee: A&A **347**, 123 (1999)
39. S. S. Kim, D. F. Figer, H. M. Lee, M. Morris: ApJ **545**, 301 (2000)
40. L. C. Ho, & A. V. Filippenko: ApJ, **472**, 600 (1996)
41. S. S. Kim, M. Morris, H. M. Lee,: ApJ **525**, 228 (1999)
42. F. Eisenhauer, A. Quirrenbach, H. Zinnecker, R. Genzel: ApJ **498**, 278 (1998)

The IMF of Starbursts Near and Far

Jorge Melnick

European Southern Observatory, La Silla, Chile

Abstract. There is good evidence that above a minimum mass, that may be a function of environment, the IMF of Starburst clusters, near and far, is well represented by a single power-law of slope very close to the Salpeter value. This contribution presents an overview of some of the observations supporting this affirmation.

1   The Nearest Starbursts

The nearest Galactic starbursts are the clusters close to the Galactic centre, discussed by Don Figer in this conference, and NGC 3603, in the region of Carina, described by Frank Eisenhauer in these proceedings. In this section I will review the observations of the nearest *extragalactic* starburst, 30 Doradus (NGC 2070), in the Large Magellanic Cloud.

1.1   The 30 Doradus Cluster

30 Dor (NGC 2070) is the ionizing cluster of the Tarantula nebula that is easily visible with the naked eye from any dark site in the southern hemisphere. The cluster has a radius of about 20 pc and contains about 1500 OB stars and more than 8000 stars more massive than $3M_\odot$. Although it is more distant than its Galactic counterparts, 30 Dor is significantly less reddened and this turns out to be an important advantage even in the near-IR.

Two effects plague the determination of the IMF in very young clusters: stochastic effects, which affect the upper end of the IMF where most clusters have only a few stars [2], and the *magnitude limit effect* [9] which affects stars near the completeness limit of the photometry. This effect is particularly important in starbursts, where the reddening varies significantly from star to star and hence the completeness limit is a function of stellar mass.

The simulations of Elmegreen [2] show that in order to achieve an accuracy of $\sim 5\%$ in the determination of the IMF slope, a cluster must have more than 1000 stars, so stochastic effects are minor in 30 Dor. The effect of reddening variations (magnitude limit bias) can be corrected if one knows the reddening to each star. A very careful analysis of the data that accounts for this and other systematic effects has been performed by Selman et al. [9]. This investigation shows three distinct peaks in the star-formation history of the cluster, occurring at 5, 2.5, and ≤ 1.5 million years ago. Above $3M_\odot$ the IMF of the cluster is consistent with a single power-law of slope $\alpha = -2.37$ (the Salpeter law has a slope $\alpha_s = -2.35$) with an error of $\sim 2\%$. When the NTT data of [9] are combined with HST data for the central part of the cluster, the resulting slope is -2.25 with an error of about 2.5%.

Figure 1 shows the combined NTT+HST IMF for M>3M$_\odot$, together with the best fitting line. The effect of magnitude limit, and completeness corrections are also shown in this figure. Siriani et al. [12] claimed that the IMF of 30 Dor flattens significantly below $\sim$3M$_\odot$, but their data have not been corrected for magnitude limit bias.

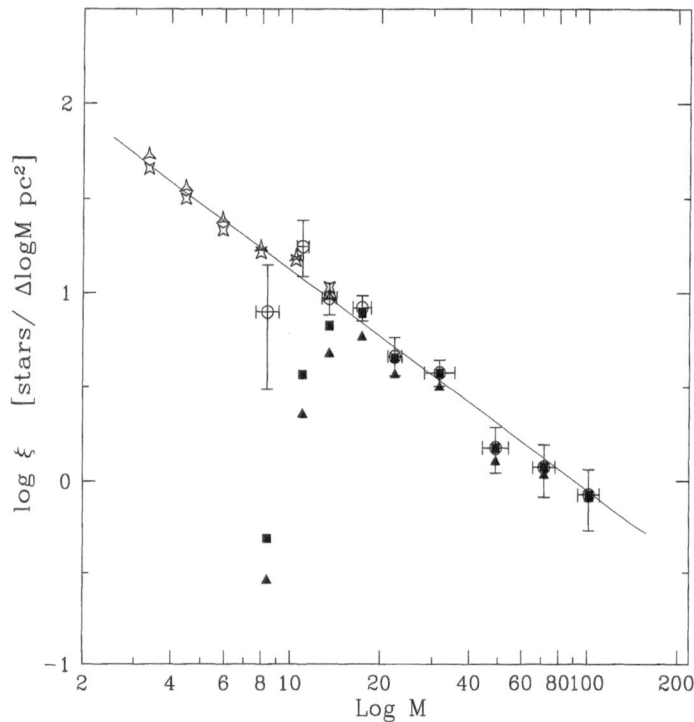

Fig. 1. The combined NTT+HST IMF of 30 Doradus. Filled squares show the raw data and filled triangles the data corrected for completeness. The open symbols with error bars present the data corrected for completeness and magnitude limit bias. The open symbols without error bars for the lowest mass bins are the published HST counts [3,4].

1.2 Comparison with Other Nearby Clusters

Massey [5] presented a compilation of IMF slopes for a sample of bright clusters and associations in the Galaxy and the Magellanic clouds. Taking stochastic effects into account, these values are consistent with a constant slope very close to Salpeter (but see [10] for a divergent opinion). Interestingly enough, the Arches cluster near the Galactic center, and NGC 3603 both appear to have IMF slopes significantly flatter than Salpeter. This has been used as evidence against the universality of the IMF slope (see contributions by Figer and Eisenhauer in these proceedings). These data, however, may be significantly affected by systematics including magnitude limit bias and/or PSF fitting, so these claims are still debatable.

2 Starbursts Near, Far, and Very Far

HII galaxies are in many senses ideal objects to study the systematic properties of starbursts near and far. These (blue-compact) dwarf galaxies (BCDs) are dominated by one or more very young and strong bursts of star formation so their integrated spectra are very similar to those of Giant HII regions in Local Group galaxies (eg. 30 Dor; NGC604). A comprehensive review of the morphological and spectral properties of HII Galaxies has recently been presented [13,6].

Figure 2 shows a histogram of bolometric luminosities of HII galaxies, computed using the observed Balmer line luminosities and the models of Rieke (private communication), at a wide redshift range from $z \sim 0$ to $z > 3$. HII galaxies span more than 4 decades in Bolometric luminosity. Note that the most luminous objects (that are also the most distant) reach luminosities comparable to those of ULIRGs!

A method has been proposed to constrain the IMF of HII galaxies using their integrated properties [1,7]. The method is based on the correlation between

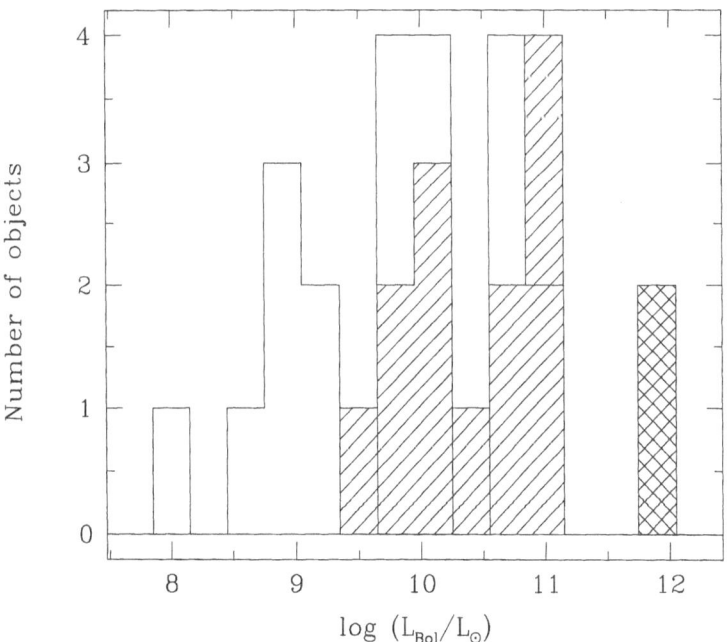

Fig. 2. Histogram of bolometric luminosities for HII galaxies near (white, $z < 0.1$), far (simple shading, $0.1 < z < 1.0$), and very far (cross shading, $z > 3$).

electron temperature and Oxygen abundance for HII regions. This relation depends of the effective temperature of the ionizing radiation and on the ionization parameter. For single starbursts, these parameters are completely determined by the total mass, the age, and the IMF of the burst [7]. Figure 3 shows the temperature-abundance plot for a sample of HII galaxies with the best available observations taken from the literature. The lines represent the starburst models of Stasinska and Leitherer [11] for a Salpeter IMF between $0.1M_\odot$ and $100M_\odot$, two different total masses, and three different ages as shown in the Figure.

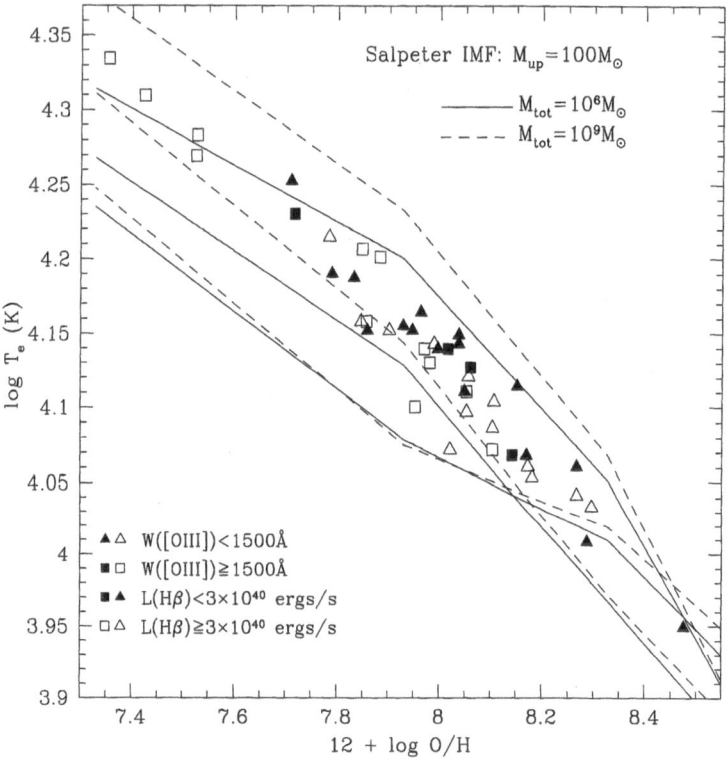

Fig. 3. The electron temperature – Oxygen abundance diagram for nearby HII galaxies. Different cuts of ages and burst masses, as indicated by W([OIII]) and L(Hβ)are given. The lines represent the models of Stasinska and Leitherer [11] for 2 different total masses (10^6 and $10^9 M_\odot$), and three different ages (1, 3, and 5Myrs) of the starbursts.

The plot shows that the observations are consistent with the models for a Salpeter IMF and total masses $\sim 10^6 M_\odot$. All objects appear to have ages between 1 and 3 Myrs, but there is a mass-age degeneracy. This degeneracy can be removed using parameters like the equivalent width of the [OIII]λ5007 line, W([OIII]), that is a sensitive age indicator, and the H_β luminosity, L(Hβ), that correlates with the total mass of ionizing stars [11]. Figure 3 shows a good consistency between models and observations when the objects are sorted by age and luminosity. A very close look at the plot shows some discrepancies that are most likely due to differences in the aperture sizes used to measure abundances, temperatures, equivalent widths, and luminosities.

It may be safely concluded, therefore, that there is no evidence for systematic variations in the IMF of HII Galaxies as a function of luminosity (mass) or metallicity. So it is reasonable to infer that the IMF of distant HII galaxies, for which no abundances and temperatures are yet available, will also be well represented by a Salpeter law above a certain minimum mass that could, in principle, be a function of mass and/or metallicity as suggested by the observations of [12].

3 Conclusions

Elmegreen ([2] and references therein) showed that the IMF resulting from randomly sampling the mass spectrum of fractal molecular clouds is a power-law of slope very close to the Salpeter value, independently of the detailed physics of star formation (at least above a certain minimum mass that in Elmegreen's theory corresponds to the thermal Jeans mass). Melnick and Selman [8] have argued that the Fractal structure of molecular clouds can be understood within the formal framework of Complexity theory, although they did not present a detailed 'toy model' to substantiate their proposal. However, if indeed the structure of molecular clouds is the result of self-organized critical processes, then it would be natural to expect the IMF to be a nearly universal function.

Moreover, since the Fractal structure of molecular clouds is a well established observational fact, there is a solid observational and theoretical basis to conclude that, above a certain minimum mass, the IMF of starbursts, near and far, is well represented by the Salpeter law.

References

1. A. Campbell, R. Terlevich, J. Melnick: MNRAS **223**, 811 (1986)
2. B.G. Elmegreen: ApJ **515**, 323 (1999)
3. D. Hunter, et al.: ApJ **448**, 179 (1995)
4. D. Hunter, et al.: ApJ **459**, L27 (1996)
5. P. Massey: in *The Stellar Mass Function*, ASP Conf Ser. vol. 142, eds. G. Gilmore & D. Howell, (San Francisco, ASP 1998), p. 17
6. J. Melnick, R. Terlevich, E. Terlevich: MNRAS **311**, 629 (2000)
7. J. Melnick: in *Evolution of Galaxies*, 10th IAU/EPC regional meeting, ed. J. Palous, (1987)
8. J. Melnick, F. Selman: in *Cosmic Evolution and Galaxy Formation*, ASP Conf Ser vol. 150, eds. J. Franco, E. Terlevich, O. López-Cruz, I. Aretxaga, in press (1999)
9. F. Selman, J. Melnick, G. Bosch, R. Terlevich: A&A **347**, 532 (1999)
10. J. Scalo: in *The Stellar Mass Function*, ASP Conf Ser vol. 142, eds. G. Gilmore & D. Howell, (San Francisco, ASP 1998) p. 201
11. G. Stasinska, C. Leitherer: ApJS **107**, 661 (1996); Pub. Astron. Inst. Czech. Acad. Sci., No. 89, Prague, p. 111.
12. M. Sirianni: ApJ **533**, 203 (2000)
13. E. Telles, J. Melnick, R. Terlevich: MNRAS **288**, 78 (1997)

Evidence in Favour of IMF Variations

Frank Eisenhauer

Max-Planck-Institut für extraterrestrische Physik,
Giessenbachstrasse, 85741 Garching, Germany

Abstract. The stellar initial mass function (IMF) determines the relative number of stars born at a given mass. Despite the tremendous effort to establish a universal IMF, the astronomical literature offers a wealth of diverse evidence for IMF variations. This review was prepared for a controversial debate at the conference "Starbursts – Near and Far" at Ringberg Castle, 2000, and gives a one-sided portrayal in favour of IMF variations. I will summarise the empirical evidence that the IMF varies with time, with environment, and for all stellar masses. While I see no obvious systematic trend in most regions of our Galaxy, there is at least an indication that the IMF is biased towards more massive stars in the early universe and in starbursts.

1 Introduction

The masses of stars span at least 3 orders of magnitude, from approximately 100 solar masses for the highest mass stars to the hydrogen burning limit of 0.1 solar masses. The relative numbers of stars born at a given mass is described by the initial mass function (IMF). This distribution is extremely important for many fields in astronomy, from the theory of star formation to the interpretation of integrated properties of galaxies at the highest redshift. Although many investigators have measured the IMF in a variety of star forming regions, we have not yet been able to come up with a final conclusion about the universality of the IMF. Specifically, Scalo's article [38] for the recent conference "The Stellar Initial Mass Function" [7] has revived the discussion, and many scientists — including the author — have not decided on their final opinion. Nevertheless the conference organisers have chosen to split the review on the universality of the IMF to keep the discussion lively and controversial. *This contribution focuses strictly on the empirical evidence in favour of IMF variations.* I will not try to balance any arguments, but leave the critical comparison with the evidence in favour of a universal IMF [8] to the reader.

The postulation of a universal IMF — in its strongest form — states that the IMF is and has always been the same in all regions of star formation everywhere; the frequencies of initial stellar masses in any unbiased sample are always drawn from the same statistical distribution. I will present evidence against this postulation with respect to "is and has always been" as well as to "everywhere", and I will also argue against this postulation being valid even in star forming regions of our own Galaxy.

Although challenging the concept of a universal IMF, I refer to a reference IMF for comparison. This IMF (small inset in Fig. 1) is the field star IMF defined

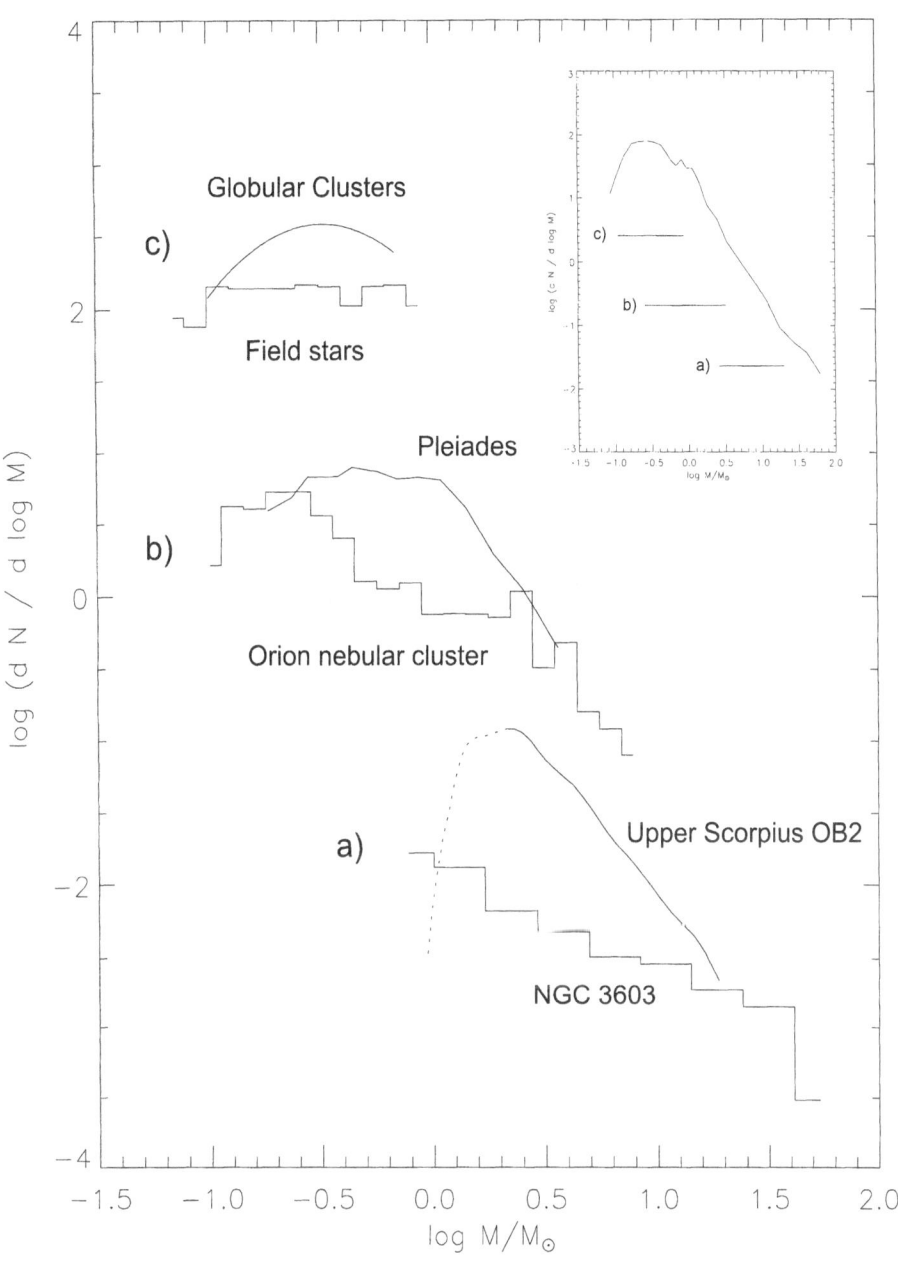

Fig. 1. Evidence for IMF variations: This figure shows a comparison of the IMF of several selected Galactic regions. Variations of the IMF are evident for all stellar masses: a) The exponent of the IMF differs by more than one in the high mass star forming regions NGC 3603 [4] and Upper Scorpius OB2 [2]. b) Compared to the Pleiades [26], the IMF of the Orion nebular cluster [10] is deficit in stars with masses between 0.5 – 1.25 $M_\odot$. c) The low-mass IMF in globular clusters is fitted best by a log-normal distribution [28], but the field stars follow a power law in the same mass interval [30]. The small inset shows the field star IMF of Scalo [37] for comparison.

by Scalo [37] in 1986, because its qualitative behaviour seems to be typical in many respects: (1) For high stellar masses the IMF can be described by a power law $\frac{d\,N}{d\,log\,M} \propto M^\Gamma$. The exponent is referred to as slope of the IMF. In the notation of this article the value for the Salpeter [35] IMF is -1.35. This power law results from scale free star formation processes, for example dominated by turbulent pressure. (2) The IMF is flat — the slope of the power law approximately 0 — for masses around 0.5 $M_\odot$, indicating a characteristic stellar mass, for example the thermal Jeans mass. (3) The IMF declines for low stellar masses and may be described by a log-normal distribution, indicating a large number of independent parameters in the star formation process. However, I will present evidence that the IMF differs in various regions from this reference IMF in shape (e.g. log-normal versus power law and turnover versus continuous), in slope (e.g. bias towards high or low masses) and in characteristic mass.

2   The Early Universe

If the IMF depends on any environmental conditions, one would expect the largest deviations for the most extreme environments. The early Universe is such an environment. There are two very compelling indications that the IMF was different in the early Universe: First, no metal-free stars and only few very-low-metallicity stars have ever been found [1,3], although low-mass stars, which have formed in the metal-poor early Universe, should have survived the last 15 billion years. Only ≈ 500 stars are known with $[Fe/H] \approx \frac{1}{300}\ [Fe/H]_\odot$ — the lowest metallicity observed in Galactic globular clusters —, and only ≈ 100 stars with $[Fe/H] \approx \frac{1}{1000}\ [Fe/H]_\odot$. No single star with primordial metallicity $[Fe/H] \approx \frac{1}{1000000}\ [Fe/H]_\odot$ has ever been observed. Second, the solar neighbourhood is deficient in metal-poor stars [20]. This discrepancy with the field star IMF is known as the "classical G-dwarf problem". Because the lifetime of these stars is longer than the age of the Universe, we would expect to find many more such stars, if low-mass stars have formed with the same frequency in the early days of our Galaxy as they do now. As outlined by Larson [14], this evidence finds its natural explanation in a time dependent IMF with a bias towards massive stars in the early Universe. Such a varying IMF is favoured for other reasons as well. For example, the gas of galaxy clusters is very hot and contains a large mass of heavy elements, which can not be explained from the proportionally few massive stars of the present day IMF. Also the observed strong evolution of the cosmic luminosity density with redshift is easier to explain with a top-heavy IMF. And microlensing experiments indicate that a significant fraction of dark matter is in the form of stellar remnants, which can only be produced from massive progenitors. See Larson's article [14] for a more detailed explanation and references.

3   Starbursts and Extragalactic Star Forming Regions

3.1   Starburst Galaxies

Starburst galaxies are another extreme environment for star formation. Their star formation rate is several magnitudes larger than in our own Galaxy. So if the thermal Jeans mass is of any importance for a characteristic stellar mass, the heating from the unusually strong star formation should give preference to more massive stars.

The starburst galaxy M 82 has been the prime example for a top heavy IMF for 20 years [31]. As the closest starburst galaxy with a distance of only 3.2 Mpc, no other starburst galaxy has been observed in such detail. The appearance of this galaxy is dominated by its nuclear starburst, which has probably been induced by a close encounter with M 81 about 100 Myr ago. The basic properties of the nuclear starburst are [24]: Bolometric luminosity $> 5.0 \times 10^{10}$ $L_\odot$, absolute K magnitude < -22.5, ionising radiation $L_{Lyc} > 10^{54}$ $photons/s$, total mass in the starburst $< 2.5 \times 10^8$ $M_\odot$, CO index > 0.21 and supernova rate ≈ 0.1 yr^{-1}. Rieke et al. [31,32] have pioneered the detailed modelling of this starburst galaxy, and have come repeatedly to the conclusion that stars with masses below a few $M_\odot$ form much less often in M 82 than in the solar neighbourhood. The argument for a top heavy IMF is basically that the infrared brightness — tracing the bolometric luminosity — is too high for the (dynamically measured) stellar mass, if one assumes a IMF similar to the field star IMF of Scalo [37]. The ionising radiation, the CO index and the supernova rate provide the necessary constraints for the age of the stellar population.

Although recent modelling [36,6] with a more complex spatial structure and temporal evolution can not exclude a Salpeter power-law IMF extending down to subsolar masses, the starburst galaxy M 82 remains one of the most cited examples for IMF variations.

3.2   Extragalactic Super Star Clusters

The same technique can also be applied to extragalactic super star clusters. Spectral diagnostics, such as hydrogen emission lines and stellar absorption features, enable us to date the starburst. Dynamical mass estimates together with infrared observations provide a reliable mass-to-light ratio, and these together constrain the shape of the IMF. Compared with starburst galaxies, individual super star clusters offer the big advantage of a simple spatial (only one cluster) and temporal (only one burst) structure. Nevertheless these clusters can be looked upon as the building blocks of starburst galaxies, and should thus trace the IMF for rather extreme environmental conditions, too. However, the IMF in starburst clusters is not necessarily biased towards massive stars.

The first super star clusters with dynamical mass estimates were NGC 1569A and NGC 1705-1 [12]. With a mass of 1.1×10^6 $M_\odot$ (NGC 1569A) and 2.7×10^5 $M_\odot$ (NGC 1705-1), these clusters are both gravitationally bound and could evolve into globular clusters [12]. However, recent evolutionary synthesis modelling of the mass-to-light ratio [39] revealed that the IMF must be very different in the two regions. While the IMF in NGC 1596A is steep (slope $\approx$ -1.5) and extends to below 0.1 $M_\odot$, the IMF in NGC 1705-1 is either flat (slope > -1) or truncated at a lower mass limit of 1 – 3 $M_\odot$.

A similar conclusion has been found for the Antennae [25]. The new generation of infrared spectrometers has allowed the dynamical mass determination of young embedded super star clusters for the first time. The authors [25] have compared each mass measurement with the results with the stellar synthesis model Starburst99 [15], that fit best the observed NIR-luminosity, Brγ, and the CO and Ca absorption features. The measurements in two of the observed clusters are consistent with a Salpeter IMF extending to subsolar masses. However, in their cluster #2 [43], such a power law IMF with a slope of -1.35 would include all the dynamical mass ($\approx 1.6 \times 10^6$ $M_\odot$) in stars more massive than 1 $M_\odot$. Therefore no mass is left for subsolar mass stars. This cluster seem to have formed proportionally more massive stars.

In conclusion, some of the measured mass-to-light ratios in super star clusters support the theoretical arguments for a top heavy IMF in starbursts, and thus systematic variations in the IMF, but the same kind of measurements also point to random variations in the IMF from cluster to cluster, even those in similar environments.

3.3   Extragalactic Star Formation Complexes

Unfortunately there are only a few dynamical mass measurements for star clusters. Statistics on IMF variations for a larger number of extragalactic star forming regions has to rely on integrated photometric and spectroscopic properties.

A recent example for such a statistical approach is [34], who have included 105 extragalactic star formation complexes for which UBVR photometry, Lyman continuum flux, metallicity and extinction measurements were available. These properties trace essentially the high mass stellar content, and the results are restricted to the IMF for stars more massive than 10 $M_\odot$. The models include the slope of the IMF, the maximum stellar mass and the age of the cluster. The star formation history is assumed either as a delta burst or as continuous. The average slope of the IMF was found to be -1.42. The standard deviation in the measured distribution is 0.91, and the authors estimate the standard error of their method to be 0.51. Therefore the measured distribution of slopes is meaningful [34]. Because the total number of analysed star formation complexes is rather large, the formal probability of measuring such a large standard deviation is basically negligible ($< 10^{-10}$), if one assumes a universal IMF.

Another investigation [23] has concentrated on 17 star formation regions in a sample of blue compact and irregular galaxies, using the integrated properties from UV, optical, FIR and radio measurements. This analysis uses the equivalent widths of Hβ, Si IV, and Ca IV to constrain the relative number of O3-O8 stars to B0-B3 stars, and thus the slope of the high mass IMF for stars more massive than 10 $M_\odot$. The evolutionary tracks of starburst regions with three modelled IMF slopes of 0.0, -1.35, -2.0 are clearly separated in a $W(H\beta)$ over $W(Si\,IV)/W(C\,IV)$ diagram. The derived slope for the starburst regions varies between 0 (3 regions), -1.35 (8 regions) and -2.0 (5 regions). Despite the variation in the derived slopes of the IMF, the authors interpret their result as in supporting a universal IMF, specifically because they see no trend with metallicity [23]. However, the scatter in the diagram is much larger than the average error bars, and I would interpret the same diagram as evidence for a varying IMF. In addition, the authors also report that objects showing flatter IMFs are always small compact star forming regions, that four of the galaxies with optical continuum dominated by a previous generation of stars have IMF slopes close to -2, and that a previous burst of star formation could have hampered the formation of lower mass stars.

4   IMF Variations in the Galaxy

A direct measurement of the IMF from star counts is still restricted to our Galaxy and the most nearby galaxies. While we can only compare the relative number of high- to low-mass stars or constrain the slope of the high-mass IMF in the early Universe and in extragalactic starburst regions, we are able to trace the IMF down to substellar masses in nearby star forming regions. Figure 1 shows a comparison of several IMFs measured in the Galaxy. We find variations in the slope and/or the shape of the IMF for all mass ranges.

4.1   The Most Massive Star Forming Regions

The most massive stars found in the Galaxy have about 100 $M_\odot$. Most of these stars have been found in either very compact star clusters or rather loose OB associations. The observations of these regions have improved significantly in the last few years with the HIPPARCOS mission, the Hubble Space Telescope (HST) and adaptive optics assisted telescopes.

For example, HIPPARCOS has revealed 178 new members of the Upper Scorpius OB2 association [2]. Only 91 stars were previously known in this association. Figure 1 shows the preliminary IMF. With a slope of $\approx$ -1.9, this IMF resembles very well Scalo's [37] field star IMF. I have compared this high-mass IMF with the IMF in NGC 3603, the most massive visible HII region in our Galaxy [4]. Adaptive optics assisted observations have revealed more than 800 stars in the central parsec of this cluster. Although the IMF shows no turnover or truncation down to < 1 $M_\odot$, its slope is only $\approx$ -0.7 for stars with masses between 3 and 30 $M_\odot$. A very similar result was found with the help of HST in the Arches cluster near the Galactic centre [5]. The slope of the IMF for stars more massive than 10 $M_\odot$ is $\approx$ -0.65, much shallower than the reference field star IMF, too. The Arches cluster and NGC 3603 are two of the most massive young clusters in the Galaxy, and each cluster has a few 1000 $M_\odot$ in O-stars. Therefore the difference between the field star IMF and the IMF in NGC 3603 and the Arches cluster can not be explained by small number statistics. Such an argument is often used to invalidate the evidence for IMF variations from star counts [13]. The finding in these two clusters supports the hypothesis that the IMF in starburst regions is biased towards massive stars. However, the deficit of low-mass stars is caused by a shallow IMF, and does not result from a truncated IMF, which has been indicated in M 82 [32].

4.2 Associations and Open Clusters

The discussion about a universal IMF was revived by Scalo [38] in 1997, at a time, when the majority of the participants of a conference on the stellar initial mass function was arguing for a universal IMF [7]. He made his case for IMF variations basically from a statistical interpretation of the measured slopes of the IMF in Galactic and Magellanic cloud associations and open clusters. Scalo plotted the slope of the IMF of 61 clusters against the average logarithmic stellar mass considered in the measurements. Although the typical error in the slopes is $0.1 - 0.4$, the spread in the diagram above ≈ 1 $M_\odot$ is so large that the author saw no basis for adopting some average value.

A serious argument against such evidence for IMF variations is that the results for different clusters have been obtained from different authors using different techniques. Therefore the IMF variations may have been mimicked by systematic errors, which have not been addressed carefully enough. I have thus repeated a similar χ^2 analysis for this conference to estimate how large the systematic errors would have to be, if the IMF were to be universal. My compilation includes the slopes of the IMFs for 51 clusters with intermediate- to high-mass star formation from 7 publications. The publications have been selected such that each paper includes more than 4 clusters, so that I could carry out statistics for each author and technique separately. Table 1 summarises the IMF properties of the different publications. Clusters with suspicious IMF measurements (for example NGC 436 in [29], which has probably undergone dynamical evolution), and results without error estimates (for example NGC 7235 in [21]) or with a different mass range (for example NGC 376 and N 24 in [9]) have not been included in the statistics. For the ease of interpretation, I have converted the reduced χ^2 values into a significance level [11], which is the probability of measuring such a high or higher χ^2, if the IMF would have a universal slope equal to the weighted mean slope in this subsample. Even at a significance level of only 5 %, we have to reject the hypothesis of a universal IMF in 5 of 7 cases. The statistics for the whole sample is even worse. The standard deviation in the slope distribution is 1.9 times the mean error in the measurements of the slopes. The likelihood for such a large discrepancy is basically negligible ($< 10^{-10}$) for a universal IMF. If the IMF is indeed universal, the authors must have underestimated their errors by a factor of two. The systematic errors would have to be even larger if we include the field stars of the Magellanic clouds, which show a particularly steep IMF with a slope of $\approx$ -3 to -4 [22].

A very intriguing example with very different IMFs at intermediate stellar masses are the Pleiades and the Orion nebular cluster. These clusters have been the target of some of the most careful recent measurements of the IMF [26,10]. The IMFs of these two clusters are also displayed in Fig. 1. In contrast to the

Table 1. IMF variations in 57 clusters with intermediate- to high-mass star formation.

Ref.	Number of clusters	Weighted mean slope	Standard deviation in slopes	Mean error in slope measurements	Reduced χ^2	Significance
[40]	4	1.43	0.34	0.15	5.15	0.15 %
[33]	5	1.27	0.80	0.33	14.82	≈ 0 [a]
[27]	6	1.22	0.31	0.30	2.92	1.2 %
[21]	12	1.12	0.44	0.34	1.78	5.3 %
[22]	5	1.27	0.18	0.16	1.64	16.2 %
[9]	12	1.92	0.54	0.34	3.22	0.021 %
[29]	7	1.42	0.26	0.15	18.08	≈ 0 [a]
all	51	1.52	0.53	0.28	6.35	≈ 0 [a]

[a] The significance level is $< 10^{-10}$.

IMF of the Pleiades, which follows the field star IMF, the IMF in Orion shows a regime of "missing" stellar masses between $0.5 - 1.25\ M_\odot$. Other examples are NGC 6231 and NGC 2264. The IMFs in these clusters have been measured in both cases with the same technique — UBVRI and Hα photometry — and by the same authors [41,42], therefore minimising systematic variations. Nevertheless, the IMF in NGC 2264 rises continously down to below $0.6\ M_\odot$, whereas the IMF decreases abruptly below $2.5\ M_\odot$ in NGC 6231.

4.3   Low-Mass Stars in Young Clusters, Globular Clusters and the Field

If we assume a Salpeter [35] power-law IMF with a slope of -1.35 extending down to a lower mass limit of $0.1\ M_\odot$, 96 % of all stars would have a mass smaller than $1\ M_\odot$, and 55 % of the stellar mass would be included in stars less massive than the sun. However, these low-mass stars are very faint, and many such stars have not been detected before the large ground based infrared surveys and HST. These observations have also revealed IMF variations in many regions.

To illustrate these variations, I will again compare different regions with the field star population. Figure 1 shows a recent measurement [30] of the field star IMF for low-mass stars. It was derived from the Deep Near-Infrared Survey (DENIS) and the 2 Micron All-Sky Survey (2MASS). In the $0.1 - 1\ M_\odot$ mass range, this IMF can be represented by a power-law mass function with a slope of ≈ 0.1. The statistical uncertainties are ≈ 0.13. A similar IMF has been observed in the young star forming regions IC 348, ρ Ophiuchi, and the Trapezium [16–18]. Their IMF is flat or slowly rising — slope $\gtrsim 0$ — from the brown dwarf regime to $0.6 - 1\ M_\odot$, where it rolls over to a power-law with a slope of ≈ -1.7. It is important to note that the IMF is not log-normal [18].

In contrast, recent HST observations indicate that the IMF of globular clusters can be described by a log-normal function with a peak near $0.3\ M_\odot$ [28].

Figure 1 shows the best fit for a sample of 12 globular clusters. The authors [28] explicitly exclude a single power-law IMF in the $0.1 - 0.6\ M_\odot$ range. Therefore the IMF of globular clusters seems to be fundamentally different from the IMF in young star forming regions and the field.

A meaningful interpretation of the IMF measurements for masses below the hydrogen burning limit is increasingly more difficult, because the available data are sparse. However, observations of young star forming regions indicate IMF variations even in the mass regime of brown dwarfs. The Taurus star forming region, for example, is significantly deficient in objects below $0.1\ M_\odot$ compared to the Trapezium [19]. If the IMF of both objects is normalised by the number of stars between $0.1 - 1\ M_\odot$, then ≈ 13 brown dwarfs with masses $> 0.02\ M_\odot$ are found in the Trapezium, but only one in Taurus.

5   Conclusions

Variations of the stellar initial mass function have been reported for all masses and in a large variety of stellar populations. We find evidence for IMF variations in the early Universe, in starburst galaxies and extragalactic star forming regions, in Galactic star clusters and associations, and in the field star population. Neither the shape nor the slope and the characteristic mass of the IMF seem to be excluded from these variations. There is some indication that the IMF is systematically biased towards more massive stars in the early Universe and in starbursts. However, I see no obvious systematic trend in those regions where the IMF could be constructed from direct star counts. If the IMF is indeed universal, it will be very difficult to prove this postulation empirically, because every individual measurement of a different IMF has to be invalidated. Only a consistent theory of star formation, with clear and testable predictions, will finally convince in the scientific world.

References

1. T.C. Beers: 'Observational Contraints on the Nature of the First Stars — Final Comments'. In: *The First Stars, MPA/ESO workshop, Garching, Germany, August 4-6, 1999*, ed. by A. Weiss, T.G. Abel, V. Hill (Springer, Heidelberg, 2000), pp. 336–342
2. A.G.A. Brown: 'The Initial Mass Function in Nearby OB Associations'. In: *The Stellar Initial Mass Function, 38th Herstmonceux Conference, Cambridge, UK, July 14-18, 1997*, ed. by G. Gilmore, D. Howell (ASP, San Francisco, 1998), ASP Conference Series, 142, pp. 45-59
3. C. Chiosi: 'Evolution of Pop III Stars'. In: *The First Stars, MPA/ESO workshop, Garching, Germany, August 4-6, 1999*, ed. by A. Weiss, T.G. Abel, V. Hill (Springer, Heidelberg, 2000), pp. 95–110
4. F. Eisenhauer, A. Quirrenbach, H. Zinnecker, R. Genzel: ApJ **498**, 278 (1998)
5. D.F. Figer, S.S. Kim, M. Morris, E. Serabyn, R.M. Rich, I.S. McLean: ApJ **525**, 750 (1999)

6. N. Förster-Schreiber: Near-infrared imaging spectroscopy and mid-infrared spectroscopy of M82: revealing the nature of star formation activity in the archetypal starburst galaxy. Ph.D. thesis, Ludwig-Maximilians-Universität, München (1998)
7. G. Gilmore, D. Howell (eds.): *The Stellar Initial Mass Function, 38th Herstmonceux Conference, Cambridge, UK, July 14-18, 1997* (ASP, San Francisco, 1998), ASP Conference Series, 142
8. G. Gilmore: these proceedings
9. R.J. Hill, B.F. Madore, W.L. Freedman: ApJ **429**, 192 (1994)
10. L.A. Hillenbrand: AJ **113**, 1733 (1997)
11. ISO Information Centre: *Statistical methods, Handbook on international standards for statistical methods* (International Standards Organization, Geneva, 1979)
12. L.C. Ho, A.V. Filippenko: ApJ **466**, L83 (1996)
13. P. Kroupa: MNRAS, accepted (2000)
14. R.B. Larson: MNRAS **301**, 569 (1998)
15. C. Leitherer, D. Schaerer, J.D. Goldader, R.M.G. Delgado, C. Robert, D.F. Kune, D.F. de Mello, D. Devost, T.M. Heckman: ApJS **123**, 3 (1999)
16. K.L. Luhman, G.H. Rieke, C.J. Lada, E.A. Lada: ApJ **508**, 346 (1998)
17. K.L. Luhman, G.H. Rieke: ApJ **525**, 440 (1999)
18. K.L. Luhman, G.H. Rieke, E.T. Young, A.S. Cotera, H. Chen, M.J. Rieke, H. Schneider, R.I. Thompson: ApJ **540**, 1016 (2000)
19. K.L. Luhman: ApJ **544**, 1044 (2000)
20. A. Martinelli, F. Matteucci: A&A **353**, 269 (2000)
21. P. Massey, K.E. Johnson, K. DeGioia-Eastwood: ApJ **454**, 151 (1995)
22. P. Massey, C.C. Lang, K. Degioia-Eastwood, C. Garmany: ApJ **438**, 188 (1995)
23. J.M. Mas-Hesse, D. Kunth: A&A **349**, 765 (1999)
24. K.K. McLeod, G.H. Rieke, M.J. Rieke, D.M. Kelly: ApJ **412**, 111 (1993)
25. S. Mengel: these proceedings
26. H. Meusinger, E. Schilbach, J. Souchay: A&A **312**, 833 (1996)
27. M.S. Oey: ApJ **465**, 231 (1996)
28. F. Paresce, G. De Marchi: ApJ **534**, 870 (2000)
29. R.L. Phelps, K.A. Janes: AJ **106**, 1870 (1993)
30. I.N. Reid, J.D. Kirkpatrick, J. Liebert, A. Burrows, J.E. Gizis, A. Burgasser, C.C. Dahn, D. Monet, R. Cutri, C.A. Beichman, M. Skrutskie: ApJ **521**, 613 (1999)
31. G.H. Rieke, M.J. Lebofsky, R.I. Thompson, F.J. Low, A.T. Tokunaga: ApJ **238**, 24 (1980)
32. G.H. Rieke, K. Loken, M.J. Rieke, P. Tamblyn: ApJ **412**, 99 (1993)
33. R. Sagar, W.K. Griffiths: MNRAS **299**, 777 (1998)
34. F. Sakhibov, M. Smirnov: A&A **354**, 802 (2000)
35. E.E. Salpeter: ApJ **121**, 161 (1955)
36. S. Satyapal, D.M. Watson, J.L. Pipher, W.J. Forrest, M.A. Greenhouse, H.A. Smith, J. Fischer, C.E. Woodward: ApJ bf 483, 148 (1997)
37. J.M. Scalo: Fundamentals of Cosmic Physics **11**, 1 (1986)
38. J.M. Scalo: 'The IMF Revisited: A Case for Variations'. In: *The Stellar Initial Mass Function, 38th Herstmonceux Conference, Cambridge, UK, July 14-18, 1997*, ed. by G. Gilmore, D. Howell (ASP, San Francisco, 1998), ASP Conference Series, 142, pp. 201-236
39. A. Sternberg: ApJ **506**, 721 (1998)
40. A. Subramaniam, R. Sagar: AJ **117**, 937 (1999)
41. H. Sung, M.S. Bessel, S.-W. Lee: AJ **114**, 2644 (1997)
42. H. Sung, M.S. Bessel, S.-W. Lee: AJ **115**, 734 (1998)
43. B.C. Whitmore, Q. Zhang, C. Leitherer, S.M. Fall, F. Schweizer, B.W. Miller: AJ **118**, 1551 (1999)

Evidence Supporting the Universality of the IMF

Gerry Gilmore

Institute of Astronomy,
Madingley Road, Cambridge CB3 0HA, UK

Abstract. The stellar initial mass function (IMF) is an underlying distribution function which determines many important observables, from the number of ionizing photons in a population of some age and metallicity, through the creation rate of various chemical elements, to the mass to light ratio of a system. This significance, together with the empirical difficulty to determine the IMF robustly, and the near complete lack of any robust theoretical predictor, has allowed investigators freedom to treat the IMF as a continuously variable parameterisation of astrophysicists' ignorance of complexity. An ability to vary a parameter in a model is not the same as a true variation in a physical system. A more instructive approach is to use available data to constrain possible variations, and thereby to allow identification of those other aspects of an observed system whose understanding can be improved. Ideally, the most sensitive physical variable, or its parameterisation, should be the best constrained. A fundamental null hypothesis, which we defend here, is that the IMF is a universal invariant function, so that all apparent variations may be ascribed to other variables, and to irreducible statistical sampling fluctuations.

1 Introduction

In the accompanying article, Eisenhauer [4] both introduces this subject and provides a summary of many indirect evidences that the IMF is seen to vary. Much of his evidence is based on the review of John Scalo [14] in the recent conference "The Stellar Initial Mass Function" [6]. Scalo's review, while compelling, is in disagreement with the remainder of the contributions to that volume. Thus, it is worth noting immediately that the concensus among expert opinion has changed dramatically in recent years, as data have improved. Previously it was assumed that the IMF was variable, and the task of observation was to quantify those changes. One hoped to identify some dominant parameter (metallicity, environment,..), whose identification would in turn lead star formation theory towards reality. The holy grail of explaining (baryonic) dark matter in the universe was a further motivation in the early days, though now accepted to be irrelevant to IMF studies: even in an extreme interpretation of the data, (and lots of such analyses exist) the stellar IMF is irrelevant to the dominant problem of dark matter.

The postulation of a universal IMF — in its strongest form — states that the IMF is and has always been the same in all regions of star formation everywhere; the frequencies of initial stellar masses in any unbiased sample are always drawn from the same statistical distribution function.

A decade ago, the scientific null hypothesis was that the IMF should vary, based on naive theory, and did vary, based on extant analyses of observations. A second major open problem, identification of the 'missing mass', was certainly relevant to determination of the number of low mass stars. Applying the long-standing scientific rule that any explanation of one problem which incidentally explains a second is progress, quantifying IMF variations became a reasonable default experiment. As the data improved however – the star count data has extended faintward by some 20 magnitudes in the red/infrared during the last two decades – and as an appreciation of the physical complexity of star forming regions observed only in integrated light arose, and, most importantly, as the number of high-quality very detailed HST studies of star clusters increased, the evidence in favour of apparent IMF variations has decreased. The unexpected current result is that any variation in the IMF is startlingly small. The observational situation has changed so much that the conservative assumption has now inverted.

That is, the conservative assumption is now that the IMF is invariant, and that apparent variations are due to unrecognised complexities in the physical situation being analysed, and the statistics of small samples, both convolved with recognised difficulties in the conversion of observables – ionisation state, metallicity, age range, stellar colour, ... into fundamental parameters, such as mass. This null hypothesis is clearly extreme: zero intrinsic variation in the IMF seems ab initio implausible, given the diversity of physical conditions in which stars form. However, the observations point to a situation in which variation in the IMF is a second-order effect.

Here we consider two complementary conservative questions: first, is there any irrefutable evidence that the IMF is variable, and secondly, do we know what the IMF is? I conclude that there is indeed no robust evidence for variation, while there remains considerable (systematic) uncertainty in quantifying the underlying (invariant) IMF.

2   The IMF as a Function of Time

Was the IMF different in the early Universe than it is today? A variety of indications suggest such changes, primarily indirectly from attempts to model the chemical evolution of the IGM, to explain galaxy cluster abundances, and to explain the lack of very metal-poor stars in the Galactic halo. All of these suffer inevitable uncertainty from the many simplyfying approximations involved in modelling gas flows, accretion histories, outflows, and so on.

2.1   Chemical Abundance Constraints

Although absolute chemical abundances remain poorly understood, some much less indirect constraints are available from the relative abundances of the chemical elements. The relative abundances of a subset of the elements are determined purely by the high mass stellar IMF, and the rate of star formation. The astrophysics of this analysis is described in many published papers and need not be detailed here. (An example of one of the many relevant reviews is [7]). The essential feature is that the ratio of the elements created purely in high mass stars to those elements created primarily in lower mass stars is manifestly a measure of the relative numbers of such stars, and so the slope of the IMF.

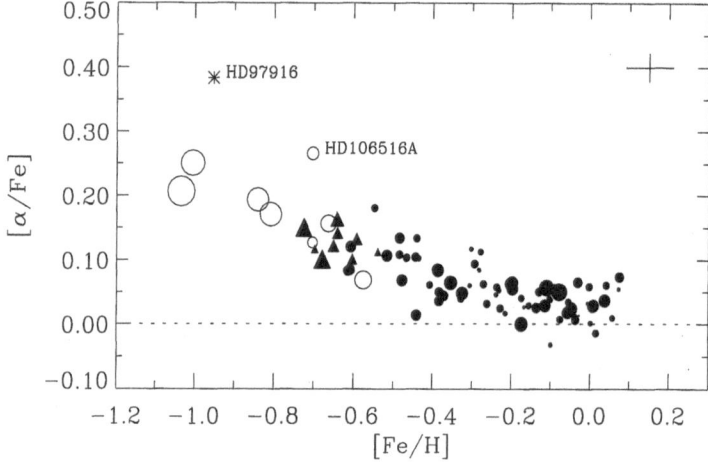

Fig. 1. The mean α (Mg, Si, Ca and Ti) abundance as a function of metallicity. The symbol size is proportional to stellar age. (From [3]).

The state of recent observational determinations of this ratio is illustrated in Fig. 1, from Chen et al. [3], who also discuss the current uncertainties in understanding elemental production sites. For present purposes however, the essential result is that observational determinations, by many different authors, agree that metal-poor stars show a systematic overabundance of the alpha elements, created in high-mass stars, relative to the abundance of the iron-peak elements, created in lower mass stars. The absolute value of the overabundance thus provides a limit on the high-mass IMF slope at very early times. Extant modelling (eg [16]) limits any IMF variation relative to the present-day IMF to be less than about 0.3 in slope, a maximum systematic change comparable to the uncertainty in direct star-count determinations of the slope today.

Thus, modelling chemical element data provides no evidence for any systematic change in the IMF slope over the whole age of the Universe.

3   The IMF as a Function of Star Formation Rate

It is very easy to expect the IMF to vary strongly in starburst regions. The relevant observational suggestions are well reviewed by Eisenhauer [4]. As so often is the case, one may however provide theoretical expectation to favour all possible outcomes. In particular, the concept of a Jeans mass, and calculations of small perturbations from stability in a self-gravitating system, are least likely to be relevant in a starburst, where one must expect dramatic inhomogeneity and significant shock energy. Analysis of observational data on current starbursts is necessarily dependent on simplifying assumptions of the stellar age and metallicity distributions, and is inevitably very sensitive to the relative spatial distributions of stars, dust and the ISM, together with any non-thermal energy sources which may be present.

There is one simple observational test, which does not probe the most extreme high star formation rates, does probe most of the available dynamic range, but is limited to low masses. That is to compare the present-day IMF in the remnants of high and low star formation rate systems. The remnant high-rate systems are globular clusters. These are limited in duration of formation by the lack of a spread in abundances, and additionally are seen today to be forming rapidly, and to be forming in starburst galaxies. Globular clusters seem to form 10^6 stars in 1Myr, for a rate of $1 - 10M_\odot yr^{-1}$. The remnant low star formation rate systems are the dSph galaxies. The star formation rate cannot have been high in these fragile galaxies. Quantitative determinations of the rates, eg by Hernandez, Gilmore, and Valls-Gabaud [8], derive star formation rates of order 10^{-3} that in galaxy disks today, and so some 3-7 orders of magnitude below those of globular clusters. The difference in present stellar volume density is similar.

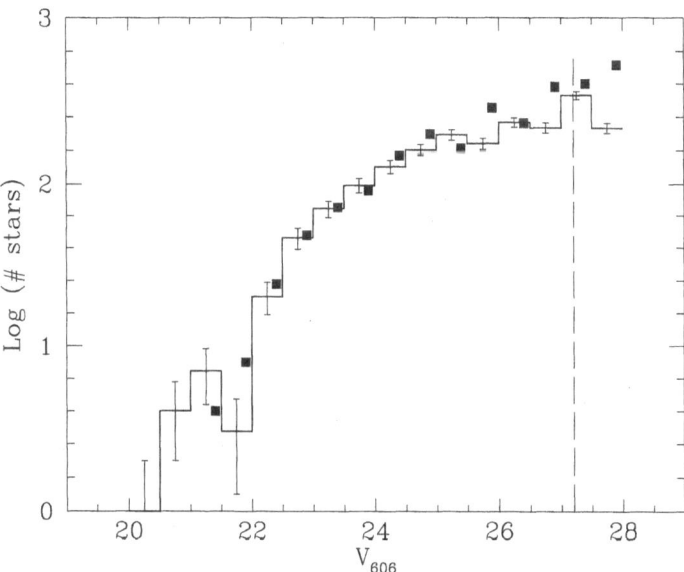

Fig. 2. Comparison between the UMi completeness-corrected luminosity functions derived from the CMD with that of the globular cluster M92 (filled squares). The vertical dashed line indicates the 50% completeness limits. The corresponding mass range is $0.8 < M/M_\odot < 0.35$. (From [5]).

A direct comparison of the stellar luminosity function in a globular cluster and dSph galaxy has recently been obtained, by Feltzing, Wyse & Gilmore [5]. They compared the luminosity functions of unevolved, low mass stars in the UMi dSph galaxy and in the globular cluster M92. Both these systems are of similar age and metallicity, so the luminosity function comparison is an IMF comparison. The systems have star formation rates and present stellar density differing by about 6 orders of magnitude, and very different mass:light ratios. The results, shown in Fig. 2, show very clearly that there is no detectable IMF difference between these systems for stars with masses in the range $0.8 < M/M_\odot < 0.35$.

4   IMF Variations Between Similar Systems?

Differential comparison between comparable systems with large numbers of stars is the most sensitive way to identify real IMF variations. The best case available, and the most studied, are the Galactic and Local Group globular clusters. These contain large numbers of stars, minimising statistical sampling errors (see below), and have little or no internal metallicity range, minimising the need for reliable stellar models (see below).

The mass-luminosity relation does of course change systematically with metallicity, generating a systematic change in luminosity function with metallicity. This is seen [15], and is to first order consistent with a constant IMF over the metallicity range from −2dex to the Solar value. At a specific metallicity one may in principle provide more exact comparisons.

The extant HST cluster luminosity function data are collected in Fig. 3, where large variations are apparent. Are these real IMF variations? Systematic changes with time away from the primordial IMF of a globular cluster are an inevitable consequence of internal dynamical evolution. Mass segregation must happen, as will stellar merging and mass-dependent mass and star loss from the cluster. To date, modelling this rich mix of physical effects has not been computationally feasible, so that highly simplified diffusion modelling (Fokker-Planck) has been necessary. Such modelling is necessarily inexact in the densest regions, where 2-body interactions are dominant. The best available calculations of this type are still unable to reproduce observed distributions of cluster properties, so considerable caution is required.

Recent advances in N-body systems, especially GRAPE, are starting to allow more realistic modelling, and hold great promise. In particular, the important effects of cluster core dynamical evolution, and external tides can be included reliably [1]. An indication of the potential importance of these effects is seen in Fig. 4, which plots the differences between cluster luminosity functions seen in Fig. 3 as a function of distance from the Galactic Plane. This distance is a crude measure of tidal forces, and suggests that these additional factors are important. Preliminary modelling from extant GRAPE facilities suggests that the range of luminosity function slopes seen is (marginally) consistent with the effects of dynamical evolutions which are feasible, assuming a universal IMF.

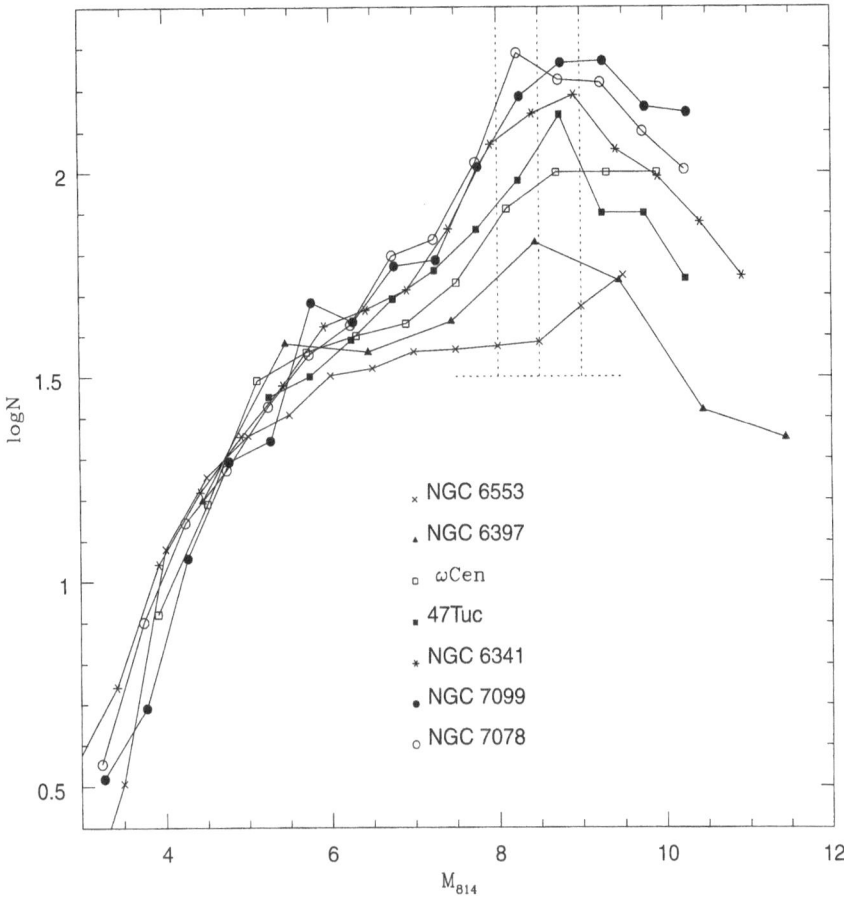

Fig. 3. A collection of luminosity functions derived from HST observations of Galactic globular clusters (Beaulieu et al. in prep). Systematic differences between clusters, even at the same metallicity, are evident. These differences are all entirely consistent with a combination of the metallicity dependence of the mass-luminosity relation, and with different dynamical evolution between clusters. In particular, the effects of the Galactic tidal field on internal cluster evolution can be as large as the differences apparent here.

Until the next generation of such models are available, it is premature to interpret the real differences in present day luminosity functions of clusters as evidence for variation in the primordial IMF.

5   Is the IMF Variable?

Determination of the IMF from star count data is a complex challenge. In addition to the obvious requirement for high-quality observational data with reliable calibrations, there are many other factors in the analysis. The most obvious among these include effects of stellar age, multiplicity, and metallicity, mass segregation, sample incompleteness, primordial and dynamical mass segregation.

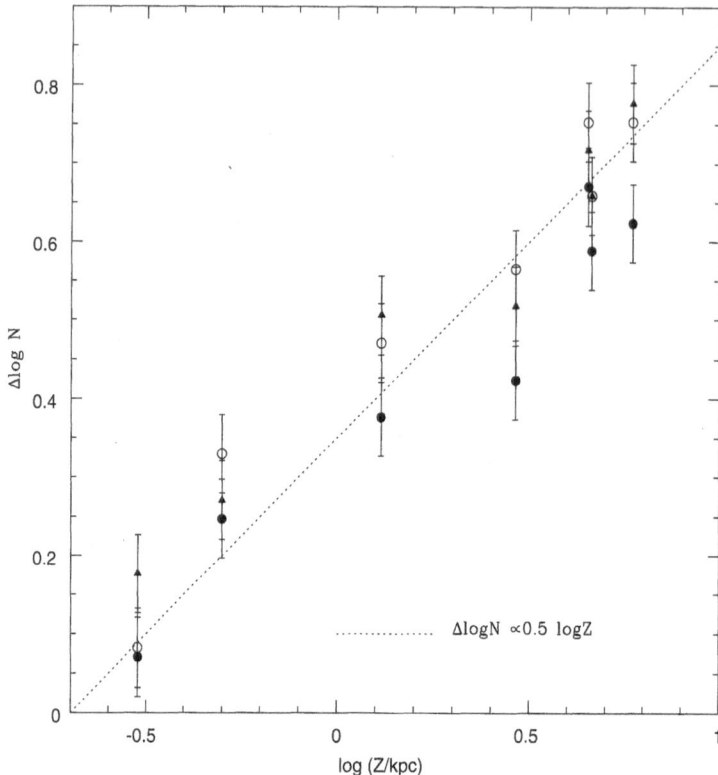

Fig. 4. The differences in the globular cluster luminosity functions shown in Fig. 3 (Δ log N stars at the LF maximum) plotted as a function of distance from the Galactic Plane. This is a crude measure of the importance of tidal forces on cluster evolution. The correlation suggest that tidal effects, which are yet to be included completely in extant models, are as significant as any real variation in the IMF between clusters.

These, plus all the many other contributions to the Malmquist bias, were considered explicitly by Kroupa, Tout & Gilmore (KTG [10]). Recently, Kroupa [9] has updated this analysis, also including the effects of sampling noise. Sampling noise is dominant in most published IMF determinations, which are studies of young open clusters with few members. Kroupa's current "best" IMF is shown in Fig. 5. Note that this IMF, which is very similar to that of KTG, increases systematically to the hydrogen burning limit.

To consider variations from this mean IMF in any observed sample, it is easiest to consider α, the gradient of this function, rather than the function itself. This is illustrated in Fig. 6, the alpha-plot.

An important point is the extent to which the scatter of observational points in the alpha-plot indicates real variation. In addition to the systematic effects considered below, there is one important effect to emphasise: statistical sampling noise, due to the finite and small number of stars in a star-forming region. This has a disproportionate effect at high masses, where the numbers are very small. The effect is shown in Fig. 7.

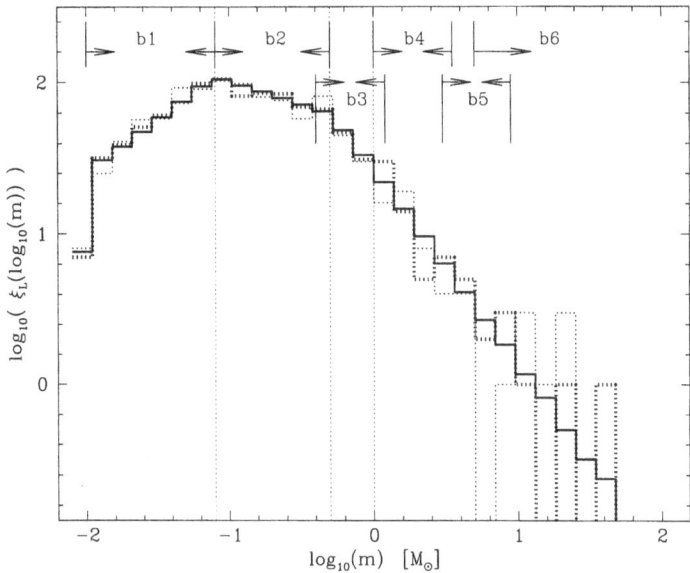

Fig. 5. The adopted logarithmic IMF for 10^6 stars (solid histogram). Two random renditions of this IMF with 10^3 stars are shown as the heavy and thin dotted histograms. The mass-ranges over which power-law functions are fitted are indicated by the arrowed six regions, while thin vertical dotted lines indicate the masses at which the fitted slope changes. (From [9]).

Overall, this analysis illustrates that sampling and observational effects, together with the many other contributions to Malmquist bias, dominate available data: any direct evidence for IMF variation is not robust.

A further very important factor is illustrated in Fig. 8: even if one has a large sample with accurate data, conversion of an observed luminosity and derived effective temperature to source mass is not a robust procedure. Extant stellar models are uncertain by an amount that is large compared to deduced changes in the IMF [2].

6   Conclusions

Variations of the stellar initial mass function have been reported for all masses and in a large variety of stellar populations. We show that in (almost) all cases the suggested variation is either dependent on simplifying assumptions made in other important parameters, or is dominated by sampling noise. Some direct evidence exists showing the IMF has been invariant over 12 Gyr, at all abundances from -2dex to at least the Solar value, in systems with stellar densities covering 6 orders of magnitude, and in systems with a range of star formation rates spanning at least 6 orders of magnitude. This remarkable and unexpected constancy, showing any real variation to be a second order, suggests the physics

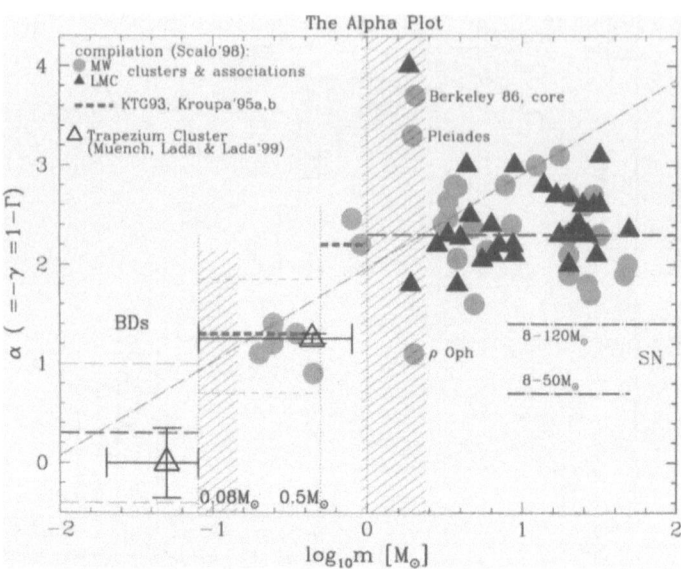

Fig. 6. The *alpha-plot*. The symbols show the compilation by Scalo [14] of determinations of α over different mass ranges for Milky-Way (MW) and Large-Magellanic-Cloud (LMC) clusters and OB associations. Unresolved multiple systems are not corrected for. The large open triangles ([12] from Orion Nebula Cluster observations, binary corrections not applied) serve to illustrate the present knowledge for $m < 0.1 M_\odot$. The horizontal long-dashed lines in the brown dwarf regime are the Galactic-field IMF of Fig. 5 with associated approximate uncertainties. For $0.08 \leq m \leq 1.0 M_\odot$ the thick short-dashed lines represent the KTG single-star IMF ([10], which has $\alpha_3 = 2.7$ for $m > 1 M_\odot$ from Scalo's [13] determination. The long-dashed lines for $m > 1 M_\odot$ show the approximate average $\alpha = 2.3$, which is adopted in the Galactic-field IMF. The Miller & Scalo [11] log-normal IMF for a constant star-formation rate and a Galactic disk age of 12 Gyr is plotted as the diagonal long-dash-dotted line. The long-dash-dotted horizontal lines labelled "SN" are those $\alpha_3 = 0.70(1.4)$ for which 50 % of the stellar (including BD) mass is in stars with $8 - 50(8 - 120) M_\odot$. The vertical dotted lines delineate the four mass ranges highlighted in Fig. 5, and the shaded areas highlight those stellar mass regions where the derivation of the IMF is additionally complicated due to unknown ages, especially for Galactic field stars: for $0.08 < m < 0.15 M_\odot$ long pre-MS contraction times make the conversion from an empirical LF to an IMF dependent on the precise knowledge of the age, while for $0.8 < m < 2.5 M_\odot$ post-main sequence evolution makes derived masses uncertain in the absence of precise age knowledge. Some published data are labelled by their star cluster names. (From [9]).

underlying the IMF is dominated locally by the central limit theorem, rather than one or a few dominant parameters.

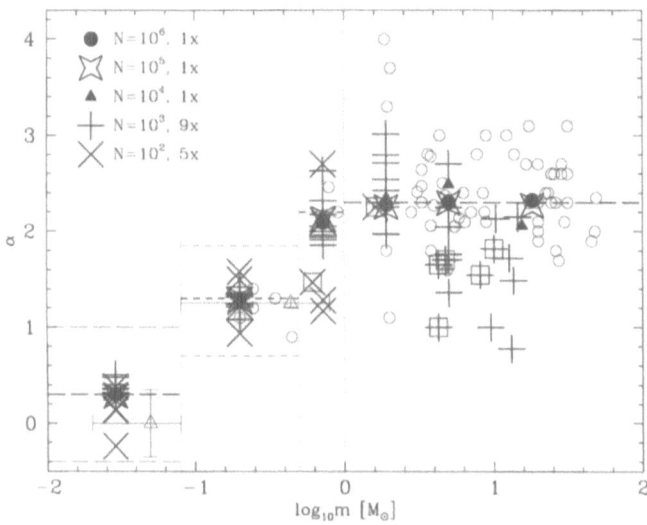

Fig. 7. Purely statistical sampling noise variation of the IMF slope α in the six mass ranges of Fig. 5 for different observed star numbers N as indicated in the key. The open circles and open triangles are as in Fig. 6. (From [9]).

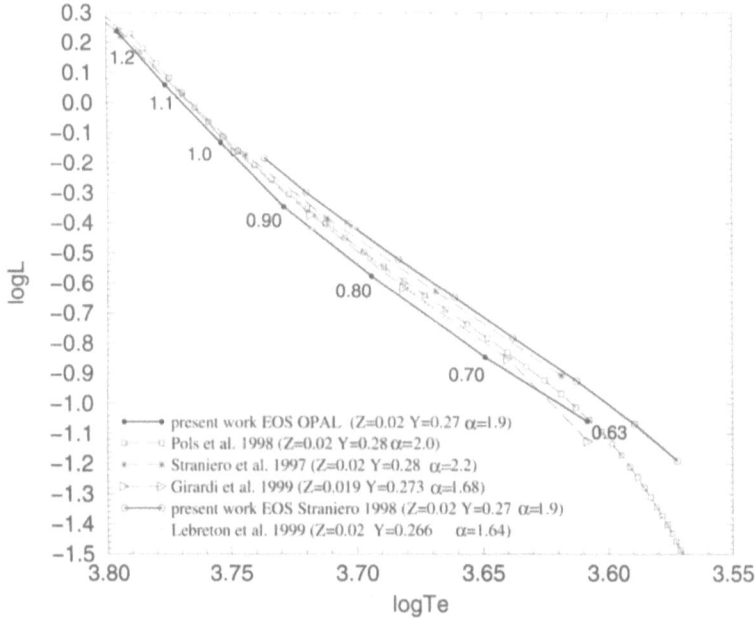

Fig. 8. Location of the ZAMS in the HR diagram, for a variety of recently published stellar models. The very considerable diversity of model parameters which provide approximately the same slope in the HR diagram illustrates the limits to which one can deduce fundamental stellar parameters, such as mass, from an observation of luminosity and temperature, given typical observational errors. (From [2]).

References

1. S. Aarseth: PASP **111,** 1333 (1999)
2. V. Castellani, S. Degl'Innocenti & P. Prada Moroni: preprint (2001)
3. Y.Q. Chen, P.E. Nissen, G. Zhao, H.W. Zhang, & T. Benoni: A&A in press (2000)
4. F. Eisenhauer: these proceedings
5. S. Feltzing, R.F.G Wyse & G. Gilmore: ApJL **516,** L17 (1999)
6. G. Gilmore, D. Howell (eds.): *The Stellar Initial Mass Function, 38th Herstmon-ceux Conference, Cambridge, UK, July 14-18, 1997* (ASP, San Francisco, 1998), ASP Conference Series, 142
7. G. Gilmore, R.F.G. Wyse & K. Kuijken: ARAA **27**, 555 (1989)
8. X. Hernandez, G. Gilmore, and D. Valls-Gabaud: MNRAS **317**, 831 (2000)
9. P. Kroupa: MNRAS, accepted (2000)
10. P. Kroupa, C. Tout, G. Gilmore: MNRAS **262,** 545 (1993)
11. G. Miller & J. Scalo: ApJS **41**, 513 (1979)
12. A. Muench, E. Lada & C. Lada: ApJ **533**, 358 (2000)
13. J.M. Scalo: Fundamentals of Cosmic Physics **11**, 1 (1986)
14. J.M. Scalo: 'The IMF Revisited: A Case for Variations'. In: *The Stellar Initial Mass Function, 38th Herstmonceux Conference, Cambridge, UK, July 14-18, 1997*, ed. by G. Gilmore, D. Howell (ASP, San Francisco, 1998), ASP Conference Series, 142, pp. 201-236
15. T. von Hippel, G. Gilmore, N. Tanvir, D. Robinson, D.H.P. Jones: AJ **112**, 192 (1996)
16. R.F.G. Wyse, G. Gilmore: AJ **104**, 144 (1992)

The Stellar Content
of Obscured Galactic Giant HII Regions

Robert D. Blum[1], Peter S. Conti[2], and Augusto Damineli[3]

[1] Cerro Tololo Interamerican Observatory, Casilla 603, La Serena, Chile
[2] JILA, University of Colorado, Campus Box 440, Boulder, CO 80309, USA
[3] IAG, University of São Paulo, Av. Miguel Stefano 4200, 04301-904, São Paulo, Brazil

Abstract. We present results of a survey of obscured Galactic Giant HII regions (GHII). Near infrared images and spectroscopy are used to identify and classify massive stars in these emergent star forming systems. Our sample has been chosen from the most luminous GHII regions in the Galaxy as indicated by radio determinations of the Lyman continuum output. Our current data have identified massive O type stars, Wolf Rayet stars, and young stellar objects at the core of a number of these very young, massive star forming regions. This large sample of GHII regions will be used to characterize and understand the massive star forming process and will be useful in building models of external, aggregate systems. The sample will also be useful as a comparison to the three known Galactic Center massive star clusters.

1   Introduction

We have begun a systematic survey of the more luminous (as determined from radio observations [10]) obscured giant H II regions in the Galaxy with an eye toward investigating their massive stellar content. J, H, and K images are used to make a broad assessment of the stellar content, and infrared spectroscopy follows, providing details of the brightest cluster members which can be used to make distance, mass, and luminosity estimates. The spectra are placed in proper context by comparison to new infrared spectral classification systems for massive stars ([6], [4]).

The present project seeks to provide a large sample of massive star clusters with which to study the young and massive stellar content in the Galaxy. This sample builds on the detailed visual studies of the Galactic OB associations, and massive star clusters NGC 3603 (in the Galaxy) and R136 (in the Large Magellanic Cloud). Finally, the investigation of a large sample of clusters in Galactic giant H II regions will be important in understanding the massive star clusters in the Galactic center ([3], [5]) which may have formed under different conditions than are typical in the disk of our Galaxy ([9]).

2   Results

Detailed results for W43, and W42, may be found in [1] and [2]. Here we present illustrative results on W31 and M17.

We have identified young stellar clusters in approximately ten GHII regions. These range from extremely young objects whose massive stars appear to be still heavily enshrouded in gas and dust (G333.6-0.2, W33) to slightly more evolved clusters whose most massive stars are clearly revealed by near infrared spectroscopy, having shed the bulk of their pre-natal circumstellar material (W31, W42, W43, G351.6-1.3, G333.1-0.4). Our sample also includes G284.3-0.3 and G333.0-0.4 for which we have incomplete data, and some new images of M17 and W49 which will build on previous work by other investigators.

Fig. 1. JHK composite image of W31. N is up and E to the left. These images were obtained on the CTIO 4m Blanco telescope. The FOV is 1.8' x 1.7'. The individual images have tip-tilt stabilized seeing of 0.5" - 0.6"and exposure times of 500s, 280s, 180s at J, H, and K, respectively.

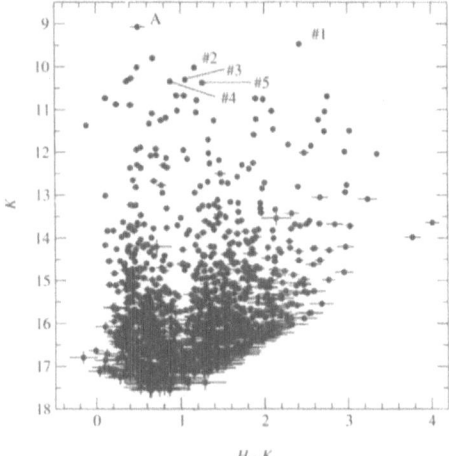

Fig. 2. K v. $H - K$ color magnitude diagram for W31. A foreground sequence is evident, topped by the brightest star, "A," which is to the N center in Fig. 1. Star 1 is a massive young stellar object (Figs. 3, 5). Such objects appear in M17 [7], W43 [1], and W42 [2]. Spectra for stars 2 - 5 are shown in Fig. 4. These stars define the upper main sequence for the cluster in W31.

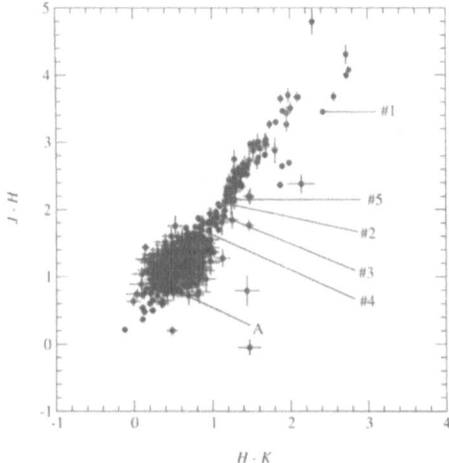

Fig. 3. $J - H$ v $H - K$ color color plot for W31. The O stars (2-5) fall along a normal reddened sequence while star 1 (and others for which we have no spectra yet) fall further to the right, exhibiting an excess emission in K, presumably due to accretion disk luminosity or other circimstellar emission.

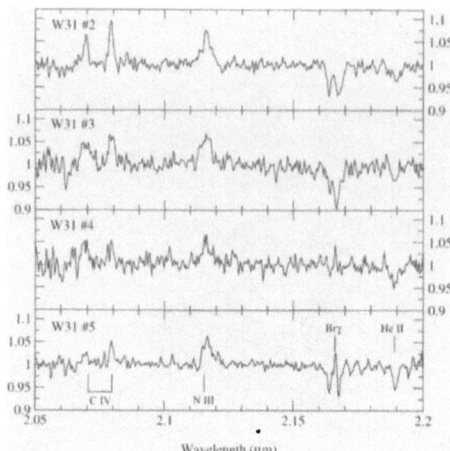

Fig. 4. Spectra of the 4 brightest stars in the W31 cluster (with no $H-K$ excess). These stars all have a k05-6 spectral type on the system of [6]. Several of the spectra exhibit residual Brγ emission superposed on the stellar absorption. This emission possibly arises in a circumstellar disk, but the difficulty in nebular background subtraction and telluric correction renders this suggestion inconclusive. Higher spectral resolution data will be required to solve the issue.

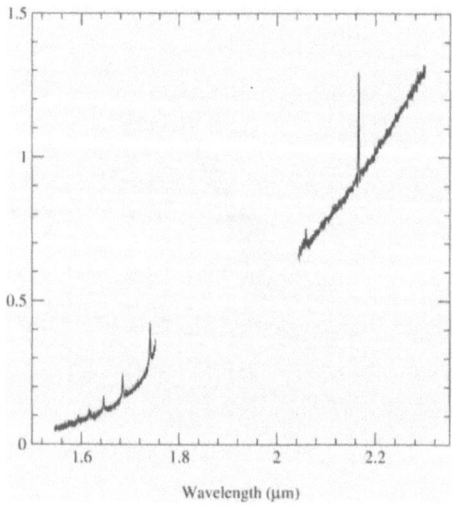

Fig. 5. H and K−band spectrum of star 1 in W31. The steep continuum and position in the color color plot (Fig. 3) suggest this object is a young stellar object. It is likely to be massive given its position in the CMD (Fig. 2). No O star features are detected in this signal to noise 70-100 spectrum. However, the compact Brackett series emission is probably circumstellar in origin.

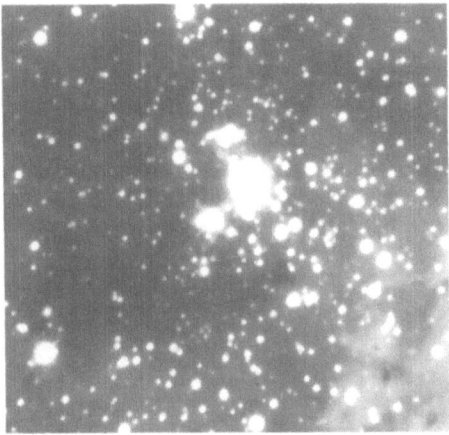

Fig. 6. *JHK* composite image of M17. N is up and E to the left. Images obtained on the CTIO Blanco 4m telescope. Tip-tilt stabilized images are 0.5", 0.7", and 0.9" at *K*, *H*, and *J*, respectively. Exposure times are 146s in *H* and *K*, and 292s in *J*. Reference [7] have identified and classified 12 O stars and YSOs in M17 including the bright central objects in this 1.8'x1.7' image.

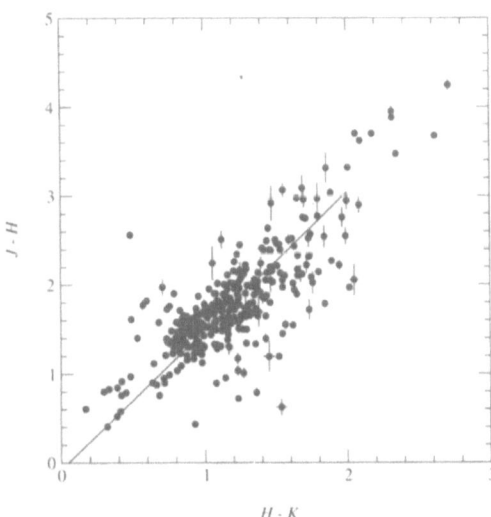

Fig. 7. Color-color plot for M17. Our higher resolution images (Fig. 6) of M17 confirm the finding of [8] that a large fraction of the M17 cluster stars exhibit excess emission. The line shows the reddened colors of an early O star from zero to 3.2 mag of extinctiont at K. M17 exhibits a relatively "flat" *K*-band luminosity function (KLF), approximately like that of the Arches Cluster. Care must be taken in converting such a KLF to a mass function.

References

1. R.D. Blum, A. Damineli, P.S. Conti: AJ **117**, 1392 (1999)
2. R.D. Blum, P.S. Conti, A. Damineli: AJ **119**, 1860 (2000)
3. A. Cotera, E.F. Erickson, S.W.J. Colgan, J.P. Simpson, D.A. Allen, M. Burton: ApJ **461**, 750 (1996)
4. D.F. Figer, I.S. McLean, F. Najarro: ApJ **486**, 420 (1997)
5. D.F. Figer, S.S. Kim, M. Morris, E. Serabyn, R.M. Rich, I.S. McLean: ApJ **575**, 750 (1999)
6. M.M. Hanson, P.S. Conti, M.J. Rieke: ApJS **107**, 420 (1996)
7. M.M. Hanson, I.D. Howarth, P.S. Conti: ApJ **489**, 698 (1997)
8. C.J. Lada, D.L. DePoy, K.M. Merrill, I. Gatley: ApJ **374**, 533 (1991)
9. M. Morris, E. Serabyn: ARAA **34**, 645 (1996)
10. L.F. Smith, P. Biermann, P.G. Mezger: A&A **66**, 65 (1978)

Part III

Star Formation in Different Environments

Star Formation in the Galactic Center and Nearby Nuclei

Mark Morris

Division of Astronomy, University of California, Los Angeles,
Los Angeles, CA 90095-1562, USA

Abstract. The abundant gas in the central molecular zone of our Galaxy is forming stars at at about the same pace, per unit mass of gas, as the disk of the Galaxy, despite the very different physical conditions in the central environment: a strong dipole magnetic field, strong tidal forces, and a large internal velocity dispersion within the abundant molecular clouds. There are at least two pronounced differences in the properties of the stars that form near the center, as compared with the disk, however. First, among the stars which form near the Galactic center, the abundance of those with higher masses is relatively large, compared to the initial mass function elsewhere. Second, an important fraction of the stars that form near the center has recently been born into massive stellar clusters. I discuss these findings and argue that the same considerations apply to most, if not all gas-rich spiral galaxies. In the mass budget of the Galactic center, star formation is apparently the dominant sink.

1 Introduction

In the potential well of a galactic nucleus, where the gas density, the energy density, and the interstellar pressure are all large relative to a galactic disk, the conditions for star formation are quite different than one finds elsewhere. Perhaps because interstellar gas in a galactic nucleus has almost nowhere else to go but into star formation, if not deeper into the well, the surface density builds up until it reaches a steady state with the star formation process. The process is presumably regulated by the rate at which gas can be brought into the central molecular zone (CMZ). The best understood example is the CMZ of our Galaxy, which shows a very pronounced gas concentration, with about 5×10^7 $M_\odot$ of molecular gas concentrated into a region of radius less than 200 pc. In this region, star formation is apparently quite active, although we see only the high-mass stars, and note that they have a propensity to form in rather spectacular clusters. In contrast, star formation is only weakly in evidence near the nucleus of M31, but in that case, it is because of the relative dearth of gas there. In this paper, we are not concerned with the mechanisms such as mergers and angular momentum loss processes which bring gas into a galactic nucleus, but rather with the question of how stars can form once the gas is present. One clear determinant of star formation seems to be the molecular cloud surface density. Other factors include the internal velocity dispersion and the magnetic field strength within clouds, the effects of tidal forces, and the large pressure of the interstellar medium. In this paper, I first discuss each of these effects and then their overall implications for the mode of star formation and the initial mass function. Then, the formation of compact, massive star clusters is considered; it seems likely that they form in relatively sudden cataclysms. Finally, I examine the mass budget of the galactic center reservoir of gas, concluding that star formation is probably the predominant mass sink.

2   Factors Playing a Role in Star Formation near a Galactic Nucleus

2.1   The Surface Density of Clouds

The surface density of molecular gas, Σ_{H_2}, is a predictor of the star formation rate, SFR, not only because the presence of more gas implies more fuel for star formation, but also because, for a fixed scale height of the gas, a greater surface density of molecular clouds implies more frequent cloud collisions – a likely trigger for star formation. In addition, the effectiveness with which star formation and young, massive stars can stimulate further star formation in nearby gas via stellar winds, collimated protostellar outflows, supernovae and radiation pressure is presumably a strong function of how much gas is nearby, $i.e.$, of Σ_{H_2}.

Overall, the dependence of SFR upon Σ_{H_2} can be parameterized in terms of the Schmidt Law, $\Sigma_{SFR} \, \alpha \, \Sigma_{H_2}^{\beta}$, where Σ_{SFR} is the surface density of star formation, and β is found empirically to be ~1.4 [18] for galactic nuclei. The SFR for the nuclear region of the Milky Way has been inferred from the combination of the total far-infrared and radio continuum luminosities [10,2]. The bolometric luminosity in the central $3° \times 2°$ is 10^9 $L_\odot$ and the 5 GHz radio flux (2500 Jy) implies a total of ~ 10^{52} Lyman continuum photons. The implied SFR is 0.3 (0.1 – 0.6) $M_\odot$ yr^{-1}, or about 10% of the SFR of the entire galaxy. Thus, the Milky Way fits reasonably well onto the Schmidt Law for galactic nuclei, if one adopts a total mass and radius of the CMZ as 5×10^7 $M_\odot$ and 200 pc, respectively. In the plot of the Schmidt law given in [18] (Fig. 7), the value of Σ_{SFR} is about a factor of 3 above the mean for its range of Σ_{H_2}, but is still within the distribution of values exhibited by other galaxies. I caution that the SFR rate used in this analysis is quite uncertain, however, both for our Galactic center and for other galactic nuclei. In particular, the conversion of radio flux and luminosity into stellar mass has been made using a standard initial mass function (IMF), whereas the true IMF in galactic nuclei probably favors massive stars ([6] and below) and may have an elevated lower mass cutoff [27]. Consequently, since massive stars dominate both the luminosity and the Lyman continuum flux, it should not be surprising if the actual SFR is lower than that presented here. If the SFR in the CMZ is only about 0.1 $M_\odot$ yr^{-1}, then it agrees better with other galaxies in the Schmidt Law plot.

2.2   Large Internal Velocity Dispersion of Clouds

The velocity dispersion measured for molecular clouds in the CMZ, even at the highest spatial resolutions obtainable, has been found to be uniformly much larger than that of clouds in the Galactic disk. Typical measured linewidths (full width at half maximum) are 25 km s^{-1} [10]. Comparable or even larger linewidths are measured toward nearby galactic nuclei, but those linewidths are usually broadened by unresolved galactic dynamics. Only in almost face-on systems like IC342 [26] can we infer that a large velocity dispersion within clouds near the nucleus is probably a general phenomenon. While molecular clouds in the CMZ are unusually warm, 50 – 80 K ([31,16] in our Galaxy, and, *e.g.*, [12,15] in others), the observed linewidths are clearly not thermal.

A variety of sources for the internal cloud velocity dispersion can be imagined, including stellar winds, cloud collisions, tidally-induced shear flows, Alvén waves, and magnetosonic waves. The strong magnetic fields found in the Galactic center lend credence to these latter two possibilities, although direct measurements of the magnetic field strengths within clouds of the CMZ population have yet to be made.

The effect of the large velocity dispersion, Δv, can be gleaned from its effect on the Jeans mass, $M_J \propto \Delta v^3 / \rho^{1/2}$. For $\Delta v = 15$ km s^{-1} and molecular hydrogen density $10^4 n_4$ cm^{-3}, $M_J = 2 \times 10^5 \ M_\odot n_4^{-1/2}$. This rather large value corresponds better with the masses of the compact young clusters near the Galactic center than it does with individual stellar masses, but it may be indicative of a propensity for relatively massive stars, compared to star formation in the much more quiescent clouds of the Galactic disk. The velocity dispersion is relevant to the Jeans mass only if the scale of the velocity fluctuations is less than the Jeans length, 13 pc $n_4^{-1/2} \times (\Delta v / 15 \ km \ s^{-1})$, a condition which is satisfied observationally.

An alternative way of describing the effect of the large velocity dispersion in CMZ clouds is to note that the pressure of the interstellar medium near the Galactic center is 2 or 3 orders of magnitude higher than that in the disk. Spergel & Blitz [41] argue that the turbulent pressure in clouds is likely to be in equilibrium with the thermal pressure of the X-ray-emitting coronal gas in which Galactic center clouds are bathed [46]. The high pressure may impede star formation if the Jeans mass is larger than the cloud clumps. In any case, the high pressure environment alters the boundary conditions for star formation, just as it does in cooling flows and elliptical galaxies [4,17].

2.3   The Magnetic Field

Current evidence indicates that the magnetic field strength within at least the inner $\sim$100 pc is on the order of a milligauss, and that the large scale field geometry is dipolar. These conclusions are based on the presence of about a dozen non-thermal radio filaments (NTFźs) which have been found there [28,29,23,22,33], several of which are shown in Fig. 1. These NTFźs, typically 50 – 100 pc long and only $\sim$0.3 pc wide, are believed to be magnetic flux tubes which are part of a pervasive field, but which happen to be illuminated by the synchrotron emission from locally injected relativistic electrons [37]. The predominant orientation of the NTFźs perpendicular to the Galactic plane, and in many cases

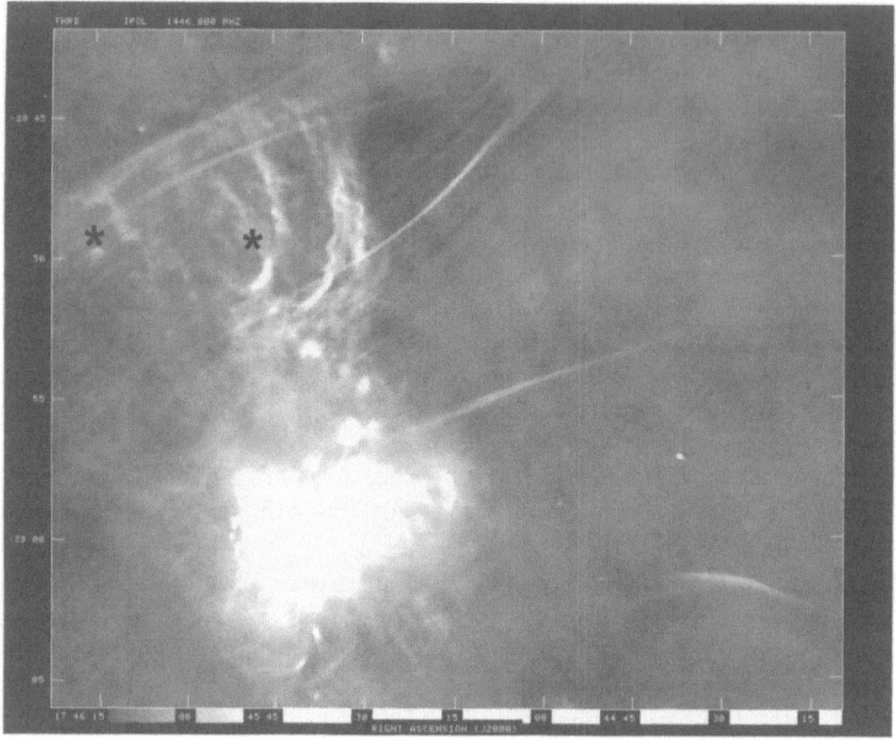

Fig. 1. Radiograph of 20cm emission intensity from the Galactic center, from [23]. The asterisks mark the locations of the Quintuplet (left) and Arches (right) clusters. The vertical extent of the region shown is 60 pc. The galactic plane is oriented in position angle -30°. The nonthermal radio filaments which presumably delineate the magnetic field are the thin structures oriented largely perpendicular to the Galactic plane. Several other sites of star formation are evident as HII regions to the north and east of the saturated SgrA complex, located at the bottom left of the figure.

passing through the plane, suggests the globally dipole geometry of the field. The curvature of the NTFźs and the progression of orientation with galactocentric distance are consistent with a divergence of the field on a few-hundred-parsec scale. The milligauss field strength is estimated from the strikingly smooth curvature of the NTFźs, in spite of the expected tumult of the interstellar medium in which they are embedded, and in spite of the large velocity dispersion and the presumably large relative motion of the clouds with which they are apparently interacting. The resistance of the NTFźs to deformation implies a rigidity which yields a lower limit to the field strength.

The origin of the central dipole field may lie in a natural and inevitable process, and may be common to all spiral galaxies with a sufficiently large gas content. The following model, originally outlined by Sofue & Fujimoto [40], and further developed by Morris [28], has recently been considered in some detail by

Chandran et al. [1]: gas flows radially inward toward the nucleus from throughout the Galaxy as it loses angular momentum by a variety of processes, including torquing and shocking by the Galactic bar, spiral density wave shocks, galactic mergers, and mass accumulation from a low-angular-momentum halo. The vertical component of the protogalactic field is therefore concentrated at the center, even allowing for ambipolar diffusion, because the rate of outward radial diffusion of this component cannot compete with the inward flux of the matter to which it is largely frozen. The horizontal component is much more complex, because, while it can indeed diffuse outward through the thickness of the galactic gas layer in a Hubble time, it can be amplified by a variety of dynamo processes, including simple galactic shear resulting from differential rotation. Shear can also tap the vertical flux to produce horizontal flux, but overall, the vertical flux is conserved. Its fate is to be concentrated toward the center until some quasi-static equilibrium is established between the dynamical pressure of incoming gas and the reverse magnetic pressure gradient of the central vertical field. Inasmuch as this model is applicable to all spiral galaxies, the implications are profound for activity in galactic nuclei: AGN activity and star formation both occur in the presence of a strong magnetic field.

How does the milligauss field affect star formation in the Milky Way? First, it is instructive to examine the magnetic Jeans mass for magnetic flux density B: $M_J(B) \alpha B^3/n^2$. Again, ignoring the anisotropy of the Jeans length in this case, $M_J(B)$ is on the order of 10^5 $M_\odot$ for B $\sim$ 1 milligauss, reflecting an approximate equipartition of magnetic and kinetic energies within clouds of the CMZ. If the clouds' velocity structure is dominated by magnetohydrodynamic waves, then this result is equivalent to that arrived at in the above discussion of the velocity dispersion. The second conclusion that one must face is that the details of the star formation process itself (angular momentum transport, etc.) are strongly affected by the field. Finally, it should not be too surprising if the relatively massive stars which form are themselves highly magnetized.

2.4 Tidal Forces

The stability of clouds near the Galactic center against tidal shear requires n > 10^7 cm^{-3} $(1.6 \ pc/r)^{1.8}$, for galactocentric radius r ([11] this applies to clouds which are more or less on circular orbits). For example, the parent cloud of the Arches cluster, which lies at $r \sim 30$ pc, would need to have had a density exceeding 5×10^4 cm^{-3} in order to remain bound. This condition selects for star formation out of denser gas than might otherwise be required, and insofar as the magnetic field is flux frozen to slightly ionized molecular gas as it evolves to become denser, giving B α $n^{2/3}$, the magnetic Jeans mass is unaffected. Also, while tidal shear may prevent a cloud in the CMZ from being gravitationally bound, the unbound gas does not leave the region. Much of the gas in the CMZ appears to be present in the form of streams of orbiting gas (*e.g.*, [39]) which, while maintaining their integrity over a few dynamical times by being pressure-bound [41], remain gravitationally unbound until accumulation of ambient material and collisions between these streams enhance their density past the point of gravitational stability.

Tidal forces can also be compressive for clouds moving inwards on predominantly radial orbits, because the clouds are moving into a converging gravitational field. In this case, the magnetic Jeans mass rises toward the center because the density and the magnetic flux density both rise as r^{-2}. Examples of approximately radially infalling gas streams can be seen in SgrA West, where the expected compression is perhaps demonstrated by the morphology of the leading tips of the infalling gas streams [30]. The dynamical time of the SgrA West features is probably too short ($\sim 10^4$ yr) to allow for star formation, but we note that star formation in radially infalling clouds located further out would give rise to stars on highly eccentric orbits. Some of the orbits of young stars in the central parsec [9] are consistent with star formation having occurred in radially infalling clouds.

2.5   Some Implications for Star Formation near a Galactic Nucleus

The ensemble of the effects described above implies a tendency for galactic nuclei to favor high mass star formation, relative to star formation further out in a galactic disk [27]. Thus, one might anticipate that the IMF is flatter in the CMZ of a galaxy than elsewhere. The existence of a lower mass cutoff to the IMF has also been considered as a possible consequence of the above effects (see discussion and references in [27]), although direct evidence of that is currently lacking. Observationally, the results for the IMF are currently mixed. For a power-law IMF, with $dN/dm \sim m^{-\alpha}$ (where $\alpha = 2.35$ is the Salpeter slope), one finds $\alpha = 1.5$ to 2.5 describing galactic nuclei. The massive, compact young clusters near the Galactic center have a relatively flat mass function, although they have already suffered a significant amount of dynamical evolution [19,20]. For further discussion of the IMF, I refer the reader to the review of Figer in this volume [6].

Although the observations are not definitively thorough, some prominent clouds in the CMZ of our Galaxy show little or no star formation. The "50 km s^{-1} cloud" is an interesting case; it harbors massive star formation along a dense ridge near its western extremity, as evidenced by the G-0.02-0.07 HII regions [3]. The compression of this ridge seems likely attributable to the expansion of the powerful SgrA East nonthermal shell source, presumably a supernova remnant (reference [24] and references therein). But the rest of this massive cloud has no known compact HII regions, IR-luminous embedded sources, H_2O masers or any other signs of star formation. By contrast, the less massive and somewhat less dense Orion Molecular Cloud has many apparent sites of star formation along its length. Thus, the "50 km s^{-1} cloud" is apparently impeded from forming stars in a piecemeal fashion, except where it has been subjected to a violent external shock. One might posit that this is the kind of powder keg that can form a starburst cluster if the shock is of sufficient strength, or if the cloud undergoes a global instability after growing to sufficient size. Another very massive molecular cloud in the CMZ, SgrB2, may now be undergoing just such a fate, as it is forming a sizeable cluster of massive stars [25,7], possibly as a result of the collision of two large clouds [13].

3   Starburst Clusters in Our Galaxy and Elsewhere

One of the remarkable findings that has occured over the past decade concerning young stars in galactic nuclei is that massive, compact young clusters are abundant where star formation in occuring at a relatively high rate. This takes its most extreme form in starburst nuclei, where super-star clusters having masses ranging up to 10^6 $M_\odot$ are widely recognized to be commonplace (e.g., [14,35,36,43]). Giant compact HII regions found in local dwarf galaxy starbursts are another indication of the formation of extremely massive, compact clusters [44,45,21].

In our Galaxy, we have the smaller, but still spectacular Arches and Quintuplet clusters [5], and the cluster of young stars occupying the central parsec (e.g., [34]). These are the most massive young clusters in our Galaxy. The first two of these are doomed to tidal disintegration on a time scale less than 10^7 years [19,20], so one may conclude, with unimpressive statistical significance, that their rate of production is on the order of one every few million years, giving a massive cluster SFR of 0.01 $M_\odot$ yr^{-1}. This estimate assumes an initial cluster mass of $\sim 2 \times 10^4$ $M_\odot$ [20]. Comparison with the overall SFR of 0.3 $M_\odot$ yr^{-1} suggests that these massive clusters may be a minor contributor to the overall SFR of our Galactic center. However, in the nuclei of starburst galaxies, the formation of clusters may be the preferred mode of star formation (e.g., [43]). An interesting hypothesis to consider is that, of the mass undergoing star formation near a galactic nucleus, the fraction which goes into massive, compact clusters is larger for more intense starbursts. The fact that starburst clusters are so much more populous, massive and compact than other known incidences of star formation suggests that the circumstances of their formation were peculiar in some fundamental way, such as catastrophic formation by extremely strong shocks or direct collision of two dense molecular clouds. These factors are presumably enhanced in high-surface-density starburst galaxies.

The compactness and high mass of the Arches cluster or any other starburst cluster raises the issue of the time scale over which the cluster must have formed. The violence implied by the formation of hundreds of O stars within a few tenths of a parsec, including protostellar jets and winds and ionized gas flows at ionization fronts, is likely to quickly shut off star formation once the process begins. Indeed, these clusters may begin formation on the scale of a Jeans mass, and then fragment hierarchically to stellar masses on a free-fall time scale. If so, then there is little room for the persistence of straggler gas clumps; the released gravitational energy rushing outward from the star formation cataclysm will commit any gas clump to immediate collapse or to a quick oblivion via ionization and Kelvin-Helmholtz instabilities. Massive starburst clusters must be quite close to coevality, so their use as probes of the IMF should be little affected by a spread in stellar ages.

4   Star Formation and The Galactic Center Mass Budget

It is instructive to review the role of star formation in the overall mass budget of interstellar gas in a galactic nucleus. Indeed, it is the dominant sink for the gas in a galactic nucleus.

Mass accretes from the disk of our Galaxy at a rate which can be estimated by noting that the molecular ring which defines the outer edge of the CMZ and contains a total of $\sim 8 \times 10^6$ $M_\odot$ [39] lies at about the location of the inner Lindblad resonance (ILR), coinciding with the innermost X1 orbit [8]. The fact that gas on this cusped orbit cannot reside there for more than about one orbital period leads to an estimate of the mass inflow rate: $\dot{M}_{ILR} = 0.4$ (0.1 - 1) $M_\odot$ yr^{-1}. No direct (i.e., kinematic) evidence for this mass inflow is yet available, but it is inevitable in the face of the processes that cause angular momentum loss by the gas [32].

With such an inflow rate, we can estimate the residence time for gas in the CMZ: $\tau \sim M_{CMZ}/\dot{M}_{ILR} \sim 5 \times 10^8$ yr, which is comparable to the time scale for angular momentum loss by clouds within the CMZ as a result of dynamical friction [42] and magnetic viscosity [28].

While these estimates are crude, the apparently close correspondence between the star formation rate and the rate of gas inflow to the CMZ is intriguing, and may imply a steady state wherein star formation is sustained at its present rate over long periods of time [38]. Other contributors to the mass budget include (see ref. [32]): (1) the rate at which bulge stars shed mass which falls directly into the CMZ: a few times 0.01 $M_\odot$ yr^{-1}, (2) accretion onto and through the circumnuclear disk (0.03 – 0.05 $M_\odot$ yr^{-1}, (3) accretion onto the central black hole ($10^{-5} - 10^{-8}$ $M_\odot$ yr^{-1}), and (4) mass outflow in a thermally-driven coronal wind (0.03 – 0.1 $M_\odot$ yr^{-1}) [46]. The latter contribution is the most significant of these; if the SFR in the CMZ of our Galaxy should be lowered from the above estimates because of the altered IMF, then the mass loss rate in the coronal wind may be comparable to the rate of star formation. Forthcoming evidence from the Chandra and XMM Observatories should help clarify this point.

References

1. B.D.G. Chandran, S.C. Cowley, M. Morris: Ap.J. **528**, 723 (2000)
2. P. Cox, R. Laureijs: 'IRAS Observations of the Galactic Center'. In IAU Symp. 136: *The Center of the Galaxy*, ed. by M. Morris (Kluwer, Dordrecht 1989), p 121
3. R.D. Ekers, J.H. Van Gorkom, U.J. Schwarz, W.M. Goss: Astron. Ap. **122** 143 (1983)
4. A.C. Fabian: Ann. Rev. Astron. Ap. **32**, 277 (1992)
5. D.F. Figer, S.S. Kim, M. Morris, E. Serabyn, R.M. Rich, I.S. McLean: Ap.J. **525**, 750 (1999)
6. D.F. Figer: in this volume.
7. R.A. Gaume, M.J. Claussen, C.G. De Pree, W.M. Goss, D.M. Mehringer: Ap.J. **449**, 663 (1995)
8. O.E. Gerhard: Rev. Mod. Astron. **5**, 174 (1992)

9. R. Genzel, C. Pichon, A. Eckart, O.E. Gerhard, T. Ott: M.N.R.A.S. **317**, 348 (2000)
10. R. Güsten: 'Gas and Dust in the Inner Few Degrees of the Galaxy'. In IAU Symp. 136: *The Center of the Galaxy*, ed. by M. Morris (Kluwer, Dordrecht 1989), p 89
11. R. Güsten, D. Downes: Astron. Ap. **87**, 6 (1980)
12. R. Güsten, E. Serabyn, C. Kasemann, A. Schinckel, G. Schneider, A. Schulz, K. Young: Ap.J. **402**, 537 (1993)
13. T. Hasegawa, F. Sato, J.B. Whiteoak, R. Miyawaki: Ap.J.Lett. **429**, L77 (1994)
14. L.C. Ho, A.V. Filippenko: Ap.J. **472**, 600 (1996)
15. P.T.P. Ho, R.N. Martin, J.L. Turner, J.M. Jackson: Ap.J. **355**, 19 (1990)
16. S. Hüttemeister, T.L. Wilson, T.M. Bania, J. Martin-Pintado: Astron. Ap. **280**, 255 (1993)
17. M. Jura: Ap.J. **212**, 634 (1977)
18. R.C. Kennicut, Jr.: Ann. Rev. Astron. Ap. **36**, 189 (1998)
19. S.S. Kim, M. Morris, H.-M. Lee: Ap.J. **525**, 228 (1999)
20. S.S. Kim, D.F. Figer, H.-M. Lee, M. Morris: Ap.J. **545**, 301 (2000)
21. H.A. Kobulnicky, K.E. Johnson: Ap.J. **527**, 154 (1999).
22. C.C. Lang, K.R. Anantharamaiah, N.E. Kassim, T.J.W. Lazio: Ap.J.Lett. **521**, L41 (1999)
23. C.C. Lang, M. Morris, M., L. Echevarria: Ap.J. **526**, 727 (1999)
24. Y. Maeda et al.: Ap.J. submitted (2001)
25. D.M. Mehringer, P. Palmer, W.M. Goss, F. Yusef-Zadeh: Ap.J. **412**, 684 (1993)
26. D.S. Meier, J.L. Turner, R.L. Hurt: Ap.J. **531**, 200 (2000)
27. M. Morris: Ap.J. **408**, 496 (1993)
28. M. Morris: 'Magnetic Phenomena'. In *The Nuclei of Normal Galaxies: Lessons from the Galactic Center*, NATO ASI Series C: Vol. 445, ed. by R. Genzel, A.I. Harris (Kluwer, Dordrecht 1994), p 185
29. M. Morris: 'What are the Radio Filaments near the Galactic Center?'. In IAU Symp. 169: *Unsolved Problems in the Milky Way*," ed. by L. Blitz & P.J. Teuben (Kluwer: Dordrecht 1996), p 247
30. M. Morris, J.-P. Maillard: 'Kinematics of Ionized Gas in the Central Parsec of the Galaxy from High-Resolution Spectroscopy of the Brackett-γ Line'. In *Imaging the Universe in 3 Dimensions: Astrophysics with Advanced Multi-Wavelength Imaging Devices*, ed. by J. Bland-Hawthorn, W. van Breugel (2000), p 196
31. M. Morris, N. Polish, B. Zuckerman, N. Kaifu: Astron.J. **88**, 1228 (1983)
32. M. Morris, E. Serabyn: Ann. Rev. Astron. Ap. **34**, 645 (1996)
33. M. Morris, K.I. Uchida: in preparation (2001)
34. F. Najarro, A. Krabbe, R. Genzel, D. Lutz, R.P. Kudritzki, D.J. Hillier: Astron. Ap. **325**, 700 (1997)
35. R. W. O'Connell, J.S. Gallagher, D.A. Hunter: Ap.J. **433**, 65 (1994)
36. R. W. O'Connell, J.S. Gallagher, D.A. Hunter, W.N. Colley: Ap.J.Lett. **446**, L1 (1995)
37. E. Serabyn, M. Morris: Ap.J.Lett. **424**, L91 (1994)
38. E. Serabyn, M. Morris: Nature **382**, 602 (1996)
39. Y. Sofue: P.A.S.J. **47**, 527 (1995)
40. Y. Sofue, M. Fujimoto: P.A.S.J. **39**, 843 (1987)
41. D.N. Spergel, L. Blitz: Nature **357**, 665 (1992)
42. A.A. Stark, O.E. Gerhard, J. Binney, J. Bally: M.N.R.A.S **248**, 14p (1991)
43. L.E. Tacconi-Garman, A. Sternberg, A. Eckart: Astron. J. **112**, 918 (1996).
44. J.L. Turner, P.T.P. Ho, S.C. Beck: Astron. J. **16**, 1212 (1998).

45. J.L. Turner, S.C. Beck, P.T.P. Ho: Ap.J.Lett. **532** L109 (2000).
46. S. Yamauchi, M. Kawada, K. Koyama, H. Kunieda, Y. Tawara: Ap.J. **365**, 532 (1990)

Star Formation in Galactic Disks

Giuseppe Gavazzi

Università degli Studi di Milano - Bicocca, Milano, Italy

1 Introduction

This review is primarily aimed as a reminder to starburst people that, besides their favourite objects which are presently forming star at the exceptional rate of 100–1000 $M_\odot$/year, there exists a much broader class of "normal" galaxies at $z = 0$ which do form stars at rates ranging from 0 (E+S0) to 10 $M_\odot$/year (BCDs), with the vast majority of normal disk galaxies doing it at 1 $M_\odot$/year. Much of what is known about the star formation properties of normal galaxies can be found in the beautiful review paper by Rob Kennicutt [14]. The Kennicutt review allows me to skip over an endless number of interesting aspects, as for example how the star formation rate varies along the Hubble sequence, and to focus on few topics that remain relatively unexplored in the literature, namely: the comparative analysis of the old stellar population indicators (NIR imaging) (Section 2) with the ongoing star formation indicators (Hα imaging) (Section 3). I will also briefly discuss an application of the population synthesis method [4] to spectrophotometric data in order to constrain the star formation history (Section 4). Moreover I will focus my analysis on the significant dependence found between the star formation history and the systemic mass and show some examples of star formation history dependences on the environmental conditions (Section 5).

2 The Past Star Formation: NIR Imaging

During the last 5 years we have concentrated on extensive H band imaging campaigns using the 2m class telescopes TIRGO and Calar Alto. We have imaged approximately 1200 giant galaxies in the Coma supercluster [6,7] and in the Virgo cluster [1,2,10]. Recently we have extended the survey using the ESO/NTT and the italian TNG (Telescopio Nazionale Galileo, the twin brother of the NTT) to comprise dwarf galaxies in the Virgo cluster [12]. Examples of these recent sensitive measurements and of the modeled light profiles are given in Figs. 1 and 2. So far the Coma supercluster survey was completed to the Zwicky limit and a $\sim 70\%$ completeness was achieved at B < 15 in Virgo. Light profile decompositions were performed [11] using the classical Bulge+Disk (de Vaucouleurs + exponential) decompositions. Furthermore the model independent "light concentration index" (C_{31}) was determined.

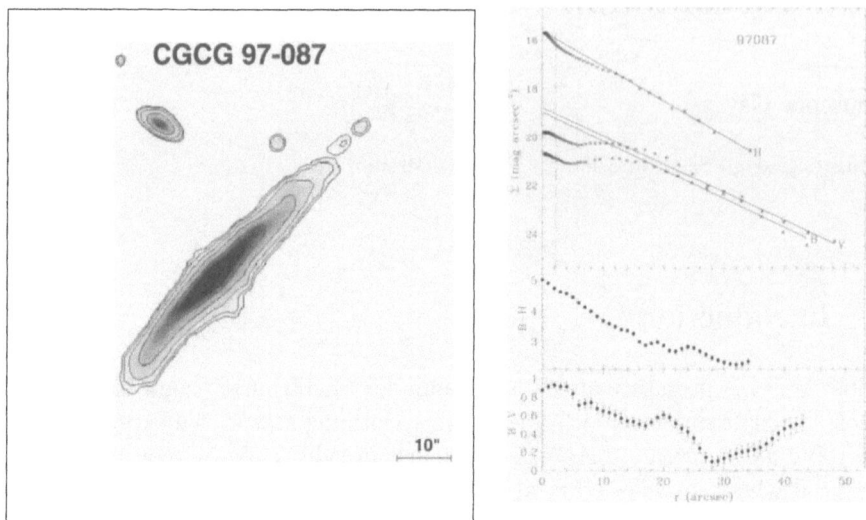

Fig. 1. The H band image of the pec galaxy CGCG-97087 in A1367 taken with the TNG (left)(contours are given from 21.0 in steps of 1.0 mag/arcsec^2) and the B, V and H light and color profiles (right). Notice the NIR central excess due to internal extinction.

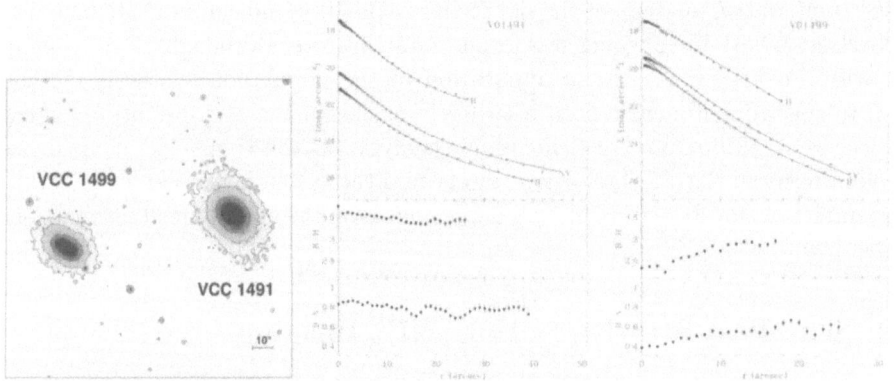

Fig. 2. The H band image of the dEs VCC1491 and VCC1499 in Virgo taken with the NTT (left) (contours are given from 21.5 in steps of 1.0 mag/arcsec^2) and their B, V and H light and color profiles (right). Notice the blue central excess in VCC1499.

2.1   The Mass-Luminosity Relation

Gavazzi et al. [8] have studied the dependence of the dynamical mass of disk galaxies on luminosity, as measured at different wavelengths. They have shown that the relation $M_{Dyn} \sim L^\alpha$ is wavelength dependent (see Fig. 3), being steeper than unity in the optical. For example in the U (B) band an increase of a factor of 10 in luminosity corresponds to a factor of 30 (20) increase in mass. Only at NIR bands (H = 1.65 μm) the M_{Dyn} vs. luminosity dependence is one of linear proportionality, i.e. the H band luminosity is a reliable tracer of the dynamical mass, with $M/L_H \sim 4.5$. This is due to both the greatly reduced dust obscuration, and to the fact that most of the galaxy mass sits in the old stellar population traced by NIR observations. In the remaining I will use H luminosity and mass as synonymous.

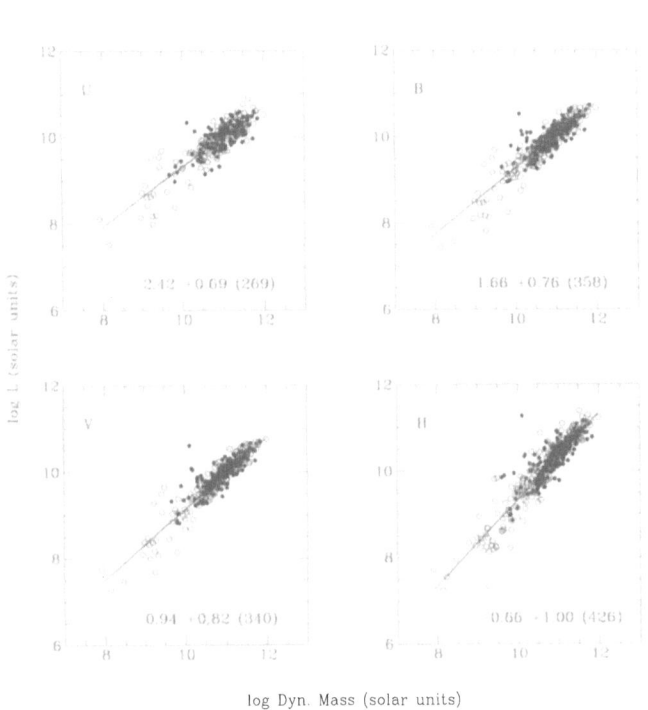

Fig. 3. The local Mass-Luminosity relation of disk galaxies at $z=0$ (from [8]). The panels carry the relations of Dynamical Mass vs. U, B, V and H total luminosity respectively. Linear fits to the observed relations are given.

2.2   The Frequency of H Band Light Decompositions

Figure 4 shows the structural properties of the old stellar population in galaxies as a function of systemic mass. Low luminosity (dwarf) galaxies have exponential profiles, while the fraction of B+D decompositions increases with luminosity. At intermediate luminosities ($L_H = 10^{10}\,L_\odot$) the two have an equal frequency (about 50%). The pure de Vaucouleurs profiles are absent below $L_H = 10^{10}\,L_\odot$ and become dominant only at the highest luminosities. The dependence of the profile decomposition on luminosity is basically morphology-independent. In fact a pattern similar to that of Fig. 4 is found subdividing the whole sample in two broad type classes: the E+S0+S0a (Early) versus the Spirals (Late), and even in two narrow Hubble type classes: Elliptical versus Sc galaxies.

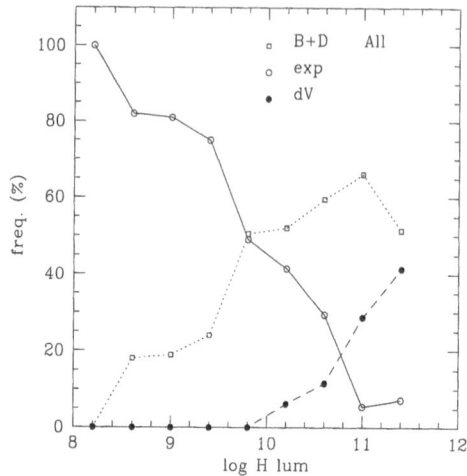

Fig. 4. The fraction of pure de Vaucouleurs, pure exponential and mixed profile decompositions as a function of the NIR luminosity.

2.3   The Photometric "Cube"

Figure 5 shows the cube defined by H luminosity, color (B-V) and the NIR light concentration index C_{31}, the fourth dimension being the morphological type (symbols). In the traditional scenario (e.g. [16]) these four parameters should not be independent of one another, i.e. the color and the concentration index should be predictable from the morphological type.

There is a strongly non-linear dependence of C_{31} on luminosity: all galaxies develop conspicuous cusps ($C_{31} > 4$) only at high luminosity, with the exception of high luminosity Scs which are bulgeless.

Galaxies obey two (separate) color-luminosity relations. Ellipticals follow a shallow relation governed by metallicity (dEs form a continuum with giant Es). Disk galaxies have a steeper color-magnitude relation that indicates a stellar population sequence with mass (see [8]).

All bulge dominated galaxies ($C_{31} > 4$) are red ($B - V > 0.8$), while the reverse is not true: $C_{31} < 4$ objects (disks) span all colors from $0.2 < B - V < 0.8$. Exceptions are the dEs which are red in spite of their low C_{31}. All dwarf galaxies (dE + Im) have low C_{31} regardless of their type, but the two type classes are segregated by color. This might result from fundamental dynamical differences between rotationally supported "cold" systems (dIrr), and "hot" systems supported by random motions (dwarfs spheroidals). The very existence of such relations among photometric parameters of the old stellar population should provide us with a strong constraint for models of galaxy evolution.

3   The Ongoing Star Formation: Hα Imaging

We have approached the study of the ongoing (10^6–10^7 yrs), massive (5–10 M$_\odot$) star formation by undertaking an extensive imaging survey of disk galaxies in the

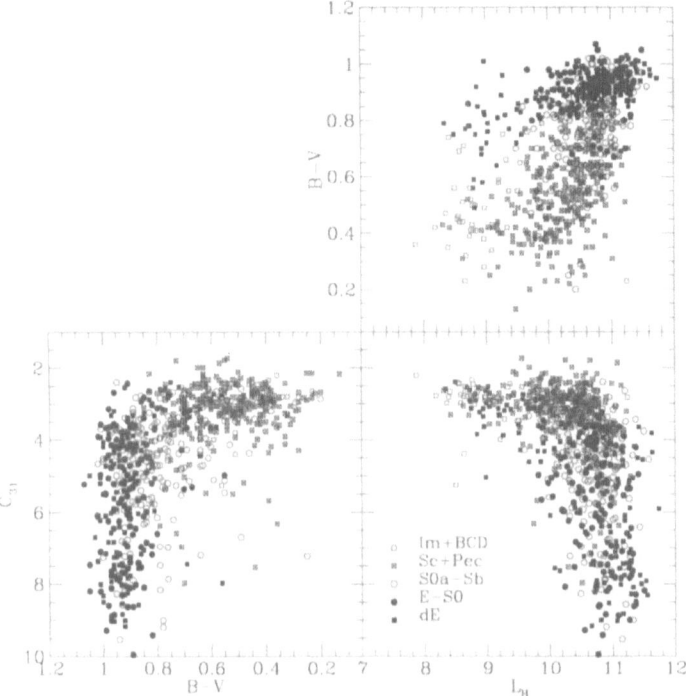

Fig. 5. The Luminosity-color-C_{31} "cube".

light of the Hα line [9,12]. We have so far imaged approximately 550 galaxies. The measurements of Coma supercluster spiral galaxies was completed up to B < 15.6 (see two examples in Fig. 6), while in Virgo we have a subsample 85% complete at B < 16.

The main result of the Hα survey is that the current star formation rate is found to anti-correlate with the total H band luminosity (see Fig. 7). Massive galaxies are presently forming stars at a rate ~ 100 times lower than dwarf galaxies. This is consistent with the prediction of a simple closed-box model (see line in Fig. 7), if it is assumed that the galaxy initial mass regulates the efficiency of collapse. Massive objects had an efficient initial collapse, producing an early total consumption of their gaseous content, while small perturbations underwent a much more slow collapse, thus their gas consumption rate was slower, and lots of gas remained available for sustaining significant star formation at $z = 0$.

4 The Star Formation History: Spectrophotometry

Since 1998 we have been undertaking a spectrophotometric survey of Virgo galaxies using the OHP 1.93m and the ESO/3.6 m telescopes, with the aim of constraining their star formation histories (see [3]). We use the drift-scanning

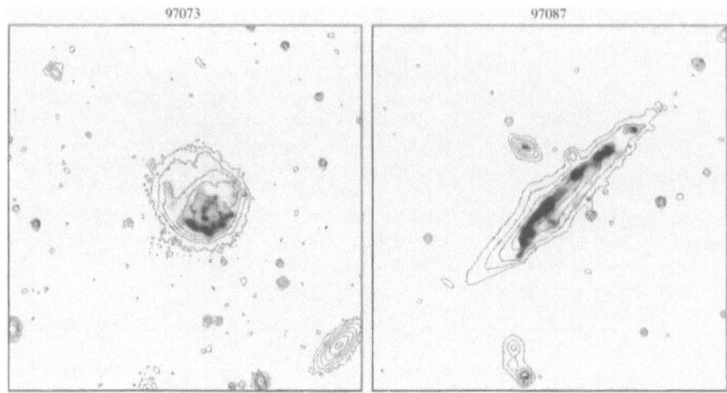

Fig. 6. Hα images of two Irr/Pec galaxies in A1367. The contours represent the underlying stellar continuum, while the Hα net flux is given by the grey scale.

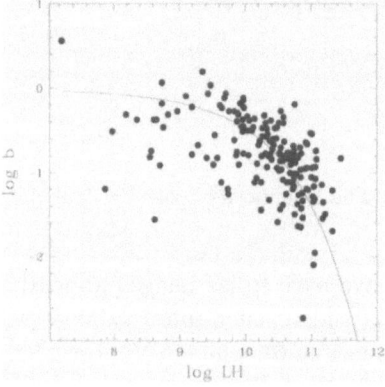

Fig. 7. The relation between the current star formation rate and the H band luminosity (mass) for the sample of isolated galaxies plus HI "healthy" (def HI < 0.2) cluster galaxies. The line represents the prediction of a simple closed-box model.

technique (the source is trailed across the long slit as in [15]) to obtain measurements representative of the whole objects, without priviledging the central regions. So far we have obtained 57 spectra trying to cover as broadly as possible a range of Hubble types (from E to BCD) and luminosity ($-16 < M_B < -20$). The spectral domain is 3800–7000 Å, at a resolution of 1000 at Hα (see Fig. 8).

We complement the optical spectroscopy with broad band photometry (from UV to NIR K) and fit the energy distributions with Bruzual & Charlot ([4]; 2000 release) synthesis population models (see Fig. 9).

Assuming a Salpeter IMF and an exponential star formation history we leave the exponential time-scale τ as the free parameter. We obtain reasonable fits as-

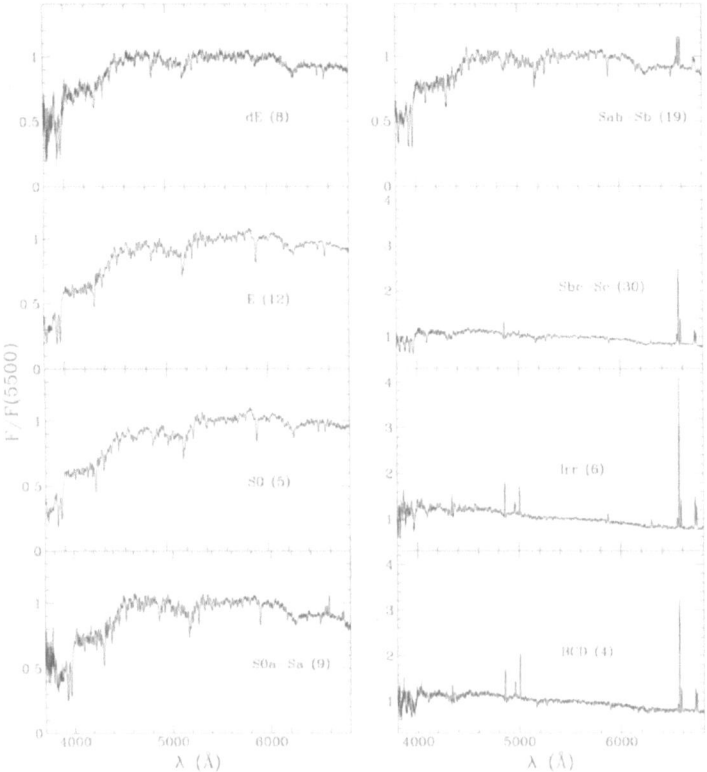

Fig. 8. Template spectra obtained combining several (number in parenthesis) spectra in bins of Hubble type. Each spectrum was reduced to the rest-frame wavelength before combination.

suming that galaxies are approximately coeval (T = 12 Gyrs) and with τ varying from 1 Gyr (quasi instantaneous burst) to 12 Gyrs (continuum star formation). Again (see also Section 3) we find a strong anti-correlation between τ and the system H luminosity (see Fig. 10), indicating that the star formation history is regulated by the total galaxy mass.

5 The Role of the Environment

To test possible environmental dependences of the star formation history of disk galaxies we compare in Fig. 11 (left panels) the present star formation rate (as obtained from Hα E.W.) in various environments (Coma supercluster versus Virgo cluster). The two are found significantly different: the average star formation rate of Virgo spirals, at any H luminosity, is a factor of ~ 10 below the corresponding SFR of spirals in the Coma region, in agreement with [13].

When galaxies with normal HI content (top-right panel of Fig. 11) are compared with HI deficient galaxies (defined by [5] as the expected HI mass from

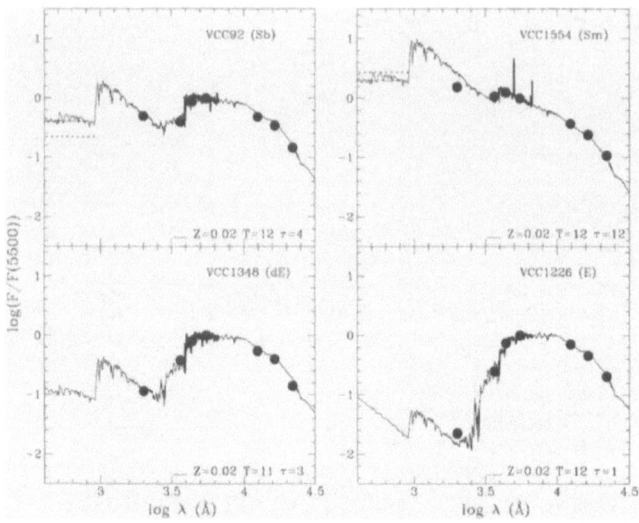

Fig. 9. Energy distributions of 4 galaxies (E, dE, Sb, Irr) fitted with Bruzual & Charlot population synthesis models. Assuming an exponential star formation history, we obtain satisfactory fits with an age $T \sim 12\,\mathrm{Gyrs}$ and $\tau = 1,\ 3,\ 4,\ 12\,\mathrm{Gyrs}$, respectively.

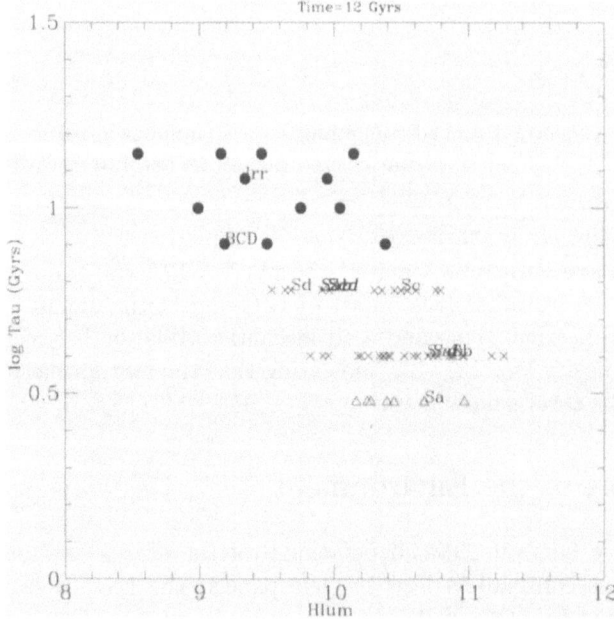

Fig. 10. The inverse relation between τ and the H band luminosity obtained by fitting the observed energy distributions with Bruzual & Charlot synthesis population models.

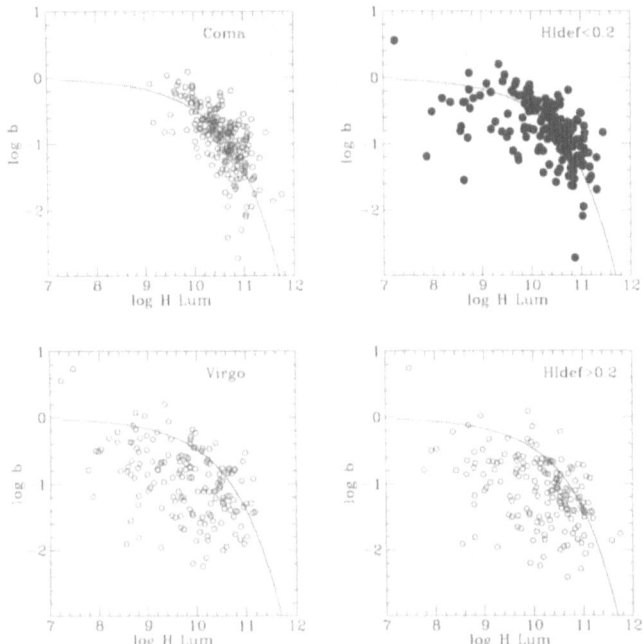

Fig. 11. The relation between the current star formation rate and the H band luminosity (mass) (same as Fig. 7) for galaxies in the Coma supercluster (top-left) and in the Virgo cluster (bottom-left). HI "healthy" (def HI < 0.2) galaxies are given in the top-right panel, while HI deficient objects are given in the bottom-right panel.

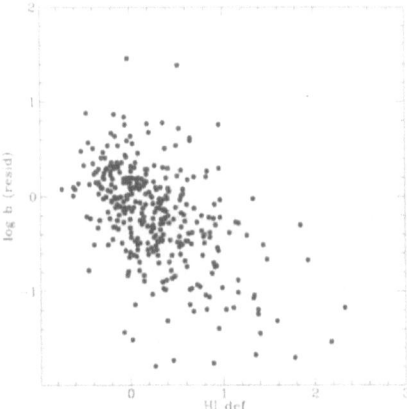

Fig. 12. The correlation between the residual Hα from the "normal" (see Figs. 7 and 11) and the HI deficiency. This figure indicates that the amount of primeval gas regulates the current star formation rate in galaxies.

unperturbed galaxies of similar size and type divided by the measured HI mass) (bottom-right panel of Fig. 11) it appears that the two have distinct star formation properties. This is best evidenced in Fig. 12 where the Hα "excess" (the difference between the measured (dots) and modeled (line) Hα E.W.) is plotted as a function of the HI deficiency. It appears that strongly HI deficient objects have star formation rates quenched by a factor of 10 with respect to HI "healthy" spirals. This quantifies the role of the environment in regulating the star formation history of galaxies and perhaps gives us a clue for understanding the morphology segregation observed in clusters of galaxies.

Acknowledgments: I wish to thank C. Bonfanti, A. Boselli, P. Franzetti, P. Pedotti, M. Scodeggio and S. Zibetti for their contribution to the present study.

References

1. A. Boselli, R. Tuffs, G. Gavazzi, H. Hippelein, D. Pierini: A&AS **121**, 507 (1997)
2. A. Boselli, G. Gavazzi, P. Franzetti, D. Pierini, M. Scodeggio: A&AS **142**, 73 (2000)
3. G. Bruzual: astro-ph 0011094 (2000)
4. G. Bruzual, S. Charlot: ApJ **405**, 538 (1993)
5. M. Haynes, R. Giovanelli, AJ, 89, 758 (1994)
6. G. Gavazzi, D. Pierini, A. Boselli, R. Tuffs: A&AS **120**, 489 (1996)
7. G. Gavazzi, D. Pierini, C. Baffa, F. Lisi, L. Hunt, A. Boselli: A&AS **120**, 521 (1996)
8. G. Gavazzi, D. Pierini, A. Boselli: A&A **312**, 397 (1996)
9. G. Gavazzi, B. Catinella, A. Boselli, L. Carrasco: AJ **115**, 1745 (1998)
10. G. Gavazzi, P. Franzetti, M. Scodeggio, A. Boselli, D. Pierini, C. Baffa, F. Lisi, L. Hunt: A&AS **142**, 65 (2000)
11. G. Gavazzi, P. Franzetti, M. Scodeggio, A. Boselli, D. Pierini: A&AS **361**, 863 (2000)
12. G. Gavazzi, S. Zibetti, A. Boselli, P. Franzetti, M. Scodeggio, S. Martocchi, submitted to A&A (2000)
13. R. Kennicutt: AJ **88**, 483 (1983)
14. R. Kennicutt: ARA&A **36**, 189 (1998)
15. R. Kennicutt: ApJS **79**, 255 (1992)
16. B. Whitmore: ApJ **278**, 61 (1984)

Nearby Dusty Starbursts

George H. Rieke

Steward Observatory, University of Arizona, Tucson, AZ 85750, USA

Abstract. Virtually all young starbursts are dusty, and hence can be understood only with a suite of techniques not strongly affected by extinction. Synthesis modeling developed around such techniques provides satisfactory fits to all the relevant observables. These models show that starbursts are short-lived and have extremely high star forming efficiency. Further advances will require improvements in the way boundary conditions such as bolometric and K-band luminosity, ionizing star effective temperature, and supernova rate are estimated and applied in the models.

1 Introduction

Virtually all young starbursts are dusty to some extent, particularly when they occur in large, metal-rich galaxies. Understanding starbursts therefore requires techniques to model their properties in dusty environments. The visible-UV region is likely to be optically thick in interstellar reddening, as indicated by a deduced extinction of about one magnitude independent of wavelength. When this situation holds, measurements penetrate only one optical depth into the starburst, and the true reddening is substantially underestimated. Consequently, the total output of the starburst is underestimated, and if there is a gradient of conditions the true nature of the starburst may escape observation. Thus, studies of dusty starbursts must emphasize spectral regions not strongly affected by reddening, particularly the infrared.

Infrared spectra of galaxies are dominated by cool giants and supergiants, rather than the range of spectral types that contribute to optical spectra. It appears that the spectral features from these stellar populations to a large extent vary *together* with metallicity, stellar temperature, and surface gravity. Even complete spectra can reveal no more than a few aspects of the stellar population. Thus, pure spectral synthesis models have little leverage to constrain the overall stellar population. Fortunately, an alternative approach based on global properties of the starburst can probe the conditions in considerable detail [22]. Boundary conditions for these models are: 1.) bolometric luminosity; 2.) luminosity in cool stars (e.g., K-band); 3.) first overtone CO band strength ($2.3\mu m$); 4.) ionizing flux; 5.) hardness of ionizing spectrum; 6.) dynamical mass; and 7.) supernova rate. The extinction is so large in many cases that even some of these infrared-oriented indicators must be corrected carefully.

In this review, I first discuss the boundary conditions for starburst models, with attention to the weaknesses in current methods to determine them. I will then summarize the general conclusions that have been reached about dusty starbursts.

2   Models for Dusty Starbursts

The various constraints discussed above are used as boundary conditions for evolutionary synthesis models. In practice, a model is generated using an assumed history of star formation and initial mass function, and the characteristics of the model are compared with those of the galaxy under study. This comparison depends on correcting the observations accurately for extinction and then matching with the boundary conditions from the models.

2.1   Extinction Levels

How far into the infrared is the extinction optically thick? The best probe is the hydrogen recombination lines, whose strengths can be readily measured out to Brα at 4.06μm, and in favorable cases in the high frequency radio regime. The relative strengths are given by Case B theory, with only a weak dependence on electron temperature. Even in galaxies with evidence for extremely heavy obscuration, comparisons of the H recombination line strengths suggest total extinctions of $A_V = 10$–50, indicating that the ISM in these galaxies becomes optically thin or nearly so by the Brα line (4.06μm) and frequently by Brγ (2.16μm) (e.g., [3,4,18]). Nonetheless, the extinction can only be treated in an approximate way, due to the complexity of the distribution of interstellar material in the starburst and to uncertainties in the extinction law (see [6,12]). Most studies of starburst reddening preferentially address galaxies with detectable ultraviolet fluxes, whose dust properties may not be representative of the dust in objects with denser interstellar media and greater extinction. In addition, the scattering properties of the dust can be important for the models, since scattered light is generally captured in the measurement beam unlike the situation for isolated stars used in determining the extinction law. Measuring the albedos of dust in external galaxies is a challenge [42].

The extinction to the stars is likely to be different from that to the gas. Fortunately, evolved stellar populations have very similar near infrared colors. In starbursts, the extinction affects these colors so strongly that the small possible intrinsic variations can be ignored. However, extinction corrections are subject to the same uncertainties due to geometry and extinction laws mentioned above.

The huge extinction values deduced from mm-wave CO emission luminosities appear not to be characteristic of these sources. There are a number of possibilities for this discrepancy. From the widths of the CO lines, Solomon et al. [30] compute dynamical masses and find them on average to be only $1/3$ the masses of molecular gas deduced from the "standard" ratio of CO luminosity to molecular gas mass. The conversion factor must be adjusted by more than a factor of three to allow for the total mass budget of the galaxy nucleus, which must include both the mass of new stars in the starburst and also the mass of old stars from the pre-existing galaxy. Thus, the mass of molecular mass is probably substantially smaller than has been assumed to date. In addition, the gas may be distributed in ways that reduce its effectiveness in overall extinction – for example, in a compact dense cloud immediately around the nucleus rather than generally distributed throughout the starburst region.

2.2 Bolometric Luminosity

To a large extent, the bolometric luminosities of dusty starbursts can be set equal to their far infrared luminosities. Some energy must escape the galaxy before being absorbed by the interstellar dust, so the far infrared value should be interpreted as a lower limit. It may be possible to quantify the typical amount of escaping luminosity through relations such as those discussed in this conference by Meurer, relating the UV luminosity of modestly obscured starbursts to their bolometric luminosities, but such estimates have not yet been applied.

2.3 Cool Star Luminosity

Determination of the K-band luminosity is subject to the problems due to the heavy extinction to the stars and thus is subject to significant uncertainties. A variety of plausible geometries for the extinction can give similar estimates of the intrinsic K-luminosity of a starburst galaxy [18]. However, frequently only a foreground screen model, perhaps point-by-point, is used. There is no clear understanding of the range of systematic errors that may arise either with foreground screens or more sophisticated models.

2.4 CO Band Strength

The composite CO band strength of a population of massive stars grows monotonically over the first 10 Myr of evolution. Although in principle use of this behavior might be confused through the CO band dependency on metallicity, it appears that the dependency on increasing numbers of red supergiants as the starburst ages is dominant. Most other strong spectral features in the near infrared have a similar behavior to the CO band, so they cannot be used easily to break the degeneracy with metallicity or to derive other information about the starburst. Another complication might arise through dilution of the CO band strength by emission by hot dust associated with the starburst. However, Ridgway et al. [21] examined this possibility and found it to be an issue only for the most extreme cases – either where there is an AGN that produces a very strong hot dust emission component, or where the galaxy is exceptionally red in the near infrared due to a combination of very high extinction and emission by hot dust. Thus, in general the CO band strength is a reliable indicator of the age of a very young starburst. Beyond 10 Myr, however, the CO band strength saturates, or even eventually decreases, so its use for this purpose becomes much more complex.

2.5 Ionizing Flux

The high extinctions make it impossible to obtain an accurate estimate of the intrinsic hydrogen recombination from optical measurements, but it appears that plausible values can be determined by dereddening the infrared line intensities – sometimes data to Brγ will suffice, and even in extreme cases data to Brα are reasonably reliable. However, starburst models depend on converting these values into true Lyman continuum strengths. This conversion requires an estimate of the portion of the Lyman continuum absorbed by interstellar dust and hence ineffective in ionizing hydrogen.

Determining the extent of absorption of ionizing flux by dust has proven problematic in Galactic HII regions, and even more so in the more remote and complex starbursts. The issue is discussed by Smith et al. [29]. They point out that an indication of the absorption efficiency could in principle be derived from the infrared excess (IRE), $L_{IR}/L_{Ly\alpha}$. If dust absorbs a large portion of the Lyman continuum luminosity, one expects a large value of the IRE. Smith et al. find that starbursts have a much smaller IRE than is typical for Galactic HII regions. Although this result might indicate a relatively small portion of the Lyman continuum is absorbed by dust in starbursts, Smith et al. prefer an explanation in terms of a larger portion of high mass stars than in Galactic HII regions. At present, all we can assume is that the Lyman continua deduced from the H recombination lines are lower limits.

2.6 Hardness of Ionizing Spectrum

Strong lines suitable for determining the effective temperature of the ionizing stars in a starburst are relatively rare even in the rich optical spectra. Often, the ratio of [OIII] 5007/Hβ is employed, or more effectively used together with [OI] 6300/Hα. However, measures that combine a forbidden and a H recombination line are subject to biases such as the effects of electron temperature and metallicity. An all-forbidden-line index proposed by Vilchez and Pagel provides improved accuracy [15], but it involves far red lines that are seldom observed. Furthermore, since it involves ratios of lines at widely separated optical wavelengths, it is influenced by reddening and is probably not applicable in the complex reddening environment in a typical starburst. Over the temperature range ~35,000 to ~40,000, the He^+/H ratio is an accurate measure, although the accessible optical HeI lines are faint and can be difficult to measure accurately [15]. In fact, in the definitive spectral atlas by Ho et al. [13], the strengths of the optical HeI lines are seldom determined.

Alternatively, effective temperatures can be measured from mid infrared fine structure line ratios [23,34]. Compared with optical forbidden lines, these lines are not strongly affected by electron temperature, and they are far less subject to extinction. On the other hand, the atomic constants are less well known. The mid infrared lines appear to give systematically lower effective temperatures than the usual optical methods [34,38]. The strong HeI line at 2.06μm has been proposed as an indicator of the fraction of He^+ (e.g, [8]), but its strength is so strongly a function of nebular conditions that it is of very limited use for this purpose [27]. The HeI line at 1.7μm is very weak, but where it can be measured it should provide the most reliable possible determination through a virtually extinction-independent measure of He$^+$/H [37]. This conversion of HeI/H to effective temperature is also nearly metallicity independent, and over the applicable temperature range should give the most accurate possible measure of the effective temperature of the ionizing stars.

Table 1 summarizes the various effective temperature determinations for the well measured galaxy He 2-10. It suggests that the combined effects of extinction and other biases do in fact result in a significant overestimate of the effective temperature from the optical indicators, and that the mid infrared fine structure lines may be more reliable than previously thought.

Table 1. Comparison of Temperature Estimates for He 2-10.

Reference	Method	Effective Temperature
[31]	[OIII] 5007/Hβ and [OI] 6300/Hα	39,000–41,000K
[35]	HeI 4471/Hβ	38,000–39,000K
[23]	mid-IR fine structure lines	35,000–37,000K
[38]	HeI 1.7μm/Br10	35,000–37,000K

As shown in Table 1, typical discrepancies between optical temperature indicators and the mid infrared fine structure lines are 40,000K or more compared to 36,000K, corresponding to a mass difference on the main sequence [25] of about 31 M$_\odot$ compared to 22 M$_\odot$. This change in maximum main sequence mass corresponds to more than a factor of two in the age of a starburst, which is a much larger uncertainty than the internal errors in starburst models.

2.7 Dynamical Mass

Dynamical masses can be measured either from emission lines or from stellar velocities determined from the 2.3μm first overtone CO bandhead. Either method must be used with some caution: the motions of emission line gas may not be dominated by gravity, or the emitting gas may have a complex distribution within the measurement beam that means it does not represent a true average gravitational potential. Stellar features may also arise from bright complexes of recently formed stars that do not represent the true average gravitational potential in the beam. There are many other well-known sources of error in modeling to derive masses, such as the underlying distribution even of a relaxed stellar population within the resolution element (r$^{1/4}$ vs. exponential, scale lengths, etc.).

Because typical dusty starbursts begin to become optically thin only near the 2.3μm CO bandhead, stellar features at significantly shorter wavelengths should be used with caution to study dynamics. One would expect that they would yield an underestimate of the peak rotational velocities (since they would only probe the foreground of the starburst). It was loudly stated at the conference that a similar problem held "by an order of magnitude" for CO bandhead-determined masses for Arp 220, but the masses from mm-wave CO line width and the CO bandhead agree to within a factor of 2–4, depending on corrections to a common beam size [28,30]. The line width from radio hydrogen recombination lines [4] is intermediate between those from the other two measures, so even in this extreme case we can argue that all methods agree with an average value to within a factor of two! In less extremely obscured starbursts, the agreement should be better and the derived dynamical masses will be dominated by systematic errors in the mass models. Given the complexities in a starburst, such errors could be significant.

2.8 Supernova Rate

In principle, the supernova rate is an important boundary condition on starbursts (e.g., [7]). Supernova rates can be estimated in a variety of ways, such as by radio emission [36] or FeII infrared line intensity [1,38]. Because the supernova rate changes dramatically in the early evolution of a starburst, such constraints can be useful. However, a more quantitative application of them requires a better connection of the supernova rate with the theory of stellar evolution than exists at present.

3 Results

Some examples of starburst modeling are to be found in [2,3,5,9–11,16,22]. A number of general conclusions have emerged from these studies. First, all the properties of bona fide starburst galaxies appear to be adequately explained by the population synthesis models, using an initial mass function that is similar at high masses to that observed in active star forming regions. It is an open question whether the IMF extends to the lowest stellar masses as observed in local star forming regions, or whether it is relatively deficient in low mass stars. The duration of the intense episode of star formation is short, of order 10 Myr or less (see also [20]). Star formation appears to be terminated by the outbreak of strong mechanical stirring of the ISM by supernova explosions. The mass budget in starbursts is very tight. Satisfying it requires that the efficiency of formation of massive stars from molecular gas is very efficient, with 50% or more of the gas being converted into stars if the IMF is like that observed in local star forming regions. This efficiency is much higher than in even the densest HII regions in the solar neighborhood. With the molecular gas masses that are consistent with the dynamical masses, the Schmidt Law, $\Sigma_{SFR} \propto \Sigma_{gas}^N$, has a large value of N, $N \sim 2$. The high star formation efficiency and large value of N suggest that star formation is triggered by cloud-cloud collisions, shocks, and winds (i.e., [26,32]), which could favor production of massive stars (e.g., [40]).

Star formation in some systems has been shown to propagate outward from the nucleus (e.g., [3,24]). However, in strongly interacting systems the pattern of star formation is far more complex, with large and powerful star forming regions in interaction zones as well as in the galaxy nuclei (e.g., [2,19,41]). Within strongly starbursting galaxies, the star formation is at least partially concentrated into immense HII regions, with Lyman luminosities an order of magnitude greater than 30 Dor, which sets the standard for the Local Group [2,3,14]). These objects appear to evolve into a population of super star clusters (e.g., [2,3,39] and references therein).

The role of interactions in initiating starbursts has been known for some time [17,22]. However, many starbursts are known not to have dramatic indications of interactions, and even in cases where the outer regions show such effects, it has sometimes been difficult to trace the influences of the interaction to the nucleus. Using infrared observations to pierce the interstellar dust, it appears that substantial advances can be made in this area. For example, NICMOS imaging of NGC 1614 shows the nucleus of a lower luminosity galaxy projected close to the main nucleus [3]. Even more interestingly, Thatte et al. [33] have discovered a probable second nucleus in the center of NGC 5236 by a large increase in velocity dispersion where there is no obvious feature in surface brightness.

4   Conclusions

The instrumentation and analysis techniques now available to study nearby dusty starbursts can provide detailed insights to the processes occurring in them. We are beginning to acquire a reasonably complete understanding of this important stage in galaxy evolution.

References

1. A. Alonso-Herrero, M.J. Rieke, G.H. Rieke: in *International Symp. On Astrophysics Research and Science Education*, ed. C. Impey, (Univ. of Notre Dame Press: South Bend, 1999), p. 213
2. A. Alonso-Herrero, G.H. Rieke, M.J. Rieke, N.Z. Scoville: ApJ **532**, 845 (2000)
3. A. Alonso-Herrero, C.W. Engelbracht, M.J. Rieke, G.H. Rieke, A.C. Quillen: ApJ, in press (2001)
4. K.R. Anatharamaiah, F. Viallefond, N.R. Mohan, W.M. Goss, J.H. Zhao: ApJ **537**, 613 (2000)
5. T. Böker, N.M. Förster-Schreiber, R. Genzel: AJ **114**, 1883 (1997)
6. D. Calzetti, A.L. Kinney, T. Storchi-Bergmann: ApJ **458**, 132 (1996)
7. J.S. Doane, W.G. Mathews: ApJ **419**, 573 (1993)
8. R. Doyon, P.J. Puxley, R.B. Joseph: ApJ **397**, 117 (1992)
9. C.W. Engelbracht, M.J. Rieke, G.H. Rieke, W.B. Latter: ApJ **467**, 227 (1996)
10. C.W. Engelbracht, M.J. Rieke, G.H. Rieke, D.M. Kelly, J.M. Achtermann: ApJ **505**, 639 (1998)
11. N.M. Förster-Schreiber: this proceedings
12. K.D. Gordon, D. Calzetti, A.N. Witt: ApJ **487**, 625 (1997)

13. L.C. Ho, A.V. Filippenko, W.L. Sargent: ApJS **98**, 477 (1995)
14. R.C. Kennicutt: ApJ **334**, 144 (1998)
15. R.C. Kennicutt, F. Bresolin, H. French, P. Martin: ApJ **537**, 589 (2000)
16. A. Krabbe, A. Sternberg, R. Genzel: ApJ **425**, 72 (1994)
17. R.B. Larson, B.M. Tinsley: ApJ **219**, 46 (1978)
18. K.K. McLeod, G.H. Rieke, M.J. Rieke, D.M. Kelly: ApJ **412**, 111 (1993)
19. I.F. Mirabel, et al.: A&A **333**, L1 (1998)
20. G. de Paz, A. Aragon-Salamanca, J. Gallego, A. Alonso-Herrero, J. Zamorano, G. Kauffmann: MNRAS **316**, 357 (2000)
21. S.E. Ridgway, C.G. Wynn-Williams, E.E. Becklin: ApJ **428**, 609 (1994)
22. G.H. Rieke, M.J. Lebofsky, R.I. Thompson, F.J. Low, A.T. Tokunaga: ApJ **238**, 24 (1980)
23. P.F. Roche, D.K. Aitken, C.H. Smith, M. Ward: MNRAS **248**, 606 (1991)
24. S. Satyapal et al.: ApJ **483**, 148, (1997)
25. G. Schaller, D. Schaerer, G. Meynet, A. Maeder: A&AS **96**, 269 (1992)
26. N.Z. Scoville, D.B. Sanders, D.P. Clemens: ApJ **310**, L77 (1986)
27. J.C. Shields: ApJ **419**, 181 (1993)
28. L.M. Shier, M.J. Rieke, G.H. Rieke: ApJ **470**, 222, (1996)
29. D.A. Smith, T. Herter, M.P. Haynes: ApJ **494**, 150 (1998)
30. P.M. Solomon, D. Downes, S.J.E. Radford, J.W. Barrett: ApJ **478**, 144 (1997)
31. H. Sugai, Y. Taniguchi: AJ **103**, 1470 (1992)
32. Y. Taniguchi, Y. Ohyama: ApJ **508**, L13 (1998)
33. N. Thatte, M. Tecza, R. Genzel: A&A Letters, in press (2000)
34. M.D. Thornley, N.M. Förster-Schreiber, D. Lutz, R. Genzel, H.W.W. Spoon, D. Kunze, A. Sternberg: ApJ **539**, 641 (2000)
35. W.D. Vacca, P.S. Conti: ApJ **401**, 543 (1992)
36. D. Van Buren, M.A. Greenhouse: ApJ **431**, 640 (1994)
37. L. Vanzi, G.H. Rieke, C.L. Martin, J.C. Shields: ApJ **466**, 150 (1996)
38. L. Vanzi, G.H. Rieke: ApJ **479**, 694 (1997)
39. B.G. Whitmore, Q. Zhang, C. Leitherer, S.M. Fall, F. Schweizer, B.W. Miller: AJ **118**, 1551 (1991)
40. A.P. Whitworth, A.S. Bhattel, S.J. Chapman, M.J. Disney, J.A. Turner: MNRAS **268**, 291 (1994)
41. C.D. Wilson, N.Z. Scoville, S.C. Madden, V. Charmandaris: ApJ **542**, 120 (2000)
42. A.N. Witt, R.S. Lindell, D.L. Block, R. Evans: ApJ **427**, 227 (1994)

Starbursts, Dark Matter and Dwarf Galaxy Evolution

Gerhardt R. Meurer

The Johns Hopkins University, Baltimore, MD 21218, USA

Abstract. Optical and H I imaging of both dwarf irregular (dI) and Blue Compact Dwarf (BCD) galaxies reveal important clues on how dwarf galaxies evolve and their star formation is regulated. Both usually show evidence for stellar and gaseous disks. However, their total mass is dominated by dark matter. Gas rich dwarfs form with a range of disk structural properties. These have been arbitrarily separated into two classes on the basis of central surface brightness. Dwarfs with $\mu_0(B) \lesssim 22\,\mathrm{mag\,arcsec^{-2}}$ are usually classified as BCDs, while those fainter than limit are usually classified as dIs. Both classes experience bursts of star formation, but with an absolute intensity correlated with the disk surface brightness. Even in BCDs the bursts typically represent only a modest $\lesssim 1\,\mathrm{mag}$ enhancement to the B luminosity of the disk. While starbursts are observed to power significant galactic winds, the fractional ISM loss remains modest. Dark matter halos play an important role in determining dwarf galaxy morphology by setting the equilibrium surface brightness of the disk.

1 Introduction

Despite their morphological differences dwarf irregular (dI), Blue Compact Dwarf (BCD), and dwarf elliptical (dE) galaxies have similar optical structures – their radial profiles are exponential, at least at large radii (e.g. [2,3,13]). Are there evolutionary connections between these morphologies? One scenario expounded by Davies & Phillips [4] starts with an initial dI galaxy; if its ISM manages to concentrate at the center of the galaxy a tremendous starburst occurs resulting in a BCD morphology. This starburst powers a galactic wind (e.g. [6,12]). If the wind is strong enough all of the ISM is expelled resulting in a dE morphology. If some ISM remains, the system fades back into a dI, and undergoes a few more dI $\Leftrightarrow$ BCD transitions before eventually expelling all of its ISM to become a dE. Here I will address the validity of this scenario. In Sec. 2, I compare the optical structure of dIs and BCDs; Sec. 3 details the H I structure and dynamics of two BCDs: NGC 1705 ($D = 6.2\,\mathrm{Mpc}$) and NGC 2915 ($D = 3.1\,\mathrm{Mpc}$), and compares them to dI galaxies; and Sec. 4 synthesizes the optical and radio results to form a new scenario where Dark Matter (DM) plays a dominant role in determining the morphology of gas rich dwarfs. Here I adopt $H_0 = 75\,\mathrm{km\,s^{-1}\,Mpc^{-1}}$.

2 The Optical Structure and Classification of Gas Rich Dwarfs

The exponential profile portion of BCDs and dIs probably signifies the presence of a rotating disk, which are certainly typical of the gas distributions in these systems [5,24]. In addition to the disk component, BCDs usually have a blue high surface brightness excess of light in their centers, rich in H II emission and young star clusters [13,15]. I will call this blue excess the starburst, since it is responsible for the starburst characteristics of BCDs. The integrated colors of this component, after subtracting off the disk, indicate that it must be due to a young stellar population with an age $\sim 10\,\mathrm{Myr}$ if instantaneous burst models are adopted or $\sim 100\,\mathrm{Myr}$ if constant star formation rate models are adopted [14]. The strong Hα fluxes are more consistent with the constant star formation rate models (because $\sim 10\,\mathrm{Myr}$ old instantaneous bursts are no longer ionizing). In either case this is much less than the Hubble time, confirming the starburst nature of this component. In comparison, the colors of the disk are typically like those of stellar populations forming continuously over a Hubble time (i.e. like dI galaxies, cf. [21]), or a bit redder suggesting an inactive population.

Surface brightness profile fitting provides a means to determine both the relative strength of the starburst, and the structural properties of the disk (see [14] for details). The outer portions of the profile are fitted with an exponential, yielding the extrapolated central surface brightness μ_0, and scale length α^{-1} of the disk. The burst and disk are separated by assuming that the disk remains exponential all the way into the center. The strongest starbursts are about twice as bright as their hosts. Hence, while starbursts can outshine the host disk they are nevertheless modest $\lesssim 1$ mag enhancements to the total B flux of BCDs. Typical flux enhancements are only a few tens of percent. About 20% of BCDs have exponential profiles all the way into their cores, hence they show no structural evidence for a starburst. The mass contribution of the starburst is even smaller than the flux contribution, typically $\lesssim 5\%$. These are not the ~ 6 mag starburst enhancements proposed to explain the excess of faint blue galaxies at moderate redshifts [1].

Figure 1 compares the disk parameters μ_0 and α^{-1} of both BCDs and dI galaxies. Note that μ_0 does not include the contribution of the starburst core. While there is some overlap, we see that BCD distribution is offset from that of dIs particularly in μ_0 which typically is $2.5\,\mathrm{mag\,arcsec}^{-2}$ more intense in BCDs than in dIs. Structurally BCD disks are very different from those of dIs.

The absence of BCDs on the left half of Fig. 1 is puzzling. Does this mean that dI galaxies do not experience starbursts? Examination of surface brightness and color profiles [21] reveal several dIs with an exponential profile, and a higher surface brightness blue excess, structural evidence for starbursts in dIs. The episodic star forming nature of dIs is well demonstrated using color-magnitude diagrams of the nearest ones (e.g. [8]). However the observed central surface brightness of the bursting dIs *including the light of this central excess* is typically $\mu(B) \gtrsim 22\,\mathrm{mag\,arcsec}^{-1}$, much fainter than the central regions of BCDs. While dIs do experience starbursts, they are pathetic and not usually recognized as

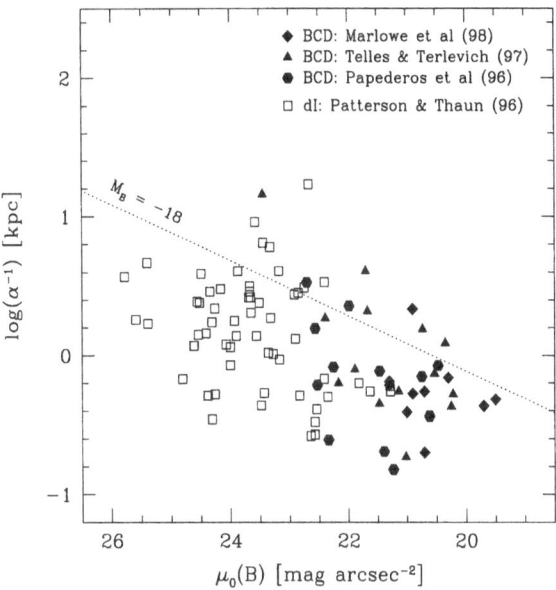

Fig. 1. Exponential disk parameters for samples of BCD and dI galaxies [14,20,21,23].

such since they are not *intense* enough. We see that there are both BCDs and dIs with central starbursts, as well as cases of both types that are exponential all the way into their centers. The separation into two classes appears to be an arbitrary segregation by central surface brightness of the underlying disk at $\mu_0(B) \approx 22\,\mathrm{mag\,arcsec}^{-2}$.

3   H I Structure and Dynamics of BCDs

Compared to dIs there are not that many H I imaging studies of BCDs; in part due to their small numbers and usual compact angular sizes. Two that have been well imaged in H I are NGC 2915 [16], and NGC 1705 [18].

Both galaxies show evidence of star formation churning up the neutral ISM. In NGC 2915, enhancements in the velocity dispersion correspond well to Hα bubbles and peculiar star formation knots. However it does not appear that H I is being ejected from the system. In the center of the galaxy, where star formation is the most vigorous, $\sigma_{\mathrm{HI}} \approx 40\,\mathrm{km\,s}^{-1}$ which is the same as the one dimensional velocity dispersion derived for the DM particles. Hence, star formation appears to be maintaining the central H I in virial equilibrium with the DM halo. This suggests that DM plays a role in the feedback process: if the starburst energizes H I to have σ_{HI} much larger than the halo velocity dispersion, then neutral ISM is thrown into the halo (or beyond) halting star formation.

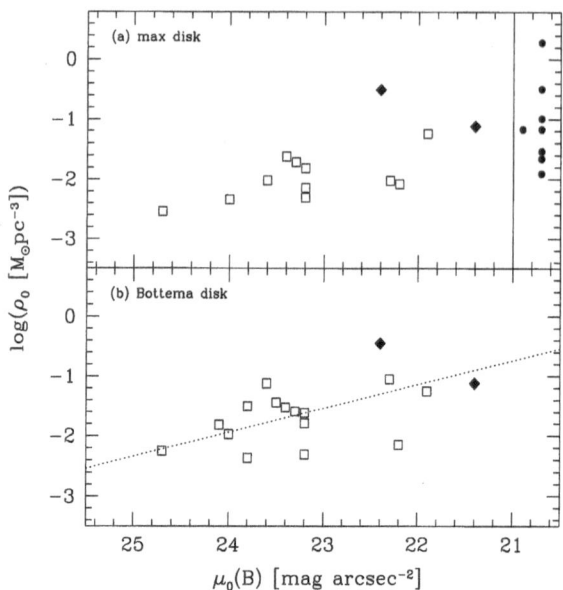

Fig. 2. DM halo central density ρ_0 plotted against disk central surface brightness. Open squares are from de Blok & McGaugh [5], while diamonds represent NGC 1705 and NGC 2915. The top panel shows the results for maximimum disk model fits, while the bottom panel shows Bottema disk fits. The circles on the right side of the top panel mark crude estimates of ρ_0 in 12 BCDs with published and unpublished RCs. The dotted line, at bottom, is a fit to the Bottema disk results with the relationship $\log(\rho_0) = 0.4 \log(\mu_0) + \text{Constant}$.

NGC 1705 displays a strong galactic wind in Hα. There is a spur of H I emission obliquely jutting out from its H I disk that appears to be a neutral ISM extension of this wind. If so, then NGC 1705 has ejected about 8% of its neutral ISM at a mass loss rate at least comparable to the star formation rate. Nevertheless, even in this BCD with one of the most spectacular Hα outflows, the majority of the ISM is retained in a disk. This starburst is incapable of totally blowing away the ISM.

Although both galaxies have some kinematic irregularities, their dominant structures are extended rotating disks which are strongly centrally peaked. These are typical properties of BCDs imaged in H I [22,24]. The disk of NGC 2915 is so extended that it has the H I appearance of a late type barred spiral. Similar galaxies include IC 10 [25] and NGC 4449 [9].

The rotation curves of both NGC 1705 and NGC 2915 show a fairly steep rise over the optical face of the galaxy which then becomes flat out to the edge of the H I distribution. They are the first BCDs with mass model fits to their rotation curves. The mass models include (1) a stellar distribution whose mass to light ratio is given by either a maximum disk model or by the optical colors (the "Bottema disk" model [5]); (2) the neutral ISM distribution; and (3) a dark matter halo. This halo is taken to be a pseudo-isothermal sphere whose free parameters are the central density ρ_0 and the core radius R_c. From these, the asymptotic rotational velocity and halo velocity dispersion can be calculated [11]. In both galaxies, DM dominates the mass distribution, even within the optical radius of the galaxy. In comparison, the stellar component has a mass equal to or *less than* the neutral gas disk.

Overall, the global dynamics of BCDs appear to be similar to dIs: they are dominated by rotating disks with normal looking RCs. A distinction between the two types is seen when the DM halo densities ρ_0 are compared, as shown in Fig. 2. Central densities found by maximum disk and Bottema disk fits are shown in separate panels. The comparison sample is taken from de Blok & Mc-Gaugh [5], and includes only galaxies with $M_B > -18$ mag. This comparison shows that NGC 1705 and NGC 2915 have two of the highest ρ_0 measurements of any dwarf galaxies. In order to check that these galaxies are typical, I crudely estimated ρ_0 from the central velocity gradient for 12 BCDs with published or unpublished RCs, and plotted them as circles at arbitrary μ_0 in the top panel of Fig. 2. These estimates are upper limits, since the contribution of the baryonic components to the velocity gradients have not been removed. Nevertheless, the comparison indicates that NGC 1705 and NGC 2915 have normal ρ_0 for BCDs. Figure 2 shows a weak but noticeable correlation between $\log(\rho_0)$ and $\mu_0(B)$, with higher surface brightness disks corresponding to higher ρ_0 halos. This result was first noted in dIs by de Blok & McGaugh [5].

4   Evolutionary Connections

The correlation in Fig. 2 can readily be explained by considering the response of a self gravitating disk immersed in a dominant DM halo core of constant density ρ_0, i.e. where the rotation curve is linearly rising. If the disk is maintained at constant stability parameter and the star formation rate per unit area scales with the gas density divided by the dynamical time [10], then it is straightforward to show that the surface brightness should scale with ρ_0 [18,19]. This is consistent with the observed correlation, as shown by the dotted line in Fig. 2. A similar correlation between surface brightness and ρ_0 holds in the center of larger starburst galaxies [17]. However for them it is normal baryonic matter that dominates ρ_0 rather than DM. In essence, the central mass density determines the equilibrium star formation rate of the embedded disk.

Following from the discussion in Sec. 2, the disk central intensity largely determines whether a dwarf galaxy is classified as a BCD or dI. The optical size of dwarfs seems to be limited to the core radius R_c. Hence, both DM halo parameters are important in governing the morphology of gas rich dwarfs.

Can there be evolution between dI and BCD classes? While some evolution in ρ_0 may be allowed, it is unlikely that there can be enough to change a typical dI into a typical BCD. That would require a 2.5 mag arcsec^{-2} change in μ_0, or equivalently, a factor of ten change in ρ_0. To do this with a mass loss or gain would require a 55% change in mass if the expansion or contraction is homologous [19]. The problem is that there isn't that much *baryonic* mass in a dwarf galaxy. To effect this large a change would require DM loss or gain. This is not feasible if DM is non-dissipative and feels only the force of gravity, as is usually assumed. I conclude that there is probably little dI ⇔ BCD evolution.

If the ISM were removed from a dI or BCD, it could still plausibly evolve into a dE galaxy. However, as noted in Sec. 3 even in a dwarf galaxy undergoing a strong starburst with a spectacular galactic wind (NGC 1705), the fractional loss of the ISM is modest. If this is typical, it would take on the order of 10 bursts to expel all the ISM from a BCD. The bursts aren't strong enough, and the ISM distributions are too flattened to allow a single burst expulsion of the ISM [7]. The demographics of dwarf galaxy morphologies point to an environmental component to their evolution. Gas rich dwarfs are found in low density environments where the frequency of external starburst triggers is low. They survive easily. The clock runs faster (more frequent triggers) in clusters, and in addition ram pressure stripping would accelerate the removal of gas from dwarfs, while tidal truncation of DM halos would assist galactic wind losses. Hence it is not surprising that gas poor dEs are found more often in clusters than the field.

5   Conclusions

We are now at a position to re-evaluate the Davies & Phillips [4] scenario for dwarf galaxy evolution. The mechanisms they invoke have clearly been verified. Dwarf galaxies do experience starbursts and these can expel some of the ISM. Mass expulsion can rival or surpass lock up into stars in regulating the gas content of dwarfs. However, the results of any single burst are not so severe. Cataclysmic bursts are not common at the present epoch, and the milder bursts that are observed may not be sufficient to change a galaxy's morphological classification. The morphology of a dwarf galaxy is largely set by its enveloping dark halo, and is relatively impervious to starbursts.

Acknowledgements: I thank the organizers for asking me to talk on this subject. Since this was done as a last minute replacement for Daniel Kunth, I was not able to prepare anything new. Instead this talk is based on my 1998 Moriond presentation [19]. I thank my collaborators on the original projects this work is based on: Sylvie Beaulieu, Claude Carignan, Ken Freeman, Tim Heckman, Neil Killeen, Amanda Marlowe, and Lister Staveley-Smith.

References

1. A. Babul, & H.C. Ferguson: ApJ **458**, 100 (1996)
2. G.D. Bothun, J.R. Mould, N. Caldwell, & H.T. MacGillivray: AJ **92**, 1007 (1986)
3. N. Caldwell, & G. Bothun: AJ **94**, 1126 (1987)
4. J.I. Davies, & S. Phillips: MNRAS **233**, 553 (1988)
5. W.J.G. de Blok, & S.S. McGaugh: MNRAS **290**, 533 (1997)
6. A. Dekel, & J. Silk: ApJ **303**, 39 (1986)

7. D.S. De Young, & T.M. Heckman: ApJ **431**, 598 (1994)
8. J.S. Gallagher III, E. Tolstoy, R.C. Dohm-Palmer, E.D. Skillman, A.A. Cole, J.G. Hoessel, A. Saha, & M. Mateo: AJ **115**, 1869 (1998)
9. D.A. Hunter, E. Wilcots, H. van Woerden, J.S. Gallagher III, & S. Kohle: ApJ **459**, 47 (1998)
10. R.C. Kennicutt: ApJ **498**, 541 (1998)
11. G. Lake, R.A. Schommer, & J.H. van Gorkom: ApJ **320**, 493 (1990)
12. A.T. Marlowe, T.M. Heckman, R.F.G. Wyse, & R. Schommer: ApJ **438**, 563 (1995)
13. A.T. Marlowe, G.R. Meurer, T.M. Heckman, & R. Schommer: ApJS **112**, 285 (1997)
14. A.T. Marlowe, G.R. Meurer, & T.M. Heckman: ApJ **522**, 182 (1999)
15. G.R. Meurer, T.M. Heckman, C. Leitherer, A. Kinney, C. Robert, & D.R. Garnett: AJ **110**, 2665 (1995)
16. G.R. Meurer, C. Carignan, S.F. Beaulieu, & K.C. Freeman: AJ **111**, 1551 (1996)
17. G.R. Meurer, T.M. Heckman, M.D. Lehnert, C. Leitherer, & J. Lowenthal: AJ **114**, 54 (1997)
18. G.R. Meurer, L. Staveley-Smith, & N.E.B. Killeen: MNRAS **300**, 705 (1998)
19. G.R. Meurer: 'Starbursts, Dark Matter, and The Evolution of Dwarf Galaxies'. In *Dwarf Galaxies and Cosmology (XVIIIth Rencontre de Moriond)* eds. T.X. Thuan, C. Balkowski, V. Cayatte, J. Tran Than Van, (Editions Frontieres, Paris 1998), pp. 337–347
20. P. Papaderos, H.H. Loose, T.X. Thuan, & K.J. Fricke: A&AS **120**, 207 (1996)
21. R. Patterson, & T.X. Thuan: ApJS **107**, 103 (1996)
22. C.L. Taylor, E. Brinks, R.W. Pogge, & E.D. Skillman: 1994, AJ **107**, 971
23. E. Telles, & R. Terlevich: MNRAS **286**, 183 (1997)
24. L. van Zee, E.D. Skillman, & J.J. Salzer: AJ **116**, 1186 (1998)
25. E. Wilcots, & B.W. Miller: AJ **116**, 2363 (1998)

The Starburst-AGN Connection

Sylvain Veilleux

Department of Astronomy, University of Maryland, College Park, MD 20742, USA

Abstract. The issue of a starburst-AGN connection in local and distant galaxies is relevant for understanding galaxy formation and evolution, the star formation and metal enrichment history of the universe, the origin of the extragalactic background at low and high energies, and the origin of nuclear activity in galaxies. Here I review some of the observational evidence recently brought forward in favor of a connection between the starburst and AGN phenomena. I conclude by raising a number of questions concerning the exact nature of this connection.

1 Introduction

Since the focus of this conference is the starburst phenomenon, many of us would rather "sweep AGN activity under the rug" and only consider star formation. This would be a mistake! There is growing evidence that intense star formation and nuclear activity often come hand in hand. The apparent correlation between the mass of dormant black holes at the centers of nearby galaxies and the mass of their spheroids (e.g., [43,23,48,29]) suggests a direct link between the formation of spheroids and the growth of central black holes. Since a starburst is a natural consequence of the dissipative gaseous processes associated with spheroid formation (e.g., [4,44,53]), a starburst-AGN connection dating back to the early universe is implied by these results. Closer to us, the presence of circumnuclear starbursts in an increasing number of local AGNs (discussed in more detail in §2) also suggests a connection between the starburst and AGN phenomena.

This possible starburst-AGN connection has direct bearings on our understanding of the early universe. A large contribution from unsuspected (hidden) AGNs would complicate the deduction of the star formation history of the universe from galaxy luminosity functions (e.g., [8]). This in turn may change our current views on the history of metal enrichment and the importance of feedback processes in the early universe (e.g., [25,63,24]). Similarly, a correction needs to be applied to account for obscured AGNs as contributors to the far-infrared extragalactic background. Recent X-ray observations with *Chandra* have added substantially to the debate. The discovery that some fraction of the X-ray background appears to be produced by a population of heavily obscured AGNs (e.g., [57,3]), objects which have been largely missed in optical surveys due to extremely heavy obscuration, has clearly further increased the importance of studies of the starburst-AGN connection in distant infrared-selected galaxies.

The rest of this paper is organized as follows. In §2, I discuss recent (< 5 years approx.) results which appear to favor a starburst-AGN connection. This discussion is not meant to be an exhaustive (or even impartial) review of the recent literature on this subject; it is only meant to illustrate some of the best cases where a starburst-AGN connection indeed appears to exist. In the last section (§3), I raise several questions regarding the nature of this starburst-AGN connection.

2   Evidence for a Starburst-AGN Connection

Since the triggering mechanism for AGN activity probably depends on the luminosity of the AGN, I make a distinction in the following discussion between the nearby, low-luminosity Seyferts and Fanaroff-Riley type I (FR I) radio galaxies and the more distant and powerful quasars, Fanaroff-Riley type II (FR II) radio galaxies, and ultraluminous infrared galaxies (ULIRGs; $\log [L_{ir}/L_\odot] \geq 12$ by definition).

2.1   Low-Luminosity Regime: Seyferts, FR I Radio Galaxies

The fueling of AGNs requires mass accretion rates $dM/dt \approx 1.7 \ (0.1/\epsilon)(L/10^{46}$ ergs s^{-1}) $M_\odot$ yr^{-1}, where ϵ is the mass-to-energy conversion efficiency. A modest accretion rate of order ~ 0.01 $M_\odot$ yr^{-1} is therefore sufficient to power a Seyfert galaxy like NGC 1068. Only a small fraction of the total gas content of a typical host galaxy is therefore necessary for the fueling of these low-luminosity AGNs. A broad range of mechanisms including intrinsic processes (e.g., stellar winds and collisions; dynamical friction of giant molecular clouds againts stars; nuclear bars or spirals produced by gravitational instabilities in the disk) and external processes (e.g., "minor" galaxy interactions or mergers) may be at work in these objects. While a detailed discussion of these processes is beyond the scope of the present paper (see [14] for a recent review of the subject), suffice it to say that there is little or no observational evidence for Seyfert nuclei to occur preferentially in barred systems (e.g., [50,37,55,38]) or to have recently experienced a major interaction or merger ([28,18,15,88], although see last paragraph of this subsection). These results seem to favor minor intrinsic processes over large-scale external processes for the fueling of low-luminosity AGNs. Ejecta from a nuclear star cluster (e.g., [59,62,91]) may be all that is needed in the cases of Seyferts and other low-luminosity AGNs to keep their nuclei active. The "angular momentum problem" in feeding low-luminosity AGNs may therefore reduce to forming the dense stellar cluster in the first place. Nuclear starbursts are the prime candidates for the formation of these clusters.

Several studies have shown that the molecular material needed to fuel nuclear starbursts in Seyfert galaxies is present near the nuclei of these objects (e.g., [51,79,41,1]). But is this material forming stars? Direct evidence for recent *nuclear* star formation now exists in a number of Seyfert 2 galaxies (i.e. Seyferts without broad recombination lines). Optical and ultraviolet spectroscopy of the nuclear regions of these galaxies often reveals the signatures of young and intermediate-age stars. The stellar Ca II triplet feature at $\lambda\lambda$ 8498, 8542, 8662 in Seyfert 2s has an equivalent width similar to that in normal galaxies while the stellar Mg Ib $\lambda5175$ is often weaker [81]. This result is difficult to explain with a combination of an old stellar population and a featureless power-law continuum from an AGN. The most natural explanation is that young red supergiants contribute significantly to the continuum from the central regions. Evidence for intermediate-age (a few 100 Myrs) stars in these galaxies is also apparent in the blue part of the spectrum, where the high-order Balmer series and He I absorption lines appear to be present in more than half of the brightest Seyfert 2 galaxies (e.g., [12,34]). A few of these objects may even harbor a broad emission feature near 4680 Å, possibly the signature of a population of young (a few Myrs) Wolf-Rayet stars [33]. The ultraviolet continuum from some of the brightest UV Seyfert 2s also appears to be dominated by young stars based on the strength of absorption features typically formed in the photospheres and in the stellar winds of massive stars (e.g., [36,33]). The extended, soft, thermal X-ray emission from these objects seems to confirm these results [45]. The bolometric luminosities of these nuclear starbursts ($\sim 10^{10}$ L$_\odot$) are similar to the estimated bolometric luminosities of their obscured Seyfert 1 nuclei.

Interestingly, a distinction appears to exist between Seyfert 1s and Seyfert 2s. Seyfert 2 galaxies have long been known to present a larger far- and mid-infrared excess than Seyfert 1s (e.g., [67,20,64,49]), but most of this excess emission may be attributed to star formation in the host galaxy rather than from a nuclear starburst. The departure of the galaxy or bulge blue luminosity of Seyfert 2s from the Tully-Fisher and Faber-Jackson relationships (e.g., [89,90,58]) and the diffuse radio emission around some Seyfert nuclei [92] may have the same origin. However, recent investigations have also suggested excess *nuclear* starburst activity in Seyfert 2s relative to Seyfert 1s [34,35] and possibly a higher frequency of companions near type 2 objects (e.g, [15,19,45]). These results cannot be explained in the context of the Seyfert unification theory (which purports that Seyfert 1s and 2s are basically the same type of objects seen from different perspectives), but they may reflect an evolutionary connection between starbursts, Seyfert 2s, and Seyfert 1s.

2.2   High-Luminosity Regime: QSOs, FR II Radio Galaxies, ULIRGs

The stringent requirements on the mass accretion rates for luminous AGNs almost certainly require external processes such as "major" galaxy interactions or mergers to be involved in triggering and sustaining this high level of activity over $\sim 10^8$ years. Substantial evidence exists that the precursors to at least some powerful AGNs have indeed been gas-rich mergers. Classical double (FR II) radio galaxies have long been known to show tidal tails and other signs of interaction [76,7]. Evidence for recent or on-going galactic interactions is also seen in several quasars (e.g., [39,2,9]). Abundant molecular gas has been detected in radio galaxies and quasars (e.g., [73,71,5,6,54,75,60,61,21,22]), and many of them also show the spectroscopic signatures of recent star formation (e.g., [80,82,10]). The far-infrared excess in some of these objects may also be attributed to star formation (e.g., [66]; see [70] for another interpretation). In this merger scenario, quasars were more common in the past because of the enhanced frequency of collisions [$\propto (1+z)^{4.0\pm2.5}$; e.g., [95]] and the larger proportion of unprocessed gas. Moreover, the observed redshift cut-off for quasars ($z \approx 5$) marks the epoch at which disk systems formed.

ULIRGs may provide the clearest observational link between galaxy mergers, starbursts and powerful AGNs. Nearly all ULIRGs show strong signs of advanced tidal interactions (e.g., [69,52,56,13]). All of them are very rich in molecular gas (e.g., [77,26,27]), most of which is distributed well within the inner kpc of the galaxy (e.g., [16,11,68]). They also present a large concentration of activity in their nuclei, including strong optical emission lines characteristic of a star-bursting stellar population and in about 30% of cases, broad or high-ionization emission lines that suggest the presence of a powerful AGN coexisting with the starburst (e.g., [42,93,94,83,84,40]). Similar results are found in the near-infrared (e.g., [31,32,86,87]) and in the mid-infrared (e.g., [30,46,65,47,17]).

The fraction of AGN-dominated ULIRGs is significantly larger among objects with high infrared luminosities and warm infrared colors (e.g.[42,83,84,86,94,40]). Current results on a limited set of ULIRGs (e.g., [65]) suggest that the dominance of AGN or starburst in ULIRGs may depend on local and short-term conditions (e.g., compression of the circumnuclear interstellar medium as a function of gas content and galaxy structure, local accretion rate onto the central black hole, etc.) in addition to the global state of the merger. Still several lines of evidence suggest that *warm* ULIRGs are indeed more advanced, transition objects and that (radio quiet) QSOs correspond to the final state of the merger-induced sequence "starburst → ULIRGs → QSOs" (e.g., [78,74,96,85]).

3   Unanswered Questions

The exact nature of this starburst-AGN connection is not at all clear. Unanswered questions include:

1. Can an AGN be triggered without a burst of star formation?
2. If a SMBH is indeed present in every (massive) galaxy, can a starburst be taking place without any AGN activity?
3. Can starbursts and AGNs simply coexist without interacting with each other?
4. If not, in what way are the starbursts and AGNs interacting?
   - Is mass loss from the central stellar cluster fueling (low-luminosity) AGNs?
   - Is the molecular gas "unused" by the starburst feeding the SMBH?
5. Is there an evolutionary connection between starbursts and AGNs?
   - Which of the starburst or AGN comes first?
   - Is the merger-induced starburst → QSO model correct?
   - Is there a similar sequence in low-luminosity objects?
   - Is there a evolutionary connection between narrow and broad-line objects?
6. Is the nature of the starburst-AGN connection dependent on look-back time?

Some of these questions should be testable in the near future. For instance, in the merger-induced scenario starburst ages should increase along the sequence "starburst → cool ULIRGs → warm ULIRGs → quasars". Detailed spectroscopic studies should be able to answer this question. Increasingly sensitive techniques and instruments to detect obscured AGNs (e.g., infrared and X-ray spectroscopy from the ground and with satellites) will allow better constraints to be placed on the contribution of the AGN to the total energy output of galaxies (question #2). Questions regarding the starburst-AGN connection in the early universe will obviously be more difficult to answer. For these we probably have to wait for the next generation of ground-based telescopes and astronomical satellites to decipher the nature of the starburst-AGN connection in proto-galaxies. In the meantime, much effort should be invested in *predicting* what we should expect to see!

References

1. A. J. Baker, N. Z. Scoville: BAAS **30**, 862 (1998)
2. J. N. Bahcall, S. Kirhakos, D. H. Saxe, D. P. Schneider: ApJ **479**, 642 (1997)
3. A. Barger et al.: AJ in press (astroph/0007175) (2001)
4. J. E. Barnes, L. Hernquist: ApJ, **370**, L65 (1991)
5. R. Barvainis, D. Alloin, R. Antonucci: ApJ **337**, L65 (1989)
6. R. Barvainis, R. Antonucci, T. Hurt, P. Coleman, H. P. Reuter: ApJ **451**, L9 (1995)
7. S. A. Baum, T. M. Heckman, W. van Breugel: ApJ **389**, 208 (1992)
8. A. W. Blain et al.: MNRAS **302**, 632 (1999)
9. P. J. Boyce et al.: MNRAS **298**, 121 (1998)
10. M. S. Brotherton et al.: ApJ **520**, 87 (1999)
11. P. M. Bryant, N. Z. Scoville: AJ **117**, 2632 (1999)
12. R. Cid Fernandes, R. Terlevich: MNRAS **272**, 423 (1995)
13. D. L. Clements, W. J. Sutherland, R. G. McMahon, W. Saunders: MNRAS **279**, 477 (1996)
14. F. Combes: Lectures given at GH Advanced Lectures on the Starburst-AGN Connection, INAOE, eds. D. Kunth, I. Aretxaga (2000)
15. M. M. de Robertis, H. K. Yee, K. Hayhoe: ApJ **496**, 93 (1998)
16. D. Downes, P. M. Solomon: ApJ **507**, 615 (1998)
17. C. C. Dudley: MNRAS **307**, 553 (1999)
18. D. Dultzin-Hacyan: in IAU Symp. #186 Galaxy Interactions at Low and High Redshift, p. 329 (1998)
19. D. Dultzin-Hacyan, Y. Krongold, I. Fuentes-Guridi, P. M. Marziani: ApJ **513**, 111 (1999)
20. D. Dultzin-Hacyan, M. Moles, J. Masegosa: A&A **206**, 95 (1988)
21. A. S. Evans, D. B. Sanders, J. A. Surace, J. M. Mazzarella: ApJ **511**, 730 (1999a)
22. A. S. Evans et al.: ApJ **521**, L107 (1999b)
23. S. M. Faber et al.: AJ **114**, 1771 (1997)
24. A. Ferrara, E. Tolstoy MNRAS **313**, 291 (2000)
25. M. Franx et al.: ApJ **486**, L75 (1997)
26. D. T. Frayer et al.: ApJ **506**, L7 (1998)
27. D. T. Frayer et al.: ApJ **514**, L13 (1999)
28. T. Fuentes-Williams, J. T. Stocke: AJ **96**, 1235 (1988)

29. K. Gebhardt et al.: ApJ **539**, L13 (2000)
30. R. Genzel et al.: ApJ **498**, 579 (1998)
31. J. D. Goldader et al.: ApJ **444**, 97 (1995)
32. J. D. Goldader et al.: ApJS **108**, 449 (1997)
33. R. M. Gonzalez Delgado et al.: ApJ **505**, 174 (1998)
34. R. M. Gonzalez Delgado, T. M. Heckman, C. Leitherer: ApJ **546**, 000 (2001)
35. Q. Gu, D. Dultzin-Hacyan, J. A. de Diego: Rev. Mex. de Astron. y Astrof., **00**, 1 (astroph/0011419) (2001)
36. T. M. Heckman et al.: ApJ **482**, 114 (1997)
37. P. Heraudeau, F. Simien, G. A. Mamon: A&AS **117**, 417 (1996)
38. L. C. Ho, A. V. Filippenko, W. L. W. Sargent: ApJ **487**, 591 (1997)
39. J. B. Hutchings et al.: ApJ **429**, L1 (1994)
40. L. J. Kewley, C. A. Heisler, M. A. Dopita, S. Lumsden: ApJ in press (2001)
41. K. Kohno, R. Kawabe, K. Sakamoto, S. Ishizuki, B. Vila-Vilaro: in The Central Regions of the Galaxy and Galaxies, ed. Y. Sofue (Dordrecht: Kluwer), 239 (1998)
42. D.-C. Kim, S. Veilleux, D. B. Sanders: ApJ **508**, 627 (1998)
43. J. Kormendy, D. Richstone: ARAA **33**, 581 (1995)
44. J. Kormendy, D. B. Sanders: ApJ **390**, L53 (1992)
45. N. A. Levenson, K. A. Weaver, T. M. Heckman: ApJ in press (astroph/0012035) (2001)
46. D. Lutz et al.: ApJ **505**, L103 (1998)
47. D. Lutz, S. Veilleux, R. Genzel: ApJ **517**, L13 (1999)
48. J. Magorrian et al.: AJ **115**, 2285 (1998)
49. R. Maiolino, M. Ruiz, G. Rieke, L. Keller: ApJ **446**, 561 (1995)
50. K. K. McLeod, G. H. Rieke: ApJ **441**, 96 (1995)
51. M. Meixner, R. Puchalsky, L. Blitz, M. Wright, T. Heckman: ApJ **354**, 158 (1990)
52. J. Melnick, I. F. Mirabel: A&A **231**, L19 (1990)
53. J. C. Mihos, L. Hernquist: ApJ, **431**, L9 (1994)
54. I. F. Mirabel, D. B. Sanders, I. Kazés: ApJ **340**, L9 (1989)
55. J. S. Mulchaey, M. W. Regan: ApJ **482**, L135 (1997)
56. T. W. Murphy et al.: AJ **111**, 1025 (1996)
57. R. F. Mushotzky et al.: Nature **404**, 459 (2000)
58. C. Nelson, M. Whittle: Ap&SS **99**, 67 (1995)
59. C. Norman, N. Z. Scoville: ApJ **332**, 124 (1988)
60. K. Ohta et al.: Nature **382**, 426 (1996)
61. A. Omont et al.: A&A **315**, 10 (1996)
62. J. Perry: in Relationships between AGNs and starburst galaxies, ed. A. V. Filippenko, ASP Conference Series (ASP: San Francisco), 31, p. 169 (1992)
63. M. Pettini et al.: ApJ **528**, 96 (2000)
64. E. A. Pier, J. H. Krolik: ApJ **418**, 673 (1993)
65. D. Rigopoulou et al.: ApJ **118**, 2625 (1999)
66. M. Rowan-Robinson: MNRAS **272**, 737 (1995)
67. J. M. Rodriguez-Espinosa, R. J. Rudy, B. Jones B.: ApJ **309**, 76 (1986)
68. K. Sakamoto et al.: ApJ **514**, 68 (1999)
69. D. B. Sanders et al.: ApJ **325**, 74 (1988a)
70. D. B. Sanders et al.: ApJ **347**, 29 (1989a)
71. D. B. Sanders et al.: A&A **213**, L5 (1989b)
72. D. B. Sanders et al.: ApJ **325**, 74 (1998)
73. D. B. Sanders, N. Z. Scoville, B. T. Soifer: ApJ **335**, L1 (1988b)
74. N. Z. Scoville et al.: AJ **119**, 991 (2000)

75. N. Z. Scoville et al.: ApJ **252**, L455 (1993)
76. E. P. Smith, T. M. Heckman: ApJ **341**, 658 (1989)
77. P. M. Solomon, D. Downes, S. J. E. Radford, J. W. Barrett: ApJ **478**, 144 (1997)
78. J. A. Surace, D. B. Sanders: ApJ **512**, 162 (1999)
79. L. Tacconi, E. Schinnerer, J. F. Gallimore, R. Genzel, L. E. Tacconi-Garman, D. Downes: BAAS **29**, 1333 (1997)
80. C. N. Tadhunter, R. C. Dickson, N. A. Shaw: MNRAS **281**, 591 (1996)
81. E. Terlevich, A. I. Diaz, R. Terlevich: MNRAS **242**, 271 (1990)
82. H. D. Tran et al.: ApJ **516**, 85 (1999)
83. S. Veilleux et al.: ApJS **98**, 171 (1995)
84. S. Veilleux, D.-C. Kim, D. B. Sanders: ApJ **522**, 113 (1999a)
85. S. Veilleux, D.-C. Kim, D. B. Sanders: ApJ submitted (2001)
86. S. Veilleux, D. B. Sanders, D.-C. Kim: ApJ **484**, 92 (1997)
87. S. Veilleux, D. B. Sanders, D.-C. Kim: ApJ **522**, 139 (1999b)
88. S. N. Virani, M. M. De Robertis, M. L. VanDalfsen: AJ **120**, 1739 (2000)
89. M. Whittle: ApJ **387**, 121 (1992a)
90. M. Whittle: ApJS **79**, 49 (1992b)
91. R. J. R. Williams, A. C. Baker, J. J. Perry: MNRAS **310**, 913 (1999)
92. A. S. Wilson: A&A **206**, 41 (1988)
93. H. Wu, Z. L. Zou, X. Y. Xia, Z. G. Deng: A&AS **127**, 521 (1998a)
94. H. Wu, Z. L. Zou, X. Y. Xia, Z. G. Deng: A&AS **132**, 181 (1998b)
95. S. E. Zepf, D. C. Koo: ApJ **337**, 34 (1989)
96. Z. Zheng et al.: A&A **349**, 735 (1999)

Finding Signatures of the Youngest Starbursts

Henry A. Kobulnicky[1] and Kelsey E. Johnson[2]

[1] University of Wisconsin, Department of Astronomy, 475 N. Charter St., Madison, WI 53706, USA
[2] JILA, University of Colorado, Boulder, CO 80309, USA

Abstract. Embedded massive starclusters have recently been identified in several nearby galaxies by means of the radio-wave thermal bremsstrahlung emission from their surrounding HII regions. Energy requirements imply that these optically-obscured starclusters contain 500–1000 O-type stars, making them similar to the "super starclusters" observed in many dwarf starbursts and mergers. Based on their high free-free optical depth and visual extinctions of $A_V \gg 10$ mag, these massive "ultra-dense" HII regions (UDHIIs) are distinct signatures of the youngest, most compact super starclusters. UDHII regions may represent the earliest stages of globular cluster formation. We review the properties of presently-known UDHIIs, and we outline a pictoral evolutionary taxonomy for massive cluster formation which is analogous to the more familiar evolutionary sequence for individual stars.

1 The Youngest Stages of Massive Star Formation

Massive $(M > 10\,M_\odot)$ stars are observed predominantly in dense clusters or OB associations. They are associated with high visual extinctions and large masses of molecular gas [7]. Current theories of massive star formation are incomplete. Some propose that massive stars form by molecular cloud collapse and subsequent accretion like low-mass stars, while others require that massive stars form via mergers and accretions of stellar-mass objects in high-density $(\rho > 10^4\,\mathrm{stars/pc^3})$ molecular clumps at the bottom of the cluster gravitational potential ([6] and references therein). Within massive molecular star-forming complexes (e.g., W49 in the Milky Way), the youngest massive stars are seen only indirectly via thermal bremsstrahlung emission from their surrounding HII regions. These ultra-compact HII regions (UCHII) have sizes of several hundred A.U., densities of $> 10^4\,\mathrm{cm^{-3}}$, emission measures of $> 10^7\,\mathrm{pc\ cm^{-6}}$, visual extinctions of $A_V > 50$ mag, and typically contain 1–2 massive stars (review in [11]; [24,9,22,5]). The lifetime of UCHII regions is $\sim$10–15% of the lifetime of an O star ($\sim$500,000 yr) based on the observations that 10-15% of Galactic O stars lie obscured within dense molecular clouds [25].

Given this picture of the earliest phases of single massive stars, we might expect that the earliest phases of massive cluster formation could be modeled as a collection of several hundred UCHII regions. Identifying such objects is problematic. Many nearby galaxies contain young, massive, blue star clusters with typical ages of several Myr (see contributions by Whitmore and others in this volume), but these have been discovered predominantly at optical wavelengths, an approach which biases the current census of massive starclusters toward the least-obscured, and, therefore, older, more evolved examples (ages $\geq$2–3 Myr). Understanding of the formation and evolution of massive starclusters requires identifying objects in their formative proto-cluster stages ($\leq$ 1 Myr). These stages will be characterized by enormous visual extinctions ($A_V > 50$ mag) which mandate the use of radio–IR techniques to uncover the physics of cluster formation. Recent radio-wave studies have pinpointed the likely precursors of super starclusters still embedded in their molecular birthplaces.

2   A Case Study of the Blue Compact Galaxy Henize 2-10

The blue compact galaxy Henize 2-10 contains five radio sources with no optical counterparts and constant flux density over 10 year time baselines [13]. Figure 1 (upper left) shows a Hubble Space Telescope F555W broadband image of the central 300 pc region along with a VLA 2 cm map (lower left) and an HST Hα image. The radio sources have no optical counterparts in the F555W image, but there is a good correlation between the radio map and the Hα morphologies, suggesting that the source of the radio emission is related to the ionized gas. The radio sources have inverted spectral indices between 2 cm and 6 cm ($S_\nu \propto \nu^{+0.5 \pm 0.3}$), consistent with an optically-thick thermal bremsstrahlung origin (Fig. 1, upper right).

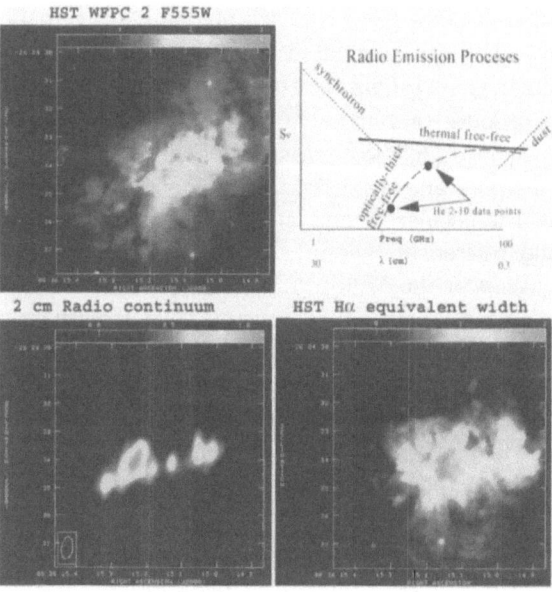

Fig. 1. Images of Henize 2-10 in optical HST F555W (top left), 2 cm radio continuum (lower left) and Hα equivalent width (lower right). The schematic at the upper right illustrates the frequency dependence of non-thermal (synchrotron emission), thermal bremsstrahlung, and optically-thick thermal bremsstrahlung emission mechanisms.

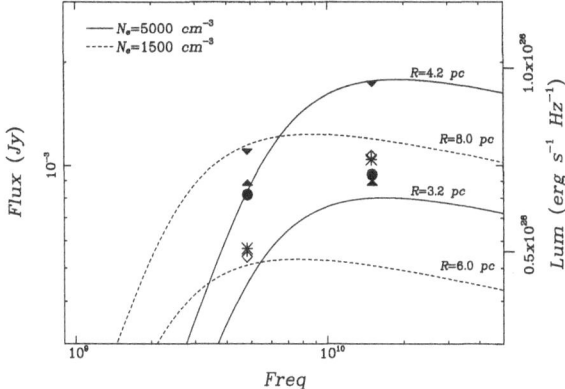

Fig. 2. VLA 6 cm (4.8 GHz) and 2 cm (14.9 GHz) fluxes and luminosities for the five radio knots in Henize 2-10. A different symbol represents data for each knot. Solid and dotted lines represent thermal bremsstrahlung plasmas modeled as spheres of radius, R, electron temperature $T_e = 6000$ K, and mean electron densities of 1500 cm^{-3} and 5000 cm^{-3}, respectively.

We model the spectral shape and luminosity of the observed sources as spheres of uniform-density plasma with an electron temperature of 6000 K. HII regions with radii between 3 pc and 8 pc, densities of 5000 cm^{-3}, and 500–1000 ionizing O7V stars are most consistent with the data. These high densities imply an overpressure compared to typical warm ionized medium pressures. Such HII regions should expand and become undetectable in the thermal radio continuum on timescales of 500,000 yr. Thus, is seems likely that the ages of these HII regions are very small, consistent with their heavily-obscured nature. For a typical Salpeter IMF extending from 0.5 to 100 M$_\odot$, the clusters contain 6×10^5 total stars, implying peak stellar densities of 5000 pc^{-3}.

Several other galaxies harbor UDHIIs discovered by their peculiar radio spectral index and high brightness temperature: one in NGC 5253 [20,19]; six in NGC 2146 [18]; 1–2 in NGC 4214 & Tololo 35 [4]. Radio recombination line studies of suggest the presence of UDHIIs in NGC 3628 and IC 694 [26].

3 Towards An Evolutionary Taxonomy for Massive Starclusters

3.1 Low Mass Stars

The evolutionary sequence of low-mass star formation has been outlined by Lada [14] and André, Ward-Thompson & Barsony [3]. Low-mass stars begin as prestellar molecular cores, and evolve through stages (Class 0 through Class III as depicted in Fig. 3) characterized by an increasing faction of infrared emission from the central star and a decreasing fraction of submillimeter emission from dust in the accretion disk.

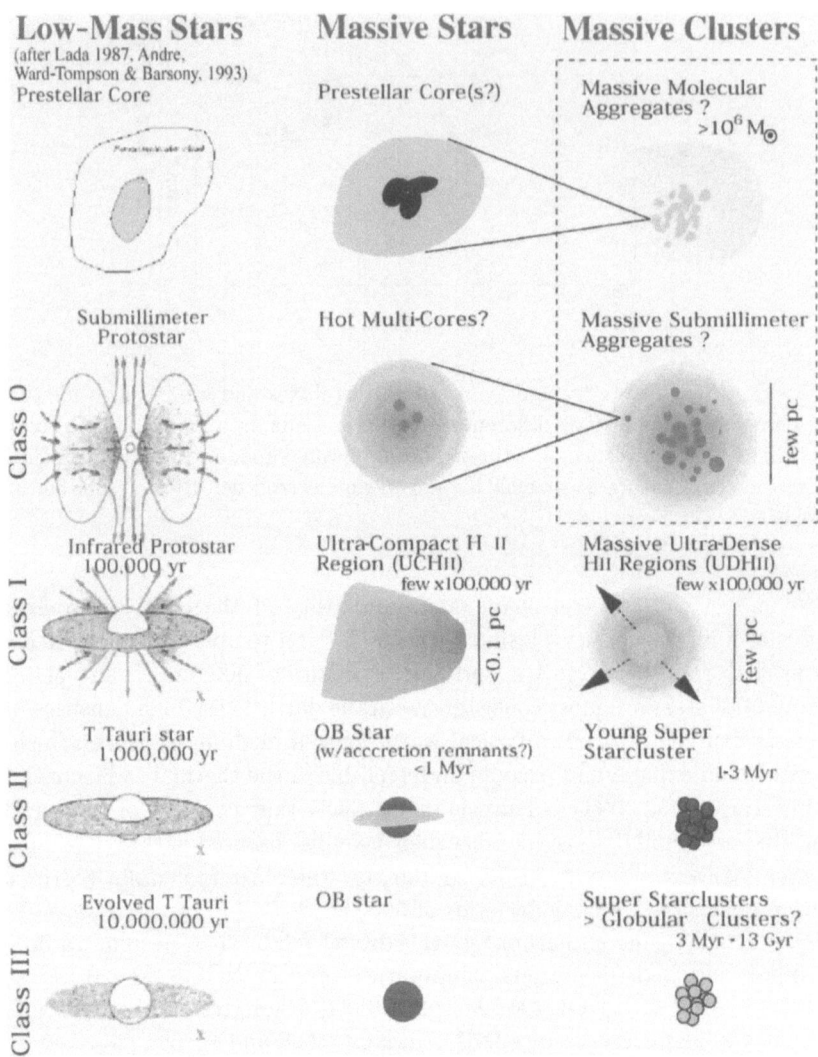

Fig. 3. A tentative, and somewhat speculative, schematic illustrating the stages of low-mass star formation (left, from [14]; [3]), massive star formation (center) and massive cluster formation (right). Ultra-dense HII regions represent the youngest phase of massive cluster formation yet identified. Dashed lines surround phases of massive cluster formation which are speculative and have no identified examples.

3.2 Massive Stars

For massive stars, no clearly defined evolutionary sequence has yet been agreed upon. In Fig. 3 (center column) we attempted to sketch an evolutionary sequence based on current literature. At the earliest times, massive stars begin as a collection of dense molecular cores (1). Because there are theoretical problems with forming stars more massive than $\sim 10\,M_\odot$ in the collapse of a single molecular core ([6] and references therein), mergers among nearly-formed proto-stars are invoked to produce the most massive objects (2). Once accretion or merging terminates and the massive star is formed, high-energy photons begin to ionize the surrounding molecular gas, producing an UCHII region (3). UCHII regions have sizes $\leq 0.1\,pc$, and often exhibit cometary morphologies. The massive star hidden within an UCHII region emerges from its natal molecular cloud through the combined actions of its stellar wind, ionizing radiation, and space velocity (4). Massive stars spend the latter 80% of their lifetimes visible as luminous O stars (5) (see review in [8]).

3.3 Massive Star Clusters

In the case of massive starclusters (Fig. 3, right column), our theoretical and observational understanding of their evolutionary phases is even less secure. Because bound massive stellar clusters contain $>$ few $\times\, 10^5\,M_\odot$ of stars in a region just a few pc in size [12,16] they must begin with the collapse and fragmentation of exceptionally compact and massive molecular structures exceeding $10^7\,M_\odot$. We term these "Massive Molecular Aggregates" in Fig. 3 (1). The star formation efficiencies in massive clusters must exceed the typical values of $\sim 0.5\%$ in Galactic SF regions [14] and 2–5% in M33 OB associations [22] in order to form the required mass of stars from typical 10^4–$10^6\,M_\odot$ Giant Molecular Clouds. Star formation efficiencies exceeding 50% appear reasonable in some young clusters (Arp 220: [2]; young Galactic clusters: [15]), and may also help increase the binding energy of the resultant cluster since there is less residual gas to be swept out of the gravitational potential [1,10]. At the center of the gravitational potential, an aggregate of thousands of warm, massive molecular cores in a space just a few pc across provides the environment for the formation of massive stars through mergers and accretion [6]. Such objects should be detectable as a collection of extremely compact subillimeter sources, so we term these "Massive Submillimeter Aggregates" (2). These first two stages are somewhat speculative schematics predicated on extrapolation of massive molecular complexes in the Milky Way where giant molecular clouds range only from 10^4-$10^6\,M_\odot$ and have sizes from 50–100 pc [17]. To our knowledge, *there have not yet been observed any molecular structures which are sufficiently massive and compact to be identified as the likely predecessors of super starclusters.* The newly-identified class of ultra dense HII regions (UDHIIs) are the massive analogs of the more familiar single-star UCHIIs, and they represent a transition phase (3) between the warm molecular and massive submillimeter aggregates and the familiar UV-bright starclusters (4,5).

Unusually compact, massive ($10^7\,M_\odot$) gravitationally-bound concentrations of molecular gas and dust on size scales of 2–6 pc could signify the genesis sites of super starclusters. However, these early phases have not yet been specifically identified. Wilson et al. [21] report the discovery of massive molecular resevoirs, termed "super giant molecular complexes", in the Antennae merger system (NGC 4038/39) which could be the source material for the "Massive Molecular Aggregates". Given the abundance of massive extragalactic starclusters, their predecessors, the "Massive Molecular Aggregates" and the "Massive Submillimeter Aggregates", should be detectable in extragalactic sources with current millimeter and upcoming sub-millimeter arrays. Identification of these precursors to massive starclusters is the first step toward understanding the dynamical conditions that produce OB associations, super starclusters, and the venerable globular cluster systems.

Acknowledgments: We are grateful for conversations with Ed Churchwell, Peter Conti, Jay Gallagher, and Jonathan Tan which have helped to clarify this presentation.

References

1. F.C. Adams: ApJ **542**, 964 (2000)
2. K.R. Anantharamaiah, F. Viallefond, N.R. Mohan, W.M. Goss, J.H. Zhao: ApJ **537**, 613 (2000)
3. P. André, D. Ward-Thompson, M. Barsony: ApJ **406**, 122 (1993)
4. S.C. Beck, J.L. Turner, O. Kovo: AJ **120**, 244 (2000)
5. E.E. Becklin, G. Neugebauer, C.G. Wynn-Williams: ApJ **182**, 7 (1973)
6. I.A. Bonnell, M.R. Bate, H. Zinnecker: MNRAS **298**, 93 (1998)
7. A.G.A. Brown, A. Blaauw, R. Hoogerwerf, J.H.J. de Bruijne, P.T. de Zeeuw, P. T.: 'OB Associations'. In: *The Origin of Stars and Planetary Systems*. ed. by C.J. Lada, N.D. Kylafis (Kluwer Academic Publishers, Dordrecht 1999) pp. 411–
8. E.B. Churchwell: A&ARv **2**, 79 (1990)
9. J.W. Dreher, K.J. Johnston, W.J. Welch, R.C. Walker: ApJ **283**, 632 (1984)
10. S.P. Goodwin: MNRAS **286**, 669 (1997)
11. H.J. Habing, F.P. Israel: ARA&A **17**, 345 (1979)
12. L.C. Ho, A.V. Filippenko: ApJ **472**, 600 (1996)
13. H.A. Kobulnicky, K.E. Johnson: ApJ **527**, 154 (1999)
14. C.L. Lada: 'Star Formation: From OB Associations to Protostars'. In *Star Forming Regions, Proceedings of IAU Symposium 115*, ed. by J. Jugaku, M. Peimbert (Reidel, Dordrecht 1987) pp. 1–17
15. E.A. Lada, N.J. Evans II, E. Falgarone: ApJ **488**, 286 (1997)
16. R.W. O'Connell, J.S. Gallagher III, D.A. Hunter: ApJ **433**, 65 (1994)
17. D.B. Sanders, N.Z. Scoville, P.M. Solomon: ApJ **289**, 373 (1985)
18. A. Tarchi, N. Neininger, A. Greve, U. Klein, U. et al.: A&A **358**, 95 (2000)
19. J.L. Turner, S.C. Beck, P.T.P. Ho: ApJ **532**, L109 (2000)
20. J.L. Turner, P.T.P. Ho, S.C. Beck: AJ **116**, 1212 (1998)
21. C.D. Wilson, N.Z. Scoville, S.C. Madden, V. Charmandaris: ApJ **542**, 120 (2000)
22. C.D. Wilson, B.C. Matthews: ApJ **455**, 125 (1995)
23. C.G. Wynn-Williams: MNRAS **151**, 397 (1971)
24. D.O. Wood, E. Churchwell: ApJS **69**, 831 (1989a)
25. D.O. Wood, E. Churchwell: ApJ **340**, 265 (1989b)
26. J.-H. Zhao, K.R. Anantharamaiah, W.M. Goss, F. Viallefond: ApJ **482**, 186 (1997)

The BIGGEST Starbursts at the HIGHEST Redshifts

Wil van Breugel[1], Michiel Reuland[1,2,3], Carlos De Breuck[4],
and Huub Röttgering[3]

[1] Univ. of California, LLNL, L-413, Livermore, CA 94550, USA
[2] Univ. of California, UC Davis, Dept. of Physics, Davis, CA 95616, USA
[3] Leiden Observatory, P.O. Box 9513, 2300 RA Leiden, The Netherlands
[4] Institut d'Astrophysique de Paris, 98bis Boulevard Arago, 75014 Paris, France

Abstract. The BIGGEST starbursts at the HIGHEST redshifts are associated with high redshift radio galaxies (HzRGs). They may well mark the beginnings of the most massive galaxies seen today. We review current evidence in support of large scale starbursts in HzRGs, and will discuss recent results of sub–mm observations.

1 Introduction

High redshift radio galaxies (HzRGs) are the likely progenitors of massive galaxies seen today. At low redshifts ($z < 1$), powerful radio sources are uniquely identified with massive elliptical galaxies. The well-behaved near–infrared 'Hubble' $K - z$ relation for such galaxies [4] then suggests that radio galaxies at very high redshifts may be associated with massive galaxies in their early stages of formation.

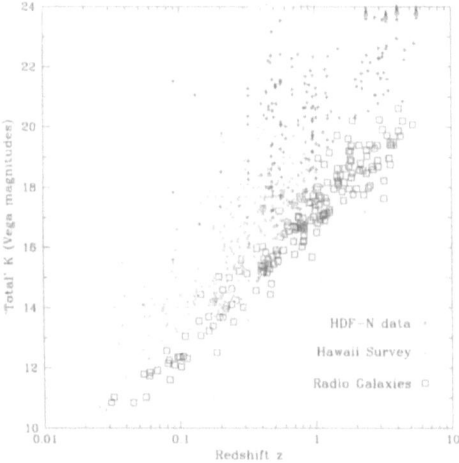

Fig. 1. The Hubble $K - z$ diagram for galaxies [4]. Radio galaxies (open squares) are the brightest and most luminous systems at all redshifts.

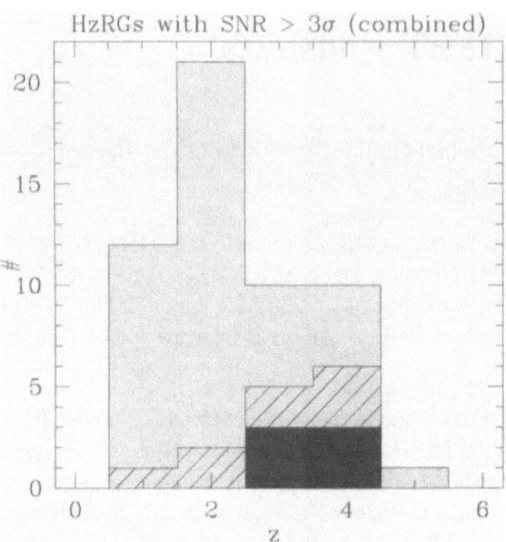

Fig. 2. Histogram of 850μm SCUBA detections with S/N > 3 (dashed) versus the total number of observed radio galaxies (grey) as a function of redshift. The data are from [1,14], with new $z > 3$ detections by [14] shown in black. At $z > 2.5$ approximately 50% of the galaxies are detected, as opposed to < 10% at $z < 2.5$.

There is increasing evidence that supports this. HST observations of HzRGs show swarms of compact knots embedded in very extended (~ 50 kpc) *rest–frame* UV continuum halos [16,13]. The sizes, UV_{rest} luminosities and star formation rates (SFRs) of these *individual* components are similar to the 'Lyman Break galaxies' (LBGs) galaxies found in field surveys at $z \sim 3-4$ [15]. In at least two HzRGs there is direct, spectroscopic evidence for massive star formation based on stellar absorption–line spectra [5,6].

Other evidence that HzRGs may be forming massive galaxies is that several appear to have huge (100 − 150 kpc), rotating (?) Ly–α halos. These include the first $z > 3$ radio galaxy discovered, B2 0902+34 (z = 3.391 [9,7,10]), 4C 41.17 (z = 3.800 [2]), 4C 1243+036 (z = 3.570 [17]), and MRC 1138−262 (z = 2.156 [11]). In standard CDM cosmogonies one could envisage that these Ly–α nebulae are evidence for accretion in large dark matter halos, signaling the formation of massive galaxies through merging of smaller starburst systems or cooling flows.

2   Dust and Molecular Gas in HzRGs

Many of the HzRGs have been detected at sub–mm continuum wavelengths [8,1,14] indicating the presence of large amounts of dust and implying star formation rates of $\sim 1000-2000$ M$_\odot$/yr. Moreover, follow–up CO mm–interferometry and continuum studies of two $z > 3.5$ HzRGs have shown that this star formation occurs over galaxy–wide scales of ~ 30 kpc [12]. The large star formation

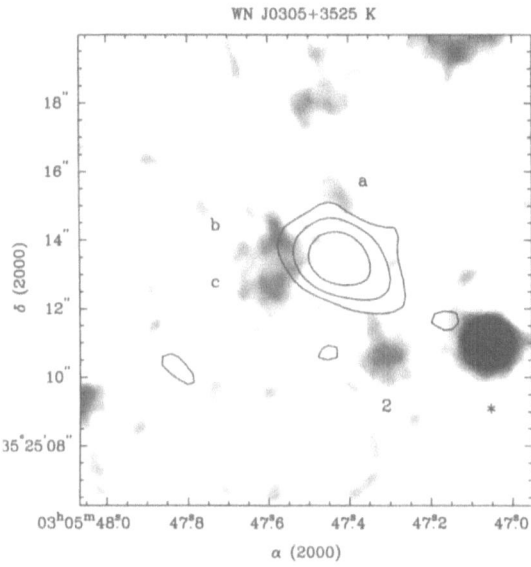

Fig. 3. Keck near–IR K–band identification of the probable $z = 4.21$ radio galaxy WN J0305+3525 [14]. The bright object to the SW is a spectroscopically confirmed star.

rates and scale sizes are strong additional evidence that HzRGs may indeed be massive forming galaxies.

Recently we have considerably enlarged the number of known HzRGs using 'Ultra Steep Spectrum' (USS) radio source selection and near–IR identification [3]. We have now begun a program to obtain JCMT/SCUBA sub–mm (850 μm) and IRAM/MAMBO (1.2 mm) observations to determine the presence of dust in these systems. Our SCUBA observations doubled the number of known $z > 3$ radio galaxies with solid ($\geq 3\sigma$) detections ([14], Fig. 2).

From these observations two important conclusions can be drawn: (i) on average the FIR–luminosity at $z > 3$ is at least a factor of 5–10 larger than at $z \sim 1$, consistent with the scenario in which the highest–z objects are massive starbursts and evolve into more quiescent objects at $z = 1$; (ii) since not all HzRGs are detected, the spread in FIR–luminosity at $z > 3$ varies greatly from object to object (e.g. $L_{FIR} > 10^{13}$ $L_\odot$, and $L_{FIR} < 10^{12}$ $L_\odot$).

3   WN J0305+3525

One of the USS sources in our sample, WN J0305+3525, has a tentative redshift of $z \sim 4.21$ and is identified with a faint multicomponent system in K–band (Fig. 3). SCUBA and MAMBO observations strongly detected this source at 850 μm and 1.2 mm respectively (Fig. 4). Assuming standard dust models and cosmological parameters these observations show that WN J0305+3525 is among

WN J0305+3525

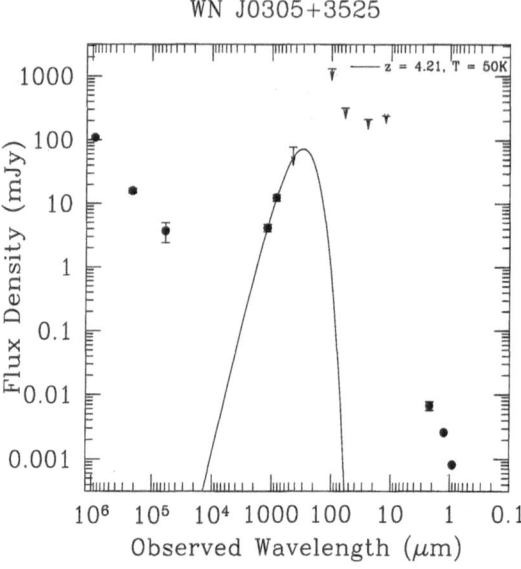

Fig. 4. Radio to Near-IR Spectral Energy Distribution of WN J0305+3525 using data from the VLA, IRAM, JCMT, IRAS and Keck. Extrapolation of the Ultra Steep radio Spectrum shows that the contribution of the non–thermal radio emission is negligible at sub–mm wavelengths. The solid line is an isothermal greybody spectrum for a conservative emissivity index $\beta = 2.0$ and dust temperature T = 50 K [12] at $z = 4.21$.

the most luminous (sub–)mm HzRGs known, with $L_{FIR} \sim 10^{13}\, L_\odot$, $M_{dust} \sim 1.5 \cdot 10^8\, M_\odot$ and SFR $\sim 1500\, M_\odot\, yr^{-1}$. WN J0305+3525 is very faint in the optical (I = 23.8), indicating that it is nearly completely obscured by dust. From our survey we have found several other such obscured HzRGs.

4   Conclusions

The evidence that HzRGs are massive forming galaxies comes from several types of observations, the key results of which are summarized below.

- Sizes
    - $UV_{rest} > 50$ kpc
    - Ly–$\alpha > 150$ kpc
    - CO ~ 30 kpc
- SFR
    - $UV_{rest} > 200\, M_\odot\, yr^{-1}$
    - $FIR_{rest} \sim 1000 - 2000\, M_\odot\, yr^{-1}$
- Luminous
    - $L_{UV} \sim 2 \times 10^{42}$ erg s^{-1} Å^{-1}
    - $L_{FIR} \sim 10^{13}\, L_\odot$

- Massive
  - $M_{\mathrm{dust}} \sim 10^8\,M_\odot$
  - $M_{\mathrm{H_2}} \sim 5 \times 10^{10} - 10^{11}\,M_\odot$

Further far–IR and (sub)–mm observations of HzRGs, in particular of highly obscured systems like WN J0305+3525, using SIRTF and future (sub)–mm interferometers will be needed to study the BIGGEST starbursts at the HIGHEST redshifts.

Acknowledgements: This work was performed under the auspices of the U.S. Department of Energy by the University of California, Lawrence Livermore National Laboratory under contract No. W-7405-Eng-48.

References

1. E.N. Archibald, J.S. Dunlop, D.H. Hughes, S. Rawlings, S.A. Eales, R.J. Ivison: MNRAS, submitted, astro-ph/0002083
2. K.C. Chambers, G.K. Miley, W.J.M. van Breugel: ApJ **363**, 21 (1990)
3. C. De Breuck, W. van Breugel, H. Röttgering, D. Stern, G. Miley, W. de Vries, S.A. Stanford, J. Kurk, R. Overzier: AJ, submitted
4. C. De Breuck, W. van Breugel, H. Röttgering, G. Miley: A&AS **143**, 303 (2000)
5. A. Dey, W. van Breugel, W.D. Vacca, R. Antonucci: ApJ **490**, 698 (1997)
6. A. Dey: in Proc. KNAW Colloq. 49, (2000), p.19
7. P. Eisenhardt, M. Dickinson: ApJ **399**, L47 (1992)
8. R.J. Ivison, J.S. Dunlop, D.H. Hughes, E.N. Archibald, J.A. Stevens, W.S. Holland, E.I. Robson, S.A. Eales, S. Rawlings, A. Dey, W.K. Gear: ApJ **494**, 211 1998)
9. S.J. Lilly: ApJ **333**, 161 (1988)
10. J.M. Martin-Mirones, E. Martinez-Gonzalez, J.I. Gonzalez-Serrano, J.L. Sanz: ApJ **440**, 191 (1995)
11. G. Miley et al.: "Watching the Birth of a Galaxy Cluster?", ESO Press Release 30 July 1999
12. P.P. Papadopoulos, H.J.A. Röttgering, P.P. van der Werf, S. Guilloteau, A. Omont, W.J.M. van Breugel, R.P.J. Tilanus: ApJ **528**, 626 (2000)
13. L. Pentericci, H.J.A. Röttgering, G.K. Miley, H. Spinrad, P.J. McCarthy, W.J.M. van Breugel, F. Macchetto: ApJ **504**, 139 (1998)
14. M. Reuland et.al., in preparation
15. C.C. Steidel, K.L. Adelberger, M. Dickinson, M. Giavalisco, M. Pettini, M. Kellogg: ApJ **492**, 428 (1998)
16. W. van Breugel et al.: in Proc. KNAW Colloq.49, (1999), p.49
17. R. van Ojik, H.J.A. Röttgering, C.L. Carilli, G.K. Miley, M.N. Bremer, F. Macchetto: A&A **313**, 25 (1996)

Star Clusters in Merging Galaxies

Bradley C. Whitmore

Space Telescope Science Institute, 3700 San Martin Dr., Baltimore, MD, 21218, USA

Abstract. Many of the most active starburst galaxies are mergers. Observations with the Hubble Space Telescope have revealed the presence of large numbers of young (typically 1 - 500 Myr), massive (10^3 - 10^7 M$_\odot$), compact (R$_{eff}$ $\approx$ 3 pc) star clusters in merging galaxies. The brightest of these clusters have all the attributes expected of protoglobular clusters, hence allowing us to study the formation of globular clusters in the local universe rather than trying to ascertain how they formed $\approx$ 14 Gyr ago. Similar clusters are also found in non-merging starbursts, barred galaxies, and even normal spiral galaxies. A compilation of the literature indicates that all these systems have similar cluster luminosity functions, with the primary difference being the normalization (i.e., roughly a tenfold increase in the number of clusters in mergers). Hence, the fact that the brightest clusters are in mergers may be largely a statistical result.

1   Introduction and a Brief History

Star formation is, arguably, the most important process in astronomy, with relevance for everything from the formation of planets, to the production of the elements in our bodies, to the detection of galaxies at the edge of the universe. Hence, it is ironic that while we have very detailed models of the structure and evolution of stars (e.g., observations fit theoretical isocrones in the HR diagram very well), we have only sketchy ideas of how stars form to begin with. An obvious approach to solving this problem is to study galaxies where lots of stars are forming, such as merging and starbursting galaxies. When we do this we find that a large fraction of the star formation is in the form of massive, compact star clusters. Understanding what triggers the formation of these clusters will go a long way toward understanding star formation in general.

A hint that young globular clusters might be formed in mergers was provided by Schweizer's [36] observations of six unresolved bluish knots in the merger remnant NGC 7252. However, with so few objects he could not be sure they were not simply field stars. Lutz [27] observed roughly a dozen blue point-like objects in the merger remnant NGC 3597, but he was not able to resolve the objects, and hence could not be certain they were not associations or giant HII regions.

The HST observations of about 60 blue compact clusters in NGC 1275 (the central cD galaxy in the Perseus cluster) by Holtzman et al. [19] was the primary catalyst in this field. Unfortunately, NGC 1275 is such a peculiar galaxy that it was not clear which of its peculiarities were responsible for the formation of the young clusters. Whitmore [45], using WFPC1 observations of the prototypical

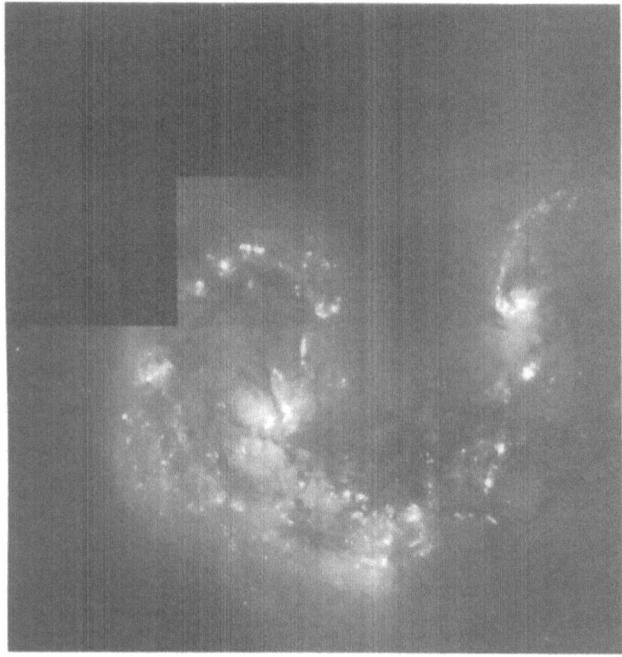

Fig. 1. Image of the Antennae Galaxies (NGC 4038/4039) from [46]

merger remnant NGC 7252 [40], found a population of about 40 blue point-like objects with luminosities and colors nearly identical to those found in NGC 1275. Whitmore and Schweizer [44] followed this up with pre-refurbishment observations of another prototypical merger, NGC 4038/4039 (the "Antennae" galaxies, see Figs. 1 and 2) with over 700 young star clusters. Subsequent observations of both these galaxies using WFPC2 (NGC 7252 - [33]; NGC 4038/4039 - [46] have increased the numbers of cluster candidates tenfold.

In total, roughly 30 different gas-rich mergers have now been observed with HST. In all cases young massive compact clusters have been observed, the brightest of which have the luminosities, colors, sizes, masses, distributions and spectra that are expected for young globular clusters (see Table 1 for a compilation, and [42]for a recent review).

2   A Case Study - The "Antennae"

Much of the early work in this field focussed on the question of whether globular clusters were being formed in mergers, since this was relevant to the question of whether spiral galaxies can merge to form elliptical galaxies. By observing merger remnants of different ages it was possible to observe the dimming and reddening of the clusters in order to put together a chronological sequence connecting the youngest mergers with the old ellipticals (e.g., [43]). However, to a large extent

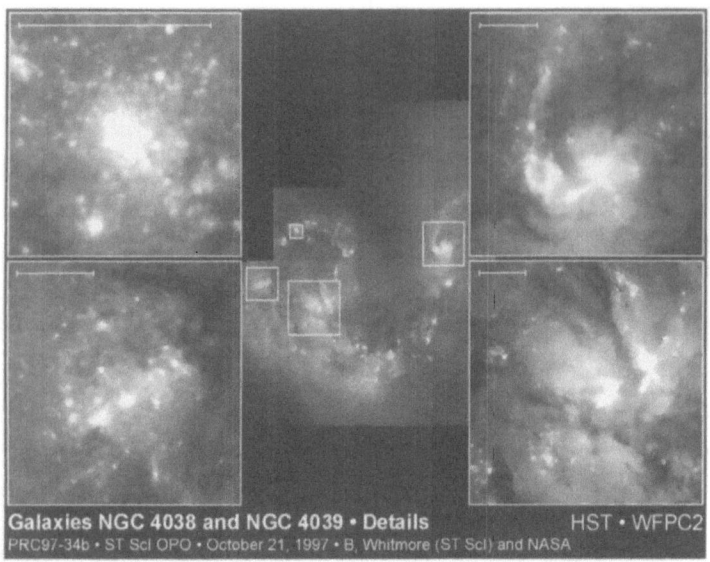

Fig. 2. Blowup of two of the brightest clusters in the Antennae (left; age estimates in the range 5 - 10 Myr) and the central regions of the two galaxies (right) from [46].

this can be done with a single galaxy, since the merging process takes several hundred million years to run to completion [32].

For example, [46] found four populations of clusters in the Antennae, the nearest and youngest of Toomre's [40] prototypical mergers. Using UBVI photometry and reddening-free Q parameters, Hα emission, and GHRS spectra of two of the brightest clusters, they find that clusters around the edge of the dust overlap region (i.e., the interface between the two colliding galaxies) appear to be the youngest, with ages < 5 Myr. Clusters in the western loop appear to be 5–10 Myr old, while clusters in the northeastern star-formation region appear to be ~100 Myr old, with a luminosity function in V that has shifted faintward by ~1.0 mag relative to the younger (0–20 Myr) clusters that dominate over most of the rest of the galaxy. A third cluster population consists of intermediate-age clusters (~500 Myr) that probably formed during the initial encounter responsible for ejecting the tails. Finally, a handful of old globular clusters from the progenitor galaxies have also been identified. Hence, it appears that we can study the entire evolution of globular clusters in this single galaxy.

The youngest clusters, essentially a fifth population in the Antennae, appear to be very red objects which [44] suggested were only now emerging from their dust cocoons. Several of these have recently been identified as strong IR sources (e.g., [41], [15], [30], [51]. In fact, the brightest IR source in the Antennae is one of these very red objects (W80), rather than the nucleus of one of the two galaxies. Wilson et al. [47] find three separate molecular clouds around W80 within a region of 1 kpc^2, and suggest that cloud-cloud collisions may play an

Table 1. Observations of Interacting Galaxies with Young Star Clusters.

Reference	Brief Description
Schweizer 1982 [36]	NGC 7252 (ground-based, 6 knots, stat. significant?)
Lutz 1991 [27]	NGC 3597 (ground-based, $\approx$10 knots, lacks resolution)
Holtzman et al. 1992 [19]	NGC 1275 (proposed "protoglobular clusters", n = 60)
Whitmore et al. 1993 [45]	NGC 7252 (prototypical merger, n = 40)
Whitmore & Schweizer '95 [44]	NGC 4038/4039 (n = 700, Antennae galaxies)
Zepf et al. 1995 [48]	NGC 1275 (ground-based spectra, .1 - 1 Gyr)
Holtzman et al. 1996 [20]	NGC 3597, NGC 6052 (mergers, not cooling flows)
Schweizer et al. 1996 [37]	NGC 3921 (102 candidate globulars, 49 "associations")
Hilker & Kissler-Patig '96 [18]	NGC 5018 (several hundred Myr to 6 Gyr)
Miller et al. 1997 [33]	NGC 7252 (n = 499, 3 pop., < 10 Myr for R < 6″)
Whitmore et al. 1997 [43]	NGC 1700, NGC 3610 (missing link with ellipticals ?)
Schweizer & Seitzer 1998 [38]	NGC 7252 (spectra, n = 8, ages, metallicities)
Brodie et al. 1998 [4]	NGC 1275 (ground-based spectra, age $\approx$ 450 Myr)
Carlson et al. 1998 [6]	NGC 1275 (n = 3000, mix of red and blue clusters)
Johnson et al. 1998 [22]	NGC 1741 (starburst, interacting, Hickson group)
Stiavelli et al. 1998 [39]	NGC 454 (5-10Myr, effects of emission on photometry)
Dinshaw et al. 1999 [9]	NGC 6090 (n = 4, NICMOS observations)
Zepf et al. 1999 [49]	NGC 3256 (n=1000, 15-20% of U light, break in LF ?)
Whitmore et al. 1999 [46]	NGC 4038/4039 (n=800 to 8000, break in LF ?)
Gallagher et al. 2000a [13]	Stephan's Quintet (n = 150, galaxies and tidal tails)
Alonso-Herrero et al. 2000 [1]	Arp 299 (ULIRG, n = 40)
Forbes & Hau 2000 [12]	NGC 3597 (ground-based, K band, α = -2)
Johnson & Conti 2000 [23]	HCG 31 (several in Hickson Compact Group 31)
Gilbert et al. 2000 [15]	NGC 4038/39 (IR spectra, ages, masses)
Mengel et al. 2000 [30]	NGC 4038/39 (IR spectra, ages, masses)
Georgakakis et al. 2000 [14]	NGC 6702 (dust-lane elliptical, 2- 5 Gyr)
Goudfrooij et al. 2000 [16]	NGC 1316 (elliptical with shells, 3 Gyr)

important role in cluster formation. However, the lack of similar morphologies for the other very red objects suggest that this may not be the universal mechanism.

To first order, the luminosity functions (hereafter LF) of young compact clusters in merging galaxies are power laws with index $\approx$ -2 (see §4). The power law index for the young clusters is markedly different than the Gaussian profile found for old globular clusters (e.g., Fig. 3 of [50]). However, various destruction mechanisms (e.g., 2-body evaporation, bulge and disk shocking, dynamical friction, stellar mass loss) should modify the distribution with time. Two-body evaporation appears to be the strongest amongst these mechanisms, destroying the fainter and more diffuse clusters first, and in certain conditions leaving a peaked distribution similar to what is seen for old globular clusters (e.g., [11]). Similarities between the luminosity functions of young clusters, and the mass functions of giant molecular clouds, have lead several groups to suggest that the progenitors of the young star clusters found in mergers are giant molecular clouds, and to propose various models for their formation (e.g., [21], [17], [37], [10]).

A closer look at the luminosity function of clusters in the Antennae by [46], using a variety of different techniques to decouple the cluster and stellar LFs, shows that the cluster LF has two power-law segments and a bend at $M_V \approx -10.4$ (≈ -11.4 after making a correction for extinction). For absolute magnitudes brighter than $M_V \approx -10.4$ the power law is steep and has an exponent of $\alpha = -2.6 \pm 0.2$, while for the range $-10.4 < M_V < -8.0$ the power law is flatter with $\alpha = -1.7 \pm 0.2$. Assuming a typical age of 10 Myr for the clusters, and 1 mag of extinction, the apparent bend in the LF corresponds to a mass $\approx 1 \times 10^5$ $M_\odot$, only slightly lower than the characteristic mass of globular clusters in the Milky Way ($\approx 2 \times 10^5$ $M_\odot$; [28]). Hence we may be seeing the precursor of what may become the peak in the globular cluster luminosity function.

3   Young Compact Clusters in Other Types of Galaxies

Young compact star clusters are also found in many starburst galaxies, but in much smaller numbers than for merging galaxies. For example, [31] found young compact clusters in nine nearby starburst galaxies using the Faint Object Camera on HST. On average, 20 % of the UV light from the galaxies comes from the clusters. They find the sizes are similar to Galactic globular clusters and the luminosity function has an index $\approx$ -2. Hence, the clusters found in the starburst galaxies appear to be similar to the clusters found in merging galaxies. Several other authors find similar examples in other starburst galaxies (see [42] for a complete list).

The case of M82, the prototypical starburst dwarf galaxy, is especially interesting. O'Connell et al. [34] find a complex of over 100 compact, luminous, "super star clusters" concentrated in the inner 100 pc of the galaxy shining though a relatively dust-free region. They estimate ages in the range from a few Myr to $\approx$ 50 Myr. [8]studied a region farther from the center of M82 where active star formation is not occurring. They estimate the ages of the clusters in this region are 20 - 100 Myr. Hence, there appears to be a range in ages for the clusters in M82, similar to results for the Antennae.

Young compact star clusters have also been found in barred galaxies (e.g., [3] and [5]), tidal tails [24], and normal spiral galaxies (e.g., [25], [26], and [7] for M33). Larsen and Richtler [26] argue that, "The cluster formation efficiency seems to depend on the SFR in a continuous way, rather than being related to any particularly violent mode of star formation". The lesson appears to be that luminous young star clusters are found whenever there is vigorous star formation. Since the ultraluminous IRAS sources are essentially all mergers [35], it is not surprising that mergers show the largest populations of young star clusters.

4   Is the Young Cluster Luminosity Function Universal?

Whitmore [42] has compiled a list of HST (and key ground-based) observations of young compact star clusters in merging, starbursting, and miscellaneous other galaxies. While it is clear that luminous young compact star clusters are produced in a wide variety of environments, they are observed in much greater number in mergers and starburst galaxies, systems where the most vigorous star formation is occurring. An important question is whether this is a statistical effect due to the lower number of clusters in galaxies with low star formation, or is it only possible to form the brightest and most massive young clusters in the chaotic systems (i.e., the mergers).

The situation may be analogous to the upper initial mass function in 30 Doradus. It was presumed that the large number of very luminous stars indicated that conditions in 30 Doradus were especially conducive for making high mass stars. However, [29] find that the IMF is normal; that the large number of massive stars is simply due to the tremendous number of stars in the system and the young age of the cluster.

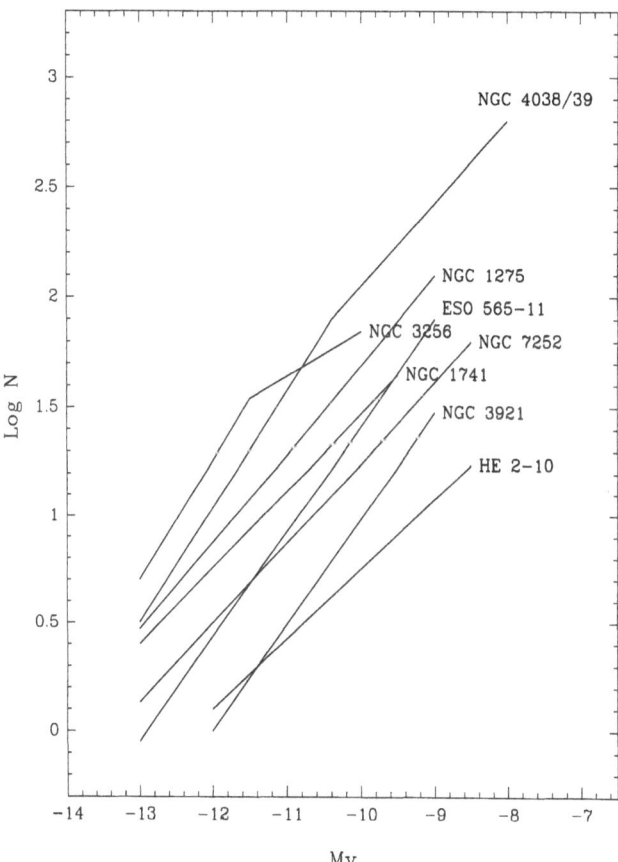

Fig. 3. Approximate luminosity functions for several merging and starbursting galaxies, normalized to have 0.25 mag bins (from [42]).

The most straightforward approach to answering this question is to examine the mass function of the clusters for a variety of galaxies. However, this is quite difficult given the large amounts of dust and the dimming caused by stellar evolution. The only galaxy where this has been attempted in detail is the Antennae [50]. However, another approach is to compare the luminosity functions, as shown in Fig. 3 (from [42]. We find that all the galaxies have luminosity functions with similar slopes, with an average power law index $\alpha = -1.93 \pm 0.06$ (uncertainty in the mean; the scatter is 0.18). The primary difference is the normalization of the luminosity function, with NGC 3256 and NGC 4038/39 having large numbers of clusters while, NGC 3921 and HE 2-10 have relatively few clusters. There is no obvious trend for a cutoff at high luminosity for the more quiescent galaxies, suggesting a universal luminosity function is a reasonable approximation at this point in time, albeit based on a small and nonuniform database.

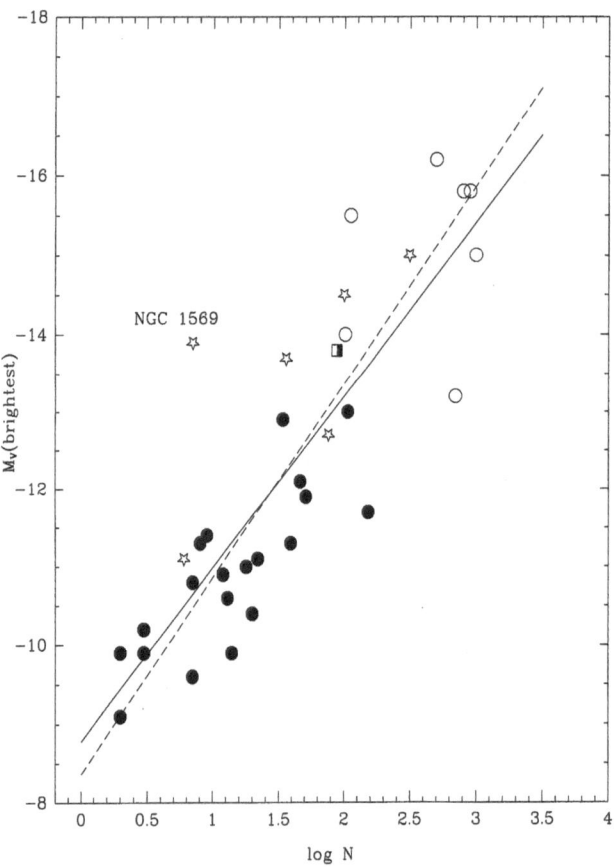

Fig. 4. Plot of the magnitude of the brightest cluster vs. the log of the number of clusters. Filled circles are spiral galaxies from [26], open circles are mergers, stars are starbursting galaxies, and the half filled square is a barred galaxy. The solid line is a best fit (excluding NGC 1569) while the dashed line is the prediction from a universal power law luminosity function with index $\alpha = -2$ (from [42]).

Such an approach is oversimplified for a number of reasons, primary amongst them being that the luminosities of the clusters vary with time. For example, a single-age burst population will evolve to the right in Fig. 3, making it difficult to determine whether the luminosity function is lower because of evolution or due to a smaller number of clusters. Other difficulties with this simplistic approach are that it assumes similar star formation histories for the various galaxies (e.g., continuous rather than sporadic bursts at different times), and ignores the fact that the faint end will probably undergo rapid evolution as the faint clusters dissolve. Nevertheless, to first order the luminosity functions appear to be remarkably similar in form.

Another approach, which allows us to increase the sample at the expense of more scatter for any particular galaxy, is to plot the magnitude of the brightest cluster vs. the number of clusters in the galaxy, as shown in Fig. 4 (from [42]). We find a clear trend between the number of clusters observed and the magnitude of the brightest cluster. The solid line is the fit to the data (excluding NGC 1569) with a slope = -2.3 ± 0.2 . The dotted line shows the trend expected if there is a universal luminosity function with α = -2 (i.e., a slope of -2.5 in Fig. 4), and the increase in the luminosity is simply due to a larger sample of clusters. Again, to first order it appears that a universal luminosity function can explain the data, even with the large scatter expected from low number statistics, non-uniform databases, differences in selection criteria, and differences in cutoff magnitudes (only those with cutoffs $\approx$ -9 have been included).

These results support the idea, originally proposed by [26], that the formation of young compact clusters does not require a violent mode of star formation. It appears that the main difference between the mergers and more quiescent galaxies is that the amplitude of the luminosity function has been increased tenfold or more, hence statistically producing larger numbers of very luminous (and presumably massive) clusters which may survive the various destructions mechanisms and become globular clusters. Hence, the conditions required to make globular clusters may exist "globally" in mergers and starburst galaxies, but only "locally" in certain regions of barred and spiral galaxies, hence the smaller numbers of clusters. Perhaps this is how many of the disk clusters in spiral galaxies like our own Milky Way are formed. Support for this idea comes from recent observations of M31 by [2] , who find that the disk globular clusters are younger than the halo clusters, with ages $\approx$ 8 Gyr.

5   A Final Thought

What fraction of all stars are formed in compact star clusters? As mentioned in §3, roughly 20 % of the UV light in merging and starbursting galaxies comes from young compact star clusters. In addition, the majority of the clusters may dissolve after $\approx$ 10 Myr, so the fraction of stars that were originally in clusters may be even higher than 20 %. The rapid destruction of most clusters would also explain the fact that in most of the recent mergers and starbursts where the clusters have been age-dated, the largest population of clusters are those with ages < 10 Myr. The alternative, that all these systems happen to be caught during a major burst, seems unlikely.

As an illustrative example, if we assume that the Antennae has been making clusters at roughly the same rate for the past 200 Myr, we can use figure 2 from [50] to show that for every 20 clusters originally formed, only 1 will survive to an age of $\approx$ 100 Myr. While this crude calculation may not be justifiable for a single galaxy, which we may be catching during a burst, once a larger sample becomes available a more careful calculation of this sort can be made.

In any case, it is clear that many of the clusters formed in mergers and star-bursting galaxies will be destroyed, and the stars will join the field population. If a large percentage of the stars in a galaxy are formed during starbursts and mergers, it is possible that the dominant mode of star formation in a galaxy is via the formation of compact star clusters.

References

1. A. Alonso-Herero, G. H. Rieke, M. J. Rieke, N. Z. Scoville: astroph/9911534 (2000)
2. P. Barmby, J. Huchra: ApJ **531**, L29 (2000)
3. A. J. Barth, L. C. Ho, A. V. Filippenko, W. L. W. Sargent: AJ **110**, 1009 (1995)
4. J. P. Brodie, L. L. Schroder, J. P. Huchra, A. C. Phillips, M. Kissler-Patig, D. F. Forbes: AJ **116**, 691 (1998)
5. R. Buta, D. A. Crocker, G. G. Byrd: AJ **118**, 2071 (1999)
6. M. N. Carlson, the WFPC2 team: AJ **115**, 1778 (1998)
7. R. Chandar, L. Bianchi, H. C. Ford, B. Salasnich: PASP **111**, 794 (1999)
8. R. de Grijs, R. W. O'Connell, J. S. Gallagher: AJ **119**, 681 (2000)
9. N. Dinshaw, A. S. Evans, H. Epps, N. Z. Scoville, M. Rieke: ApJ **525**, 702 (1999)
10. B. G. Elmegreen, Y. N. Efremov: ApJ **480**, 235 (1997)
11. S. M. Fall, Q. Zhang: in preparation (2001)
12. D. A. Forbes, G. K. T. Hau: MNRAS **312**, 703 (2000)
13. J. S. Gallagher, C. J. Homeier, C. J. Conselice: preprint (2000)
14. A. E. Georgakakis, D. A. Forbes, J. P. Brodie: astroph/0011275 (2000)
15. A. Gilbert, J. R. Graham, I. S. McLean, E. E. Becklin, J. Larkin, M. K. Wilcox, D. F. Figer, N. A. Levenson, H. I. Teplitz: this volume (2000)
16. P. Goudfrooij, J. Mack, M. Kissler-Patig, G. Meylan, D. Minniti: MNRAS, **in press** (2000)
17. W. E. Harris, R. E. Pudritz: ApJ **429**, 177 (1994)
18. M. Hilker, M. Kissler-Patig: A&A **314**, 357 (1996)
19. J. A. Holtzman, et al. (the WFPC team): AJ **103**, 691 (1992)
20. J. A. Holtzman, et al. (the WFPC2 team): AJ **112**, 416 (1996)
21. C. Jog P. Solomon ApJ **387**, 152 (1992)
22. K. E. Johnson, C. Leitherer, W. D. Vacca, P. S. Conti: AJ **117**, 1708 (1998)
23. K. E. Johnson, P. Conti: AJ **119**, 2146 (2000)
24. K. Knierman, S. C. Hunsberger, S. D. Gallagher, J. C. Charlton, B. C. Whitmore, A. Kundu, J. E. Hibbard, D. Zaritsky: in preparation (2001)
25. S. S. Larsen, T. Richtler: A&A **345**, 59 (1999)
26. S. S. Larsen, T. Richtler: A&A **354**, 836 (2000)
27. D. Lutz: A&A **245**, 31 (1991)
28. G. Mandushev, N. Spassova, A. Staneva: A&A **252**, 94 (1991)

29. P. Massey, D. Hunter: ApJ **493**, 180 (1998)
30. S. Mengel, M. D. Lehnert, N. Thatte, R. Genzel: this volume (2001)
31. G. R. Meurer, T. M. Heckman, D. Leitherer, A. Kinney, C. Robert, D. R. Garnett: AJ **110**, 2665 (1995)
32. J. C. Mihos, G. D. Bothun, D. O. Richstone: ApJ **418**, 82 (1993)
33. B. W. Miller, B. C. Whitmore, F. Schweizer, S. M. Fall: AJ **114**, 2381 (1997)
34. R. W. O'Connell, J. S. Gallagher, D. A. Hunter, W. N. Colley: ApJ **446**, L1 (1995)
35. D. B. Sanders, et al.: ApJ **325**, 74 (1998)
36. F. Schweizer: ApJ **252**, 455 (1982)
37. F. Schweizer, B. Miller, B. C. Whitmore, S. M. Fall: AJ **112**, 1839 (1996)
38. F. Schweizer, P. Seitzer: AJ **116**, 2206 (1998)
39. M. Stiavelli, N. Panagia, M. Carollo, M. Romaniello, I. Heyer, S. Gonzaga: ApJ **492**, L135 (1998)
40. A. Toomre: In: *The Evolution of Galaxies and Stellar Populations* ed. by B. M. Tinsley, R. B. Larson (Yale, New Haven 1977) pp. 401
41. L. Vigroux, et al.: A&A **315**, L93 (1996)
42. B. C. Whitmore: In: *A Decade of HST Science,* ed. by M. Livio (Cambridge University Press, Cambridge, 2001), in press.
43. B. C. Whitmore, B. W. Miller, F. Schweizer, S. M. Fall: AJ **114**, 1797 (1997)
44. B. C. Whitmore, F. Schweizer: AJ **109**, 960 (1995)
45. B. C. Whitmore, F. Schweizer, C. Leitherer, K. Borne, C. Robert: AJ **106**, 1354 (1993)
46. B. C. Whitmore, Q. Zhang, C. Leitherer, S. M. Fall, F. Schweizer, B. W. Miller: AJ **118**, 1551 (1999)
47. C. D. Wilson, N. Scoville, S. C. Madden, V. Charmandaris: ApJ **542**, 120 (2000)
48. S. E. Zepf, D. Carter, R. M. Sharples, K. M. Ashman: ApJ **445**, L19 (1995)
49. S. E. Zepf, K. M. Ashman, J. English, K. C. Freeman, R. M. Sharples: AJ **118**, 752 (1999)
50. Q. Zhang, S. M. Fall: ApJ **527**, L81 (1999)
51. Q. Zhang, S. M. Fall, B. C. Whitmore: submitted (?000)

Spectroscopy of Compact Star Clusters in NGC 4038/4039

Sabine Mengel, Matthew D. Lehnert, Niranjan Thatte, and Reinhard Genzel

Max-Planck-Institut für extraterrestrische Physik, Giessenbachstrasse
85748 Garching, Germany

Abstract. The large populations of young star clusters observed in interacting galaxies like NGC 4038/4039 ("The Antennae") are widely believed to be the progenitors of part of the globular cluster systems seen in local elliptical galaxies. For a comprehensive study of the young clusters in the Antennae we have obtained near infrared broad and narrow band images (SOFI on the NTT), integral field spectroscopy (MPE-3D at the AAT) and medium and high resolution spectroscopy (ISAAC/VLT-UT1 and UVES/VLT-UT2). We find that all of the bright star clusters are young ($< 20\,\mathrm{Myr}$), with the interaction region hosting the youngest clusters ($\sim 5\,\mathrm{Myr}$). The nuclear starbursts are older ($\sim 65\,\mathrm{Myr}$), but also show more recent star formation activity. Age variations on small spatial scales are seen throughout the merger. Cluster masses range from 10^5 to a few $\times 10^6\,\mathrm{M_\odot}$. A comparison between dynamically determined masses and those estimated from photometry in combination with starburst models suggests variations in the IMF from cluster to cluster.

1 Introduction

Over the last years, many interacting and merging galaxies have been found to host large numbers of newly formed star clusters [5,14,16,17]. It is very suggestive that they might evolve into the high metallicity part of the globular cluster population of elliptical galaxies when the merger is complete, but this hypothesis needs to be tested by determining the characteristics of individual young clusters and the cluster population as a whole. Globular clusters have typical masses of $1 \times 10^5\,\mathrm{M_\odot}$ and a mass function which is lognormal. The cluster system in the Antennae, however, has a power law luminosity function [17], and the same shape is also suggested for the mass function [19]. Masses determined from photometric data are as high as a few $\times 10^6\,\mathrm{M_\odot}$ for some of the clusters [9,19], which could represent the top end of the future globular cluster mass function. But it is also possible that some clusters will be destroyed during future evolution, which amongst other parameters depends on the relative fractions of low and high mass stars, and therefore on the initial mass function (IMF) [2]. Narrow band imaging and spectroscopy allows us to constrain the cluster ages by comparison to stellar evolutionary synthesis models. The photometric mass is determined from the observed K-band magnitude with respect to that expected for a cluster of given mass and age. It is therefore heavily dependent on model assumptions for the star formation parameters (timescale, IMF slope, limiting masses, metallicity). The dynamical mass is a more model independent estimate. From our spectroscopy in combination with HST imaging data that yield the cluster sizes, we determine the dynamical masses M_{dyn} of several clusters. A comparison with the corresponding photometric masses M_{phot} allows some conclusions about the IMF.

2 Observations, Analysis and Results

For near infrared (NIR) imaging of the Antennae we used SOFI on the NTT. The goal of these observations was an age dating and an extinction map of all the bright clusters in the merger. We selected broad band filters J, H and Ks, and the Brγ and CO (2.32 μm) narrow band filters. The field of view covered both galaxies, excluding the tails.

NIR integral field spectroscopy was performed using MPE-3D on the AAT. These data sets allow us to study the star formation history on smaller spatial scales and in more detail. Six fields ($6\rlap{.}''4 \times 6\rlap{.}''4$) were observed in the K-band, including the two nuclei and four star clusters in different regions of the merger. Data cubes were created, with 64×64 ($0\rlap{.}''2$/pix) spatial and 600 ($R \approx 1000$) spectral pixels.

Studying cluster velocity dispersions requires higher spectral resolution, which was provided by ISAAC and UVES. Two different slit positions including three young star clusters were observed with ISAAC on VLT-UT1, using the highest possible spectral resolution ($R \approx 9000$). The central wavelength was set to 2.31 μm, including the CO absorption bandheads and some continuum. Several red supergiants were observed as template stars. UVES echelle spectroscopy mainly aimed at getting high resolution ($R \approx 38,000$) spectra of the Calcium triplet absorption feature around 8500 Å. Three fields were observed, including a total of four clusters. Figure 1 shows the K-band image of NGC 4038/39, indicating the clusters selected for 3D, ISAAC and UVES observations.

The imaging and imaging spectroscopy data were used for age dating the single clusters and to derive the extinction. Very young clusters were identified by a large equivalent width in Brγ emission, whereas clusters with an emission dominated by red supergiants were traced by the equivalent width of CO absorption between 2.29–2.41 μm. The comparison with evolutionary synthesis models (Starburst99, [8]) determined the assigned age range. For starburst parameters we usually assumed an instantaneous burst, a Salpeter IMF between 1–100 M$_\odot$ and solar metallicity. Only for [W99]-2, where UVES spectra suggested a higher metallicity, did we assume $Z = 2 Z_\odot$. The ages of all the clusters detected in the SOFI images (≈ 150) are below 20 Myr. This is very remarkable since, according to models [1,12], last pericenter was around 200 Myr ago, and evidence of enhanced star formation dating back to this age could have been expected [11]. In the nuclei, the starbursts indeed started roughly 70 Myr ago. The region where the two disks overlap hosts several very extincted star clusters, and it is the site of the most recent star formation, as was suggested by ISO data [6,10,13]. Most of the star clusters there are around 5 Myr old, while the loop located to the north-west of the merger hosts clusters ~ 10 Myr old. Even though Whitmore et al. [17] see several star clusters which they assign an age of ~ 100 Myr, these

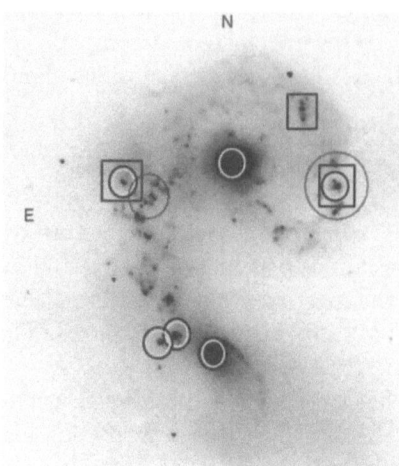

Fig. 1. Ks-band image of NGC 4038/4039, obtained with SOFI on the NTT. Overlaid are indicators of the positions observed with MPE-3D (small black and white circles), with ISAAC (large circles) and UVES (boxes).

cannot be the remains of a starburst of comparable strength to the one seen now, because otherwise they would be far more numerous. Our determination of the cluster ages shows that the triggering of the burst cannot arise from shocks driven through the galaxies as a result of the collision during pericenter, because regions at projected distances of several kpc host essentially coeval star clusters. The high gas densities in the star formation regions are more likely to be due to the gravitational response of the galaxies, which amplifies the tidal distortions and could lead to the formation of spiral structures as they are seen in NGC 4038/4039. On small spatial scales (a few $\times 100$ pc), age differences between neighbouring clusters are observed which might be caused by direct interaction, such as star formation triggered by supernovae. This is also true for the nuclei, where young clusters are seen on the background of an older starburst [9].

The extinction measurement was derived by comparing our Brγ emission map of the merger to the Hα image [17] from the HST and yielded the extinction for individual clusters, which varied between $0.3 \leq A_V \leq 5$ mag. Those clusters which had a high equivalent width in CO absorption were selected for the high resolution spectroscopy, K-band bright and obscured for ISAAC, I-band bright and unextincted for UVES spectroscopy.

The derivation of a cluster mass using a stellar velocity dispersion σ and the half light radius r_{hp} assumes a gravitationally bound cluster, the validity of the Virial Theorem, a spherically symmetric system, and that light traces mass: $M\langle v^2 \rangle \sim 0.4GM^2/r_h$ where $\langle v^2 \rangle$ is the RMS 3-dimensional stellar velocity and r_h the half mass radius. The conversion factor between the gravitational radius and the half mass/light radius depends on the cluster profile, but for a King

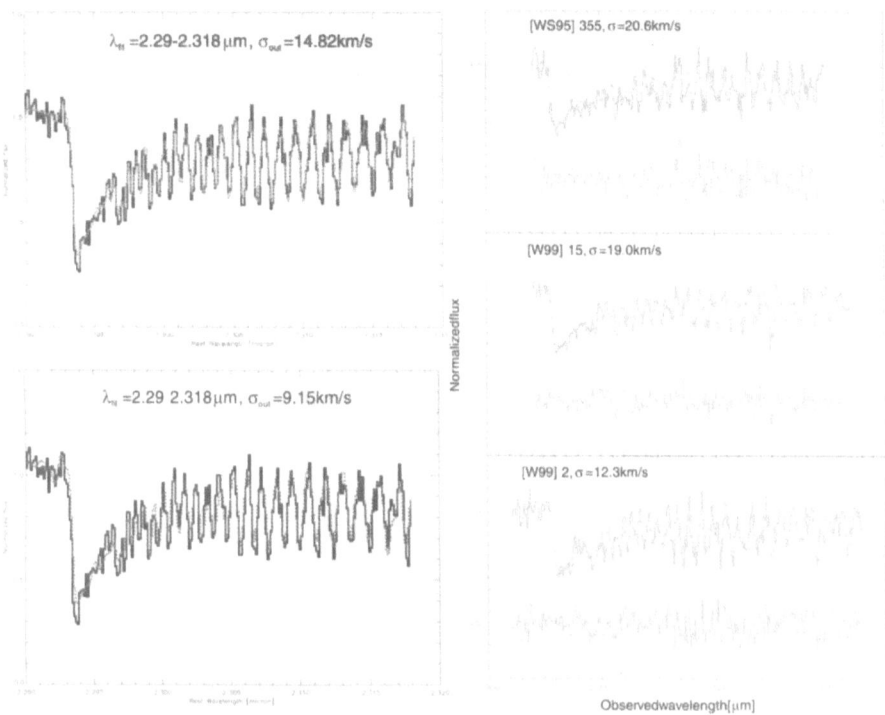

Fig. 2. For $\sigma_{\mathrm{in}} = 15$ km/s, SNR $= 15$ and a fitted wavelength range 2.29–2.318 μm, tests have been performed to determine the reliability of the results obtained for the cluster spectra. In a first step, matching templates were used, meaning that the same star was used for the creation of the simulated noisy, broadened spectrum (black, solid) and the redetermination of the velocity dispersion (template in grey, dotted). This case is shown in the top left panel. The next step was checking for the influence of template mismatch by fitting a simulated spectrum with the wrong and/or diluted template. This case is shown in the lower left panel. Note the large difference in the redetermined velocity dispersion σ_{out}. The mismatch in the tip and the rising edge of the bandhead is obvious. The right side of the figure shows the three cluster spectra obtained with ISAAC ([WS95]-355, [W99]-15 and [W99]-2 from top to bottom, solid) with their fits (dotted) and the residuals. In all cases a spectrum of a M3Iab supergiant, diluted with a continuum contribution of 14%, provided the best fit. The velocity dispersions determined from these fits are 20.6, 19.0 and 12.3 km/s, respectively.

model of moderate concentration, the factor 0.4 is a reasonable approximation. The dynamical mass is then: $M_{\mathrm{dyn}} \approx 10\sigma^2 r_{\mathrm{hp}}/G$. Since the cluster spectrum at an age of ~ 10 Myr is a superposition of red supergiants (M2I–M3I) and hot main sequence stars ($\sim$ B2V), a template spectrum is a combination of both. The relative B-star contributions were 14% in K-band and 35% in I-band. The velocity dispersion in the cluster spectra determined from comparison with these template spectra (via $\chi^2(\sigma)$ minimization) is sensitive to the selected template. We established two criteria for the best matching template: 1) the determined velocity dispersion σ needs to be relatively independent of the fitted wavelength range within the CO bandheads, and 2) the rising edge of the 2.29 μm bandhead reacts very sensitively to template mismatch and is required to be fitted well. An example of our tests is displayed in Fig. 2. In all cases a diluted spectrum of an M3Iab supergiant provided the best fit. These fits are also displayed in Fig. 2.

In principle, the comments above also apply to the UVES spectra. Several stellar absorption lines, such as the reddest line of the Calcium triplet at a rest wavelength of 8662 Å and three other weaker absorption lines were selected for the fit as it is shown in Fig. 3.

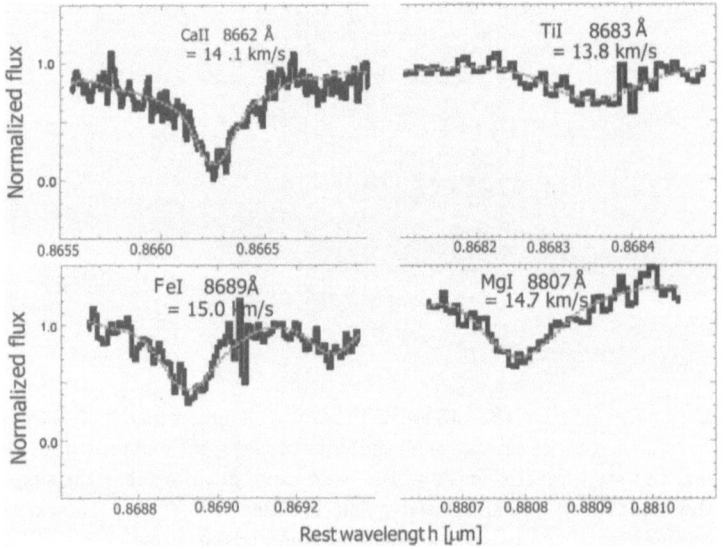

Fig. 3. Displayed are parts of the normalized spectrum of cluster [W99]-2 (black, solid), together with the fit (grey, dotted). The four pieces were fit separately, and the average value of the velocity dispersion is $\sigma = 14.2$ km/s. It is worth noting that this is the cluster which was also observed with ISAAC, and comparing the results (see Fig. 2).

Another critical parameter concerning the mass determination of the clusters is their projected half light radius r_{hp}. Even using HST, at the distance of the Antennae, clusters with $r_{hp} \approx 4$ pc are only slightly resolved, and it is even more difficult to determine the light profile model. We used the *ishape* routine developed by Larsen [7] to perform this task. *ishape* was designed to determine the sizes of marginally resolved sources with various possible light distributions. The selected light profile is convolved with the 2-dimensional point spread function and compared with the cluster image. By varying r_{hp}, χ^2 is minimized. For our clusters, moderately concentrated King profiles provided the best fit. For one cluster ([WS95]-355) we could not achieve a reliable fit and instead use the value kindly provided by Whitmore [18].

Table 1. The cluster masses as they were derived from the ISAAC and UVES spectra by comparison to the supergiant template spectrum. The data for clusters [WS95]-355 and [W99]-15 were from ISAAC, for [W99]-2 from both ISAAC and UVES ($\sigma = 12.3 \pm 1.2$ and 14.3 ± 1.0 respectively), and for [WS99]-1 from UVES. $M_{\rm vir}$ is the virial mass calculated from $M_{\rm vir}$ and $r_{\rm hp}$ (the projected half light radius determined using *ishape*; except [WS95]-355: provided by [18]). $M_{\rm phot}$ is the mass determined from the absolute extinction corrected K-band magnitude M_{K_0}, by comparison to that expected (Starburst99, [8]) for a cluster of the given age and with starburst parameters instantaneous burst, solar metallicity (except [W99]-2: twice solar) and Salpeter IMF between 1–$100\,M_\odot$. The comparison of virial and photometric mass in the last column is used for a statement on the IMF.

Cluster	Age	r_{hp}	M_{K_0}	σ	$M_{\rm vir}$	$M_{\rm phot}$	$\frac{M_{\rm vir}}{M_{\rm phot}}$
	$[10^6\,{\rm yr}]$	[pc]	[mag]	[km/s]	$[10^6\,M_\odot]$	$[10^6\,M_\odot]$	
WS95-355	8.5±0.3	4.8±0.5	-16.1±0.2	20.6±1.1	4.7±1.0	0.25±0.25	20
W99-15	8.7±0.3	3.6±0.4	-15.5±0.2	19.0±1.0	3.0±0.6	0.15±0.15	20
W99-2	6.6±0.3	4.6±0.4	-17.5±0.2	13.3±0.5	1.9±0.3	1.9±0.3	1.0
W99-1	8.1±0.5	3.6±0.4	-16.5±0.2	9.1±0.8	0.69±0.19	0.34±0.15	2.0

Table 1 summarizes our results for the mass determinations. It is obvious from the table that there are vast differences in the ratios of the two mass estimates between the four clusters if the same starburst parameters (with the exception of [W99]-2: $Z = 2\,Z_\odot$) are assumed for all of them. In all cases, $M_{\rm vir}$ is at least as high as $M_{\rm phot}$, in three cases even higher. If $M_{\rm phot}$ equals $M_{\rm vir}$ ([W99]-2), the assumed starburst model yields the same cluster mass as that determined dynamically. Since our assumed model had a lower mass cutoff at $1\,M_\odot$, this would indicate that there are either actually no stars below that mass present, or that the IMF slope, which we assumed to be Salpeter ($\alpha = 2.35$), is in reality flatter. In those cases where the observed ratio is larger than 1, the additional observed dynamical mass allows us to extrapolate the assumed IMF to masses below $1\,M_\odot$ (a factor of 2.6 would correspond to an extrapolation to $0.1\,M_\odot$ with Salpeter slope), or to steepen the IMF. For example, the results for [WS95]-355 and [W99]-15 would be compatible with an IMF with an exponent of $\alpha = 3.3$, and a lower mass cutoff of $0.1\,M_\odot$. Even though the uncertainties involved in the measurements are fairly substantial, there is an indication for IMF variations between different star clusters in NGC 4038/4039.

References

1. J.E. Barnes: ApJ **331**, 699 (1988)
2. D.F. Chernoff & M.D. Weinberg: ApJ **351**, 121 (1990)
3. A.I. Díaz, E. Terlevich & R. Terlevich: MNRAS **239**, 325 (1989)
4. L. Ho & A. Filippenko: ApJ **472**, 600 (1996)
5. J.A. Holtzman et al.: AJ **103**, 691 (1992)
6. D. Kunze, et al.: A&A **315**, L101 (1996)
7. S.S. Larsen: A&AS **139**, 393 (1999)

8. C. Leitherer, et al.: ApJS **123**, 3 (1999)
9. S. Mengel, M.D. Lehnert, N. Thatte, L.E. Tacconi-Garman, & R. Genzel: ApJ, in press, (2001)
10. I.F. Mirabel, et al.: A&A **333**, L1 (1998)
11. J.C. Mihos, L. Hernquist: ApJ **464**, 641 (1996)
12. A. Toomre & J. Toomre: ApJ **178**, 623 (1972)
13. L. Virgoux, et al.: A&A **315**, L93 (1996)
14. B.C. Whitmore, F. Schweizer, C. Leitherer, K. Borne, & C. Robert: AJ **106**, 1354 (1993)
15. B.C. Whitmore & F. Schweizer: AJ **109**, 960 (1995)
16. B.C. Whitmore, B.W. Miller, F. Schweizer, & S.M. Fall: AJ **114**, 2381 (1997)
17. B.C. Whitmore, Q. Zhang, C. Leitherer, S.M. Fall, F. Schweizer, & B.W. Miller: AJ **118**, 1551 (1999)
18. B. Whitmore: private communication (2000)
19. Q. Zhang & S.M. Fall: ApJ **527**, 81 (1999)

Near-IR Spectroscopy and Population Synthesis of Super Star Clusters in NGC 1569

Andrea M. Gilbert and James R. Graham

Astronomy Department, University of California, Berkeley, CA 94704, USA

Abstract. We present H- and K-band NIRSPEC spectroscopy of super star clusters (SSCs) in the irregular starburst galaxy NGC 1569, obtained at the Keck Observatory. We fit these photospheric spectra to NextGen model atmospheres to obtain effective spectral types of clusters, and find that the information in both H- and K-band spectra is necessary to remove degeneracy in the fits. The light of SSC B is unambiguously dominated by K0 supergiants ($T_{eff} = 4400 \pm 100$ K, $\log g = 0.5 \pm 0.5$). The double cluster SSC A has higher T_{eff} (G5) and less tightly constrained surface gravity ($\log g = 1.3 \pm 1.3$), consistent with a mixed stellar population dominated by blue Wolf-Rayet stars and red supergiants. We predict the time evolution of infrared spectra of SSCs using Starburst99 population synthesis models coupled with empirical stellar spectral libraries (at solar metallicity). The resulting model sequence allows us to assign ages of 15–18 Myr for SSC B and 18–21 Myr for SSC A.

1 Introduction

At a distance of about 2 Mpc [8], the irregular galaxy NGC 1569 is one of the nearest starbursts [7]. It is rich in ionized and neutral gas, and recently underwent a global starburst that lasted at least 100 Myr and ended 5–10 Myr ago [4]. The starburst produced numerous H II regions and young star clusters, and still drives an x-ray superwind [18,6]. One of the most notable features of the starburst is the super star clusters (SSCs) near the center of the galaxy [1,13]. They are some of the nearest and earliest-known examples of young, massive, compact star clusters. These SSCs, which may evolve into clusters resembling present-day globulars [11], have been found in all types of starburst environments, from dwarf irregulars to galaxy mergers.

2 Observations & Data Reduction

We obtained spectra and images of the SSCs in NGC 1569 on January 17, 2000 UT using the near-infrared spectrometer NIRSPEC on the Keck II telescope at the W. M. Keck Observatory. The night was photometric, with seeing of about 0.″7. The K-band image (taken with the N7 filter by NIRSPEC's slit-viewing camera) in Fig. 1 shows the brightest central SSCs of which we obtained K-band spectra (2.03–2.47 μm), along with many other clusters and stellar sources. For the brightest clusters, SSCs A and B, we also obtained H-band spectra (1.49–1.78 μm).

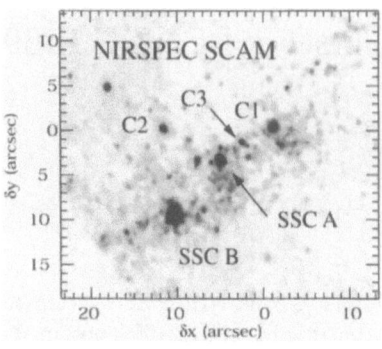

Fig. 1. NIRSPEC K-band image of NGC 1569 in 0.″7 seeing. Clusters for which spectra were obtained are labeled

Data were reduced in the usual fashion, and the optimally extracted K-band spectra are shown in Fig. 2a. The strongest features in K band are the CO first overtone bands longward of $2.295\,\mu$m. These saturated bands are found in the cool atmospheres of supergiants and giants. Their presence implies that the clusters contain stars that are at least 6 or 7 Myr old, since that much time is required before the most massive stars evolve off of the main sequence.

Stellar evolution models for a simple stellar population predict that the hot blue stars capable of driving nebular emission will have completely evolved off of the main sequence by the time red supergiants appear. Thus the conjunction of strong CO bands together with weak Brγ and HeI emission in the spectra of Clusters A and C1 suggests that they are not instantaneous bursts. In the case of cluster C1 we believe that the weak emission lines are explained by contamination from a nearby HII region [18]. The weak nebular emission from SSC A (a double cluster [2]) is likely due to the Wolf-Rayet stars known to be present there [3]. Thus the spectrum of SSC A cannot be explained with a single simple stellar population according to current population synthesis models. Cluster C2, on the other hand, shows both Brγ and HeI in absorption, which may be due to a much older population dominated by A stars and red giants.

Optimally extracted H-band spectra of SSCs A and B are shown in Fig. 2b, together with the spectrum of a K1.5 Ib supergiant from the KPNO stellar atlases [17,12]. A comparison of the star and cluster spectra reveals that most of the features in the cluster spectra are real, not noise, and that H-band photospheric spectra are remarkably rich in strong metallic and molecular features. Only the CO bands stand out in the K-band spectra, while many metal features in H band are as strong as the second overtone CO bands. Most of the features in H band are blends of lines from CO, OH, and metals such as Fe, Si, Al, and Mg.

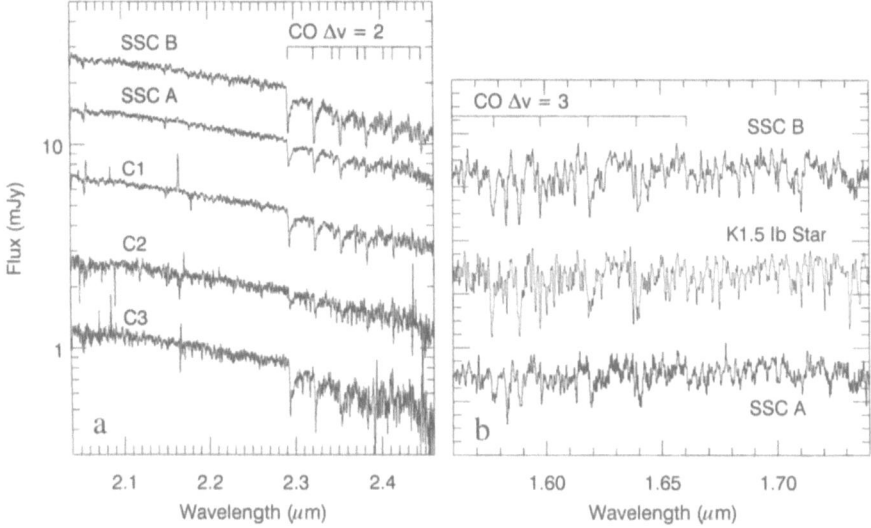

Fig. 2. NIRSPEC spectra of clusters in NGC 1569. (**a**) *K*-band spectra. Strong overtone CO bands signify the presence of red supergiants. Clusters A and C1 show weak Brγ and He I emission, while C2 shows both of these in absorption. (**b**) *H*-band spectra of SSCs A and B with slopes removed, compared with the spectrum of a K1.5 Ib supergiant [12]. Note the correspondence of features between star and cluster spectra, and the richness of the *H* band

3   The Cluster Integrated Light

The most massive, evolved members of a single-aged stellar population tend to dominate its integrated light. Thus the simplest approach to interpreting the integrated light may be to ask, what is the dominant spectral type of the cluster? We attempt to characterize a cluster by an effective spectral type by fitting its *H*- and *K*-band spectra to a grid of NextGen model atmospheres [5], in order to determine an effective temperature, surface gravity, and metallicity for each object. The NextGen atmospheres were available for spherically symmetric giant stars with a range in T_{eff} of 3000 K to 6800 K, a range in $\log g$ of 0.0 to 3.5, and metallicities of [Fe/H] = 0.0 (solar), −0.3, −0.5, and −0.7.

We first test the utility of this procedure by fitting the model atmospheres to an empirical spectrum of a star of known spectral type to evaluate the precision and accuracy with which it selects the atmospheric parameters. Figure 3a shows the resulting χ^2 contours for a fit to a solar-metallicity K1.5 Ib spectrum. The resulting parameters, T_{eff} = 4400 ± 100 K and $\log g$ = 0.0 ± 0.5, are a good match to those of the star. Both *H*- and *K*-band spectra were required in order to remove degeneracies in the fits.

Next we fit the NextGen atmospheres at metallicity [Fe/H] = −0.5 (which is closest to that of NGC 1569, [Fe/H] = −0.6 to −0.7 [3,9]) to the *H* + *K* spectra of the brightest *K*-band cluster, SSC B. The resulting χ^2 contours in Fig. 3b

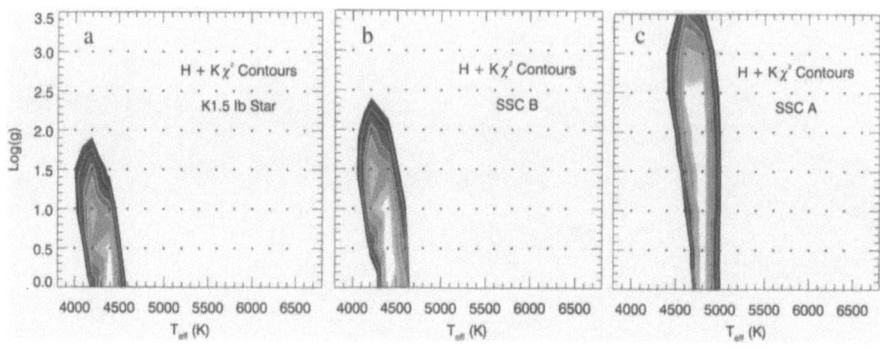

Fig. 3. χ^2 contours for fit of NextGen models for: (**a**) spectrum of a K1.5 Ib star; (**b**) SSC B; (**c**) SSC A. White filled contour indicates 1 sigma confidence level, and line contours represent 90, 95, and 99% confidence limits. Best-fit parameters for SSC B imply an effective spectral type for the cluster light of K0 supergiant. Best-fit parameters for SSC A imply an effective spectral type for the cluster light of G5 supergiant with $\log g = 1.3 \pm 1.3$. The temperature and range of $\log g$ are consistent with a mixed population of stars, either two short bursts or an extended epoch of star formation

have the same shape as those found for the star in Fig. 3a, indicating that the light of cluster B is heavily dominated by stars with a very narrow range in spectral type. The effective spectral type for SSC B is that of a K0 supergiant, with $T_{eff} = 4400 \pm 100$ K and $\log g = 0.5 \pm 0.5$.

For the double cluster, SSC A, the χ^2 contours are less tightly constrained in the fit parameters than for SSC B. Figure 3c shows them to be centered at a hotter $T_{eff} = 4800 \pm 200$ K and larger range of $\log g = 1.3 \pm 1.3$, typical for stars of types G5 I and G5 III. It is unlikely, however, that such stars dominate the cluster's emission, since optical/UV evidence indicates that hot blue Wolf-Rayet stars are present together with the red evolved stars creating the strong infrared CO bands. Thus the inferred effective spectral type determined from the stellar atmospheres may simply result from the superposition of the two distinct populations.

4   IR Spectral Population Synthesis

Since a cluster consists of stars with a range of stellar masses, luminosities, and temperatures, it is more informative to model the integrated population directly than to focus on its effective spectral type. Thus we employ the technique of population synthesis to calculate the distribution of stars in the H-R diagram of a cluster as a function of time. We then calculate the integrated spectrum of the cluster by adding up the appropriate numbers of stellar spectra – either empirical spectra or model atmospheres.

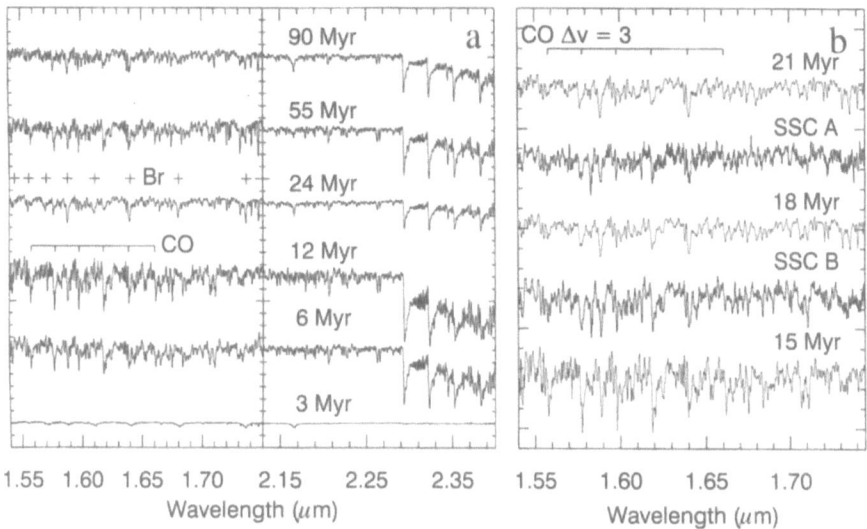

Fig. 4. (a) Sequence of model cluster spectra as a function of time, calculated using Starburst99 and KNPO stellar atlases. (b) SSC A and B H-band spectra shown in an age progression between synthetic cluster spectra (at solar metallicity) at ages of 15, 18, and 21 Myr

4.1   Models

All models in this paper were constructed using the updated evolutionary synthesis code Starburst99 [10]. The code incorporates the most recent stellar evolutionary tracks from the Geneva group at metallicities ranging from very metal-poor, to twice solar metallicity [15,16], and it has been updated to allow the use of isochrone synthesis. Starburst99 is a particular set of synthesis models which are optimized to reproduce properties of galaxies with active star formation, so it puts most of the emphasis on early evolutionary phases. Later phases, like AGB stars or white dwarfs, are covered only crudely or not at all.

Origlia et al. [14] show that low-metallicity tracks do not reproduce the CO 1.62 μm and CO 2.29 μm indices of young LMC clusters. However, if the fraction of time spent as a RSG during the core-helium phase is forced to at least 50%, and if the RSG temperature is maintained to less than 4000 K, the models agree well with the observations. Our modeling technique was modified according to this prescription (Leitherer, private communication, 2000).

In order to generate model cluster spectra, we combine the Starburst99 models with the empirical libraries of stellar spectra obtained at Kitt Peak by Wallace & Hinkle [17] and Meyer et al. [12]. For a given cluster population, we add up the spectra of component stars, and include nebular continuum emission (but not the recombination lines) based on the number of ionizing photons predicted for the cluster. Thus we generate a time series of model cluster spectra such as that shown in Fig. 4a for a 10^6 M$_\odot$ cluster with Salpeter IMF ranging from 0.1 to

$100\,M_\odot$. For the first few Myr, nebular emission powered by the hottest stars dilutes the photospheric emission from the cluster, but by an age of 6–7 Myr, the most massive stars have evolved off of the main sequence to become red supergiants, whose spectra are marked by deep CO bands.

Finally, we can place observed cluster spectra in an evolutionary sequence by fitting them to the model sequences. Figure 4b displays the H-band spectra of SSCs A and B together with the three model cluster spectra (15, 18, and 21 Myr) which most closely resemble the observations. Note the correspondence between features and the decrease in their strength with time. Since the models are for solar metallicity clusters, they presumably have stronger metal features at a given age than expected for a lower-metallicity cluster. Thus the age estimates we derive from these models will be too large during this epoch of the cluster's evolution.

5 Conclusions

We have presented new high-quality near-infrared spectra of several of the SSCs in the nearby irregular starburst, NGC 1569, and demonstrated the utility of the rich H-band spectral region for modeling stellar populations. We found that combining H- and K-band spectra removed some of the degeneracy in fitting just one band to model spectra.

We used population synthesis models together with model stellar atmospheres and empirical stellar spectra to fit for the effective spectral type of a cluster, and to generate sequences of synthetic cluster spectra to help determine the ages of observed clusters. Since the empirical libraries are only complete for solar metallicity, we are constructing models that use model atmospheres at lower metallicities that are more appropriate for systems like NGC 1569.

References

1. H. Arp, A. Sandage: AJ **90**, 1163 (1985)
2. G. De Marchi et al.: ApJ **479**, L27 (1997)
3. R.M. Gonzalez Delgado et al.: ApJ **483**, 107 (1997)
4. L. Greggio et al.: ApJ **504**, 725 (1998)
5. P.H. Hauschildt, F. Allard, E. Baron: ApJ **512**, 377 (1999)
6. T.M. Heckman et al.: ApJ **448**, 98 (1995)
7. F.P. Israel: A&A **194**, 24 (1988)
8. I.D. Karachentsev, N.A. Tikhonov: A&A **286**, 718 (1994)
9. H.A. Kobulnicky, E.D. Skillman: ApJ **489**, 636 (1997)
10. C. Leitherer et al.: ApJS **123**, 3 (1999)
11. G.R. Meurer: Nature **375**, 742 (1995)
12. M.R. Meyer, S. Edwards, K.H. Hinkle, S.E. Strom: ApJ **508**, 397 (1998)
13. R.W. O'Connell, J.S. Gallagher III, D.A. Hunter: ApJ **433**, 65 (1994)
14. L. Origlia, J.D. Goldader, C. Leitherer, D. Schaerer, E. Oliva: ApJ **514**, 96 (1999)
15. G. Schaller, D. Schaerer, G. Meynet, A. Maeder: A&AS **96**, 269 (1992)
16. D. Schaerer et al.: A&AS **102**, 339 (1993)
17. L. Wallace, K.H. Hinkle: ApJS **107**, 312 (1996)
18. W.H. Waller: ApJ **370**, 144 (1991)

The Obscured Mid-Infrared Continuum
of NGC 4418: A Dust- and Ice-Enshrouded AGN

Henrik Spoon[1], Jacqueline Keane[2], Xander Tielens[2], Dieter Lutz[3],
and Alan Moorwood[1]

[1] ESO, Karl-Schwarzschild-Strasse 2, 85748 Garching, Germany
[2] Kapteyn Astronomical Institute, Pb. 800, 9700 AV Groningen, The Netherlands
[3] MPE, Postfach 1312, 85741 Garching, Germany

Abstract. We report the detection of absorption features in the 6–8 μm region superimposed on a featureless mid-infrared continuum in NGC 4418. For several of these features this is the first detection in an external galaxy. We compare the absorption spectrum of NGC 4418 to that of embedded massive protostars and the Galactic centre, and attribute the absorption features to ice grains and to hydrogenated amorphous carbon grains. From the depth of the ice features, the central source responsible for the mid-infrared emission must be deeply enshrouded. Combined with the small size of the mid-infrared source, only an AGN can be effectively hidden within the geometry at hand.

1 Introduction

The ISO mission has considerably enhanced our knowledge of the mid-IR properties of normal, starburst, Seyfert and Ultraluminous Infrared Galaxies (ULIRGs). The spectra of most sources are dominated by ISM emission features, the most prominent of which are the well-known PAH emission bands at 6.2, 7.7, 8.6, 11.3 and 12.7 μm. The PAH features and the emission lines have been used qualitatively and quantitatively as diagnostics for the ultimate physical processes powering the galactic nuclei [11,21,25,31]. A broad absorption band due to the Si-O stretching mode in amorphous silicates, centered at 9.7 μm, is also commonly detected in galaxies. Since the center of the silicate absorption coincides with a gap between the 6.2–8.6 μm and 11.3–12.8 μm PAH complexes, it is not readily apparent whether a 9.7 μm flux minimum should be interpreted as the "trough" between PAH emission features or as strong silicate absorption, or as a combination of the two. In spectra observed towards heavily extincted Galactic lines of sight, a strong silicate feature is often accompanied by ice absorption features in the 6–8 μm region [33]. Until recently this combination had not been reported in equally extincted extragalactic sources, despite detections of ice features at shorter wavelengths [29,30]. Here we report on the strongly absorbed, ISO-PHT-S spectrum of NGC 4418, a nearby (D=29 Mpc; 1"=140pc) luminous (L_{IR}=10^{11} L$_\odot$) bright IRAS galaxy. NGC 4418 and the distant ULIRG IRAS 00183-7111 [31] are the first detections of these ice features in external galaxies.

NGC 4418 is well-known for its deep 9.7 μm silicate feature [26]. Additional evidence for strong extinction is the weakness (Hα) and absence (Hβ, Brα, Brγ)

of hydrogen recombination line emission [15,24,20,19], commonly detected in galaxies. HST-NICMOS images [28] show hardly any structure in the inner 400pc×400pc, except for large scale extinction ($\Delta A_V \sim 2$). The IRAS colors of NGC 4418 indicate that — unlike most other galaxies — the 12–100μm emission of NGC 4418 is dominated by a warm dust component, peaking shortward of 60μm. VLA radio maps at 6 and 20 cm [5,8] show NGC 4418 to be compact (70pc×50pc at most). This, as well as the presence of large quantities of warm dust, has been taken as evidence for the presence of an otherwise hidden AGN in NGC 4418. In this paper we present mid-IR spectral evidence lending further support for the presence of an AGN in NGC 4418.

2   Observations

A low resolution ($\lambda/\Delta\lambda \sim 90$) ISO-PHT-S spectrum of the central 24"×24" of NGC 4418 was obtained on 1996 July 14 as part of a project on the interstellar medium of normal galaxies [14]. The measurement was carried out in triangular chopped mode, using a chopper throw of 150". The resulting spectrum is thus free of contributions from zodiacal light. The ISO-PHT-S data were reduced using PIA 8.2. The absolute calibration is accurate to within 20%. The resulting spectrum is shown in Fig. 1.

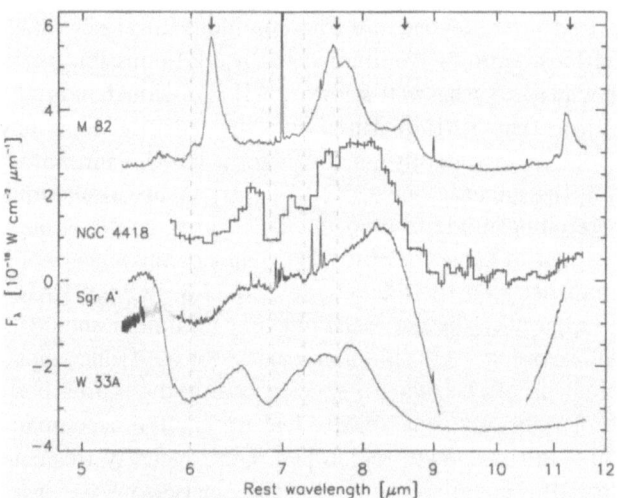

Fig. 1. A comparison of the ISO spectra of NGC 4418, M 82, the embedded massive protostar W 33A and the Galactic centre (Sgr A*). Except for NGC 4418, the spectra have been scaled and offset. We removed the strong 7μm [Ar II] line from the Sgr A* spectrum. The vertical dashed lines facilitate comparision between the four spectra with well known Galactic absorption features. The arrows mark the rest wavelengths of the 6.2, 7.7, 8.6 and 11.3μm PAH features. The zero flux levels for M 82, NGC 4418 and W 33A are indicated with horizontal dotted lines.

3   The Mid-IR Spectrum of NGC 4418

3.1   Spectral Features

The mid-IR spectrum of NGC 4418 (Fig. 1) bears little resemblance to the spectrum of (almost) any other galaxy obtained by ISO. The spectra of normal and starburst galaxies [25,14] are dominated by strong PAH emission features. Seyfert galaxies with a clear line of sight to AGN-heated hot dust, on the other hand, are dominated by a strong mid-IR continuum with PAHs barely recognizable [4]. The mid-IR spectrum of NGC 4418 is rich in features but does not resemble PAH spectra. Rather it is similar to spectra observed towards heavily extincted Galactic lines of sight, such as deeply embedded massive protostars or the Galactic centre (Fig. 1). These show no evidence for PAH emission features but do show strong absorption features. A simple criterion for the role of PAH emission and absorption features is based on the location of maxima in the 6–7 μm region: PAH spectra show the 6.2 μm emission feature, whereas absorption spectra show a peak at 6.5–6.7 μm, which is not an emission feature but a window of reduced absorption between two features.

A detailed comparison of the absorption features with Galactic templates may be able to shed further light on the origin of the extinction in NGC 4418. The absorption features were determined by fitting a local polynomial to the peaks of the 5.8–8 μm spectrum of NGC 4418. The absence of the 6.2 μm PAH feature indicates that emission features are very weak in the spectrum, and justifies this simple procedure to derive the shape and depth of the absorption features. Fig. 2 shows the resulting NGC 4418 optical depth spectrum along with the optical depth spectrum of ices towards the massive protostar W 33A [13] and the spectrum of the Galactic centre (Sgr A*), which displays absorptions due to ices as well as features due to dust in the diffuse ISM [3]. The spectrum of NGC 4418 shows absorption features at 6.0, 6.8, 7.3, 7.6, 10 and 18 μm (Table 1). The 6.0 μm feature in NGC 4418 is similar to that in the Galactic sources but with a perhaps more pronounced long wavelength wing. The 6.85 μm feature is considerably narrower than the molecular cloud feature but is similar in width to the diffuse ISM feature. The band at 7.3 μm is substantially broader than the molecular cloud and diffuse ISM features. The absorption band near 7.6 μm is similar to that observed locally.

The presence of ice along the line of sight toward NGC 4418 is suggested by the identification of the 7.6 μm band with CH_4 [1,2] and by the presence of the 6.0 μm band due to H_2O ice [17,3]. The origins of the 6.85 μm and 7.3 μm bands in NGC 4418 are unclear. Interstellar ices also show features at these wavelengths, however, their relative strengths as well as widths are markedly different in NGC 4418. The 6.85 μm and 7.3 μm band ratios are consistent with the features observed towards Sgr A*, which have been attributed to CH deformation modes in HAC-like dust grains [3]. Thus, as for the Galactic centre, both ice characteristics for shielded dense molecular cloud environments and HAC-like grain characteristics for diffuse ISM dust seem to be present along the line of sight. This conclusion could be tested through observations in the 3 μm window, which contains the strong $3\,\mu\mathrm{m}$ H_2O ice band and the $3.4\,\mu\mathrm{m}$ CH stretching modes of HAC materials. Guided by variations seen for the Galactic centre region [3], and between M 82 and NGC 1068 [30], we speculate that the relative weight of ice and HAC components may vary considerably among galaxies.

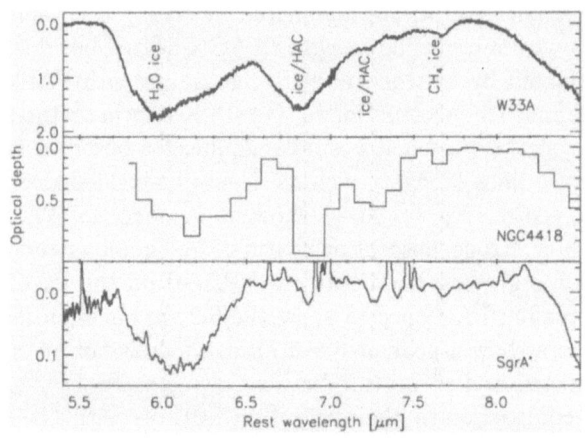

Fig. 2. The optical depth spectra for W 33A, NGC 4418, and Sgr A*. The vertical lines indicate the positions of the 6.0, 6.85, 7.3, and $7.6\,\mu\mathrm{m}$ ice and HAC absorption bands seen toward Galactic lines of sight.

3.2 Dust and Ice Column Densities

The column densities of the ice species can be derived by dividing the integrated optical depth (τ_{int}) by the molecular band strength [12,2]. Table 1 summarizes the computed column densities of CH_4 and H_2O ice. The column densities are consistent with those derived for embedded massive protostellar objects in molecular clouds [1,17]. Also shown in Table 1 are the calculated HAC column densities for the 6.85 and $7.3\,\mu\mathrm{m}$ bands, assuming the intrinsic integrated band intensities for saturated aliphatic hydrocarbons from values by Wexler [32]. These integrated intensities are stronger than other current literature values [27,10]. For NGC 4418 a substantial fraction of the carbon is locked up in HAC ($\sim$20%) as compared to the Galactic centre (a few % in Sgr A* [23]).

We consider two choices for the local mid-IR continuum in NGC 4418. The first continuum interpolates the peaks of the observed data, whereas the other assumes a stronger $18\,\mu\mathrm{m}$ silicate absorption. The effect of the different continua on the silicate feature is noticable when fitting the silicate profile by the Galactic Centre Source GCS 3, a pure absorption feature, i.e. no intrinsic emission [9]. However, the exact optical depth is still difficult to determine due to saturation of the silicate feature in NGC 4418. Adopting the first continuum results in a NGC 4418 silicate profile in which the blue wing is well matched but the red wavelength wing is poorly fitted by the GCS 3 profile. If however the second continuum is adopted, then the fit to the red wing improves while still maintaining a reasonable match to the blue wing of the $10\,\mu m$ silicate band. The $9.7\,\mu m$ and $18\,\mu m$ optical depths for the second case are $\tau_{9.7} \sim 7$ (corresponding to $A_V \sim 130$) and $\tau_{18} \sim 1.5$. These numbers, along with the derived hydrogen column density, are shown in Table 1. The apparent optical depth ratio of the $9.7/18\,\mu m$ silicate bands, regardless of adopted continuum, is significantly larger than the ratio determined by Demyk et al. [6] for two Galactic protostars. This might suggest that complex radiative transfer effects are important, which are however beyond the scope of this paper.

Table 1. Observed parameters for the NGC 4418 features.

λ [μm]	τ	τ_{int} [cm^{-1}]	carrier	N [$10^{17}\ cm^{-2}$]
6.0	0.8	87	H_2O	73
6.85	1.1	27	HAC?	~ 200
7.3	0.6	12	HAC?	~ 200
7.67	0.12	1.2	CH_4	1.6
9.7	7	—	silicates	—
18	~ 1.5	—	silicates	—
			atomic H	$\sim 2 \times 10^6$ *

* Column to the mid-IR dust source calculated
assuming Galactic gas to dust ratios. Note that
the X-ray absorbing column to the central engine
may be even higher.

4   Discussion and Conclusions

We have compared the ISO PHT-S spectrum of NGC 4418 with spectra of our template sources and found no sign of PAH emission, neither from the nucleus, nor from that part of the disk contained within the 24"×24" ISO-PHT-S aperture. Instead we found deep absorption features imposed upon a featureless mid-IR continuum. We identify the 6–$8\mu m$ absorption features with foreground ices and HAC-like grains.

The nature of the central source in NGC 4418 cannot be infered from the observed mid- to far-IR spectrum alone, given the absence of any "signposts", like the $6.2\,\mu m$ PAH emission feature ($F_{PAH} < 6 \times 10^{-20}\ W/cm^2$), and of fine structure lines ([OI], [CII] and [OIII] [22]); $F_{[NeII]} < 2.3 \times 10^{-20}\ W/cm^2$, from archival ISO-SWS data). Both a heavily enshrouded AGN or a similarly extincted nuclear starburst could be responsible for the observed continuum. Even if a starburst could be accomodated within the compact central source ($<70\,pc$ in the mid-IR [28] and 25–70 pc at 6 and 20 cm [8,16]), it would be highly unlikely to block the escape of any mid-IR starburst indicator from a region of that size. The most likely origin is therefore a heavily enshrouded AGN as suggested previously by Roche et al. [26,16,7]. The 0.1" point source (5mJy) detected with the Parkes Tidbinbilla Interferometer (PTI) at 13cm [18,19] may actually pinpoint the AGN itself. Far-IR to millimetre sizes are less well constrained but the warm IRAS colours suggest that the emission in this range also arises in the nuclear region.

Our finding of strong absorptions due to cold silicates and ices in NGC 4418 leads us to believe that the same absorptions may be present in the mid-IR spectra of other galaxies. Indeed, we have found similar absorptions in about a dozen of 225 galaxies observed spectroscopically by ISO. Spectral identifications have to be done with great care since simultaneous presence of PAH emission makes other spectra more complex than the one of NGC 4418.

Since the overall shape of the NGC 4418 6–11 μm spectrum with its maximum near 8 μm mimicks at first glance a PAH spectrum, we point out the need for high S/N spectra to clearly identify the indicators for bona fide PAH spectra or absorption dominated spectra. The most obvious discriminator is the 6.2μm PAH peak to be contrasted with the 6.5–6.7 μm pseudo-maximum in absorption spectra, which is due to a window between two absorption features. In a forthcoming paper we will address this issue for our large ISO galaxy sample.

References

1. A. C. A. Boogert, W. A. Schutte, A. G. G. M. Tielens, et al.: A&A **315**, L377 (1996)
2. A. C. A. Boogert, W. A. Schutte, F. P. Helmich, et al.: A&A **328**, 649 (1997)
3. J. E. Chiar, A. G. G. M. Tielens, D. C. B. Whittet, et al.: ApJ **537**, 749 (2000)
4. J. Clavel, B. Schulz, B. Altieri, et al.: A&A **357**, 839 (2000)
5. J. J. Condon, G. Helou, D. B. Sanders, B. T. Soifer: ApJS **73**, 359 (1990)
6. K. Demyk, A. P. Jones, E. Dartois, et al.: A&A **349**, 267 (1999)
7. C. C. Dudley, C. G. Wynn-Williams: ApJ **488**, 720 (1997)
8. S. A. Eales, E. E. Becklin, K. -W. Hodapp, et al.: ApJ **365**, 478 (1990)
9. D. F. Figer, I. S. McLean, M. Morris: ApJ **514**, 202 (1999)
10. D. G. Furton, J. W. Laiho, A. N. Witt: ApJ **526**, 752 (1999)
11. R. Genzel, D. Lutz, E. Sturm, et al.: ApJ **498**, 579 (1998)
12. P. A. Gerakines, W. A. Schutte, J. M. Greenberg, et al.: A&A **296**, 810 (1995)
13. E. L. Gibb, D. C. B. Whittet, W. A. Schutte, et al.: ApJ **536**, 347 (2000)
14. G. Helou, N. Y. Lu, M. W. Werner, S. Malhotra, et al.: ApJ **532**, L21 (2000)
15. K. Kawara, M. Nishida, M. M. Phillips: ApJ **337**, 230 (1989)
16. K. Kawara, Y. Taniguchi, N. Nakai, et al.: ApJ **365**, L1 (1990)
17. J. V. Keane, A. G. G. M. Tielens, A. C. A. Boogert, et al.: A&A, submitted (2000)
18. L. J. Kewley, C. A. Heisler, M. A. Dopita, et al., ApJ **530**, 704 (2000)
19. L. J. Kewley: priv. comm.
20. M. D. Lehnert, T. M. Heckman: ApJS **97**, 89 (1995)
21. D. Lutz, H. W. W. Spoon, D. Rigopoulou, et al.: ApJ **505**, L103 (1998)
22. S. Malhotra, G. Helou, D. Hollenbach, et al.: In: *The Universe as seen by ISO* ed. by P. Cox, M.F. Kessler (ESA-SP 1999) **427**, pp.813
23. Y. J. Pendleton, S. A. Sandford, L. J. Allamandola, et al.: ApJ **437**, 683 (1994)
24. S. E. Ridgway, C. G. Wynn-Williams: ApJ **428**, 609 (1994)
25. D. Rigopoulou, H. W. W. Spoon,R. Genzel, et al.: AJ **118**, 2625 (1999)
26. P. F. Roche, D. K., Aitken, C. H. Smith, S. D. James: MNRAS **218**, 19P (1986)
27. S. A. Sandford, L. J. Allamandola, A. G. G. M. Tielens et al.: ApJ **371**, 607 (1991)
28. N. Z. Scoville, A. S. Evans, R. Thompson, et al.: AJ **119**, 991 (2000)

29. H. W. W. Spoon, J. Koornneef, A. F. M. Moorwood, et al.: A&A **357**, 898 (2000)
30. E. Sturm, D. Lutz, D. Tran, et al.: A&A **358**, 481 (2000)
31. Q. D. Tran, D. Lutz, R. Genzel, et al.: ApJ, Submitted (2000)
32. A. S. Wexler: Applied Spec. Rev. **1**, 29 (1967)
33. D. C. B. Whittet, W. A. Schutte, A. G. G. M. Tielens, et al.: A&A **315**, L357 (1999)

Modeling Starbursts in Interacting Galaxies

Chris Mihos

Dept of Astronomy, Case Western Reserve University, Cleveland, OH, USA

Abstract. Dynamical modeling of colliding galaxies is often used to study the triggering mechanism and evolution of merger-driven starburst galaxies. Here I briefly discuss modeling techniques and results, and examine the robustness of predictions of inflow triggering and starburst properties. The properties of the models are then compared to observed interacting systems to determine the successes and failures of these dynamical models of starburst mergers. Finally, I present new data on a poststarburst "E+A" galaxy which may present difficulties for the standard picture of merger-induced starburst activity.

1 Modeling Techniques

Dynamical modeling of interaction induced starbursts is a particularly difficult prospect due to the detailed, non-linear coupling between the physical processes involved. The gravitational influence of the encounter, coupled with the self-gravitating response of the galactic disks, drives strong instabilities and inflow in the gaseous component of the galaxies. These compressions of the ISM can lead to strongly enhanced star formation rates and often to the subsequent triggering of starburst winds. These energetic winds may in turn affect the dynamics of the gas and the ability to sustain high star formation rates. The complex physical processes, operating on different physical scales, present a serious technical challenge for modeling starburst mergers. Here I briefly describe how the different processes are computed in the models.

- **Gravity:** Because of the large dynamical range and small volume filling factors associated with galaxy interactions, Lagrangian treecodes [2,11] are the optimal solution for solving Poisson's equation for the gravitational evolution of the system. The major uncertainty in the gravitational force calculation is the mass distribution of the galaxy model, particularly the nature of the dark halo (which controls the dynamical friction timescales) and the relative contribution of disk mass and dark matter in the inner scale length (which controls the global stability of the disk).
- **Gas Dynamics:** Given the Lagrangian nature of the gravitational force calculation, particle hydro techniques are largely favored over Eulerian grid codes. Both "sticky particle" [4,26] and smoothed particle hydrodynamics (SPH) [7,8,12,23,24] techniques have been used. While qualitatively both appear to give similar results, SPH models incorporate a more sophisticated

thermodynamic treatment of the gas. In particular, SPH models can include heating from shocks and from young stars, and radiative and adiabatic cooling. However, the shock-capturing capabilities of these Lagrangian techniques are limited compared to Eulerian methods.

- **Star Formation:** A variety of techniques have been explored to simulate star formation in dynamical models. The most popular are density-dependent Schmidt laws of the form $SFR \sim \rho_{gas}^n$, where $n \sim 1 - 2$ [4,23,24,26]. Some modelers have also used a requirement that the gas be Jeans unstable [8,9,20,27,31]; however, whether this condition is applicable on the physical scales involved is unclear. The use of threshold density models (i.e., the Quirk-Kennicutt criteria) have also been advocated [19]; these models may be appropriate for encounters where galaxies retain their disk dynamics, but applying them in the violent world of merging galaxies is less well-motivated.

- **Feedback:** Robust techniques for modeling the feedback from supernovae and stellar winds remain elusive. Both kinetic [23,24,26,27] and thermal [8,20] feedback have been used, in which energy is added to neighboring particles in the form of a radial "kick'" and excess heat, respectively. Both techniques have their shortcomings: adding kinetic energy assumes an efficiency at driving bulk flows which must certainly be dependent on the properties of the gas (i.e., the equation of state, the depth in the potential well, the sub-resolution filling factor, etc.), and no single parametric technique will be appropriate across all possible conditions. Adding thermal energy to the gas fails in large part because the high gas density in the models often instantly radiates the energy away (in less than a model timestep), effectively resulting in no feedback. Attempts to circumvent these problems remain highly model dependent.

- **Add-ons:** At this stage in the calculations, other physical effects can be tracked in the models, such as population synthesis models, the injection of heavy elements into the ISM and the formation of metallicity gradients, dust models, etc. However, given the uncertainties in the underlying physical models for star formation and feedback, such attempts to add increasing layers of complexity are brave indeed.

It is important to realize that the main problem with these simulations at present lies not in the lack of physical resolution. Using large supercomputers, simulations with spatial resolution of ~ 100 pc and mass resolution of $\sim 10^5$ $M_\odot$ are routinely possible, and going to even higher resolution is not difficult computationally. However, the physical description of star formation and feedback, and the computational implementation of these effects, remains the crucial stumbling block, and until we have robust methods for incorporating these effects into the models, higher resolution is meaningless at best.

2   A Comparison of Simulation Results

Given the rising degree of uncertainty as the different physical processes are included in the calculations, the skeptical observer might also ask what, if anything, are the robust results which come from these sorts of dynamical models. We now examine the results from different groups modeling a similar merger event, to examine how reproducible the main results are using somewhat different modeling techniques.

Figure 1 shows the results of our fiducial merger model [24], referred to as MH96. This merger involves two equal mass disk galaxies, with 10% of the disk mass in the form of gas. The gas is evolved using standard SPH techniques, and using an isothermal equation of state. Star formation is included using a density-dependent Schmidt law with index $n = 1.5$. Shortly after the initial collision the self-gravity of the disk amplifies the perturbation into a strong spiral arms. This nonaxisymmetric mode leads to inflow of gas into the central regions, triggering modestly increased star formation rates while the galaxies are still widely separated. Once the galaxies fall back together (due to dynamical friction from their extended dark halos), gravitational and hydrodynamical torques act to drive a second stronger phase of gas inflow and extreme central starburst activity. In

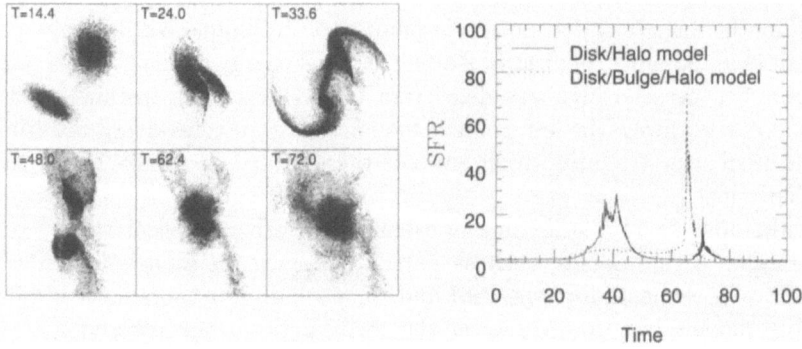

Fig. 1. Merger simulation from MH96. The left panel shows the evolution of the disk stars in the model with a central bulge. The right panel shows the evolution of the calculated star formation rate in models with and without a central bulge. Unit time is roughly 13 million years.

these models, the triggering of gaseous inflows and starburst activity is closely tied to the self-gravitating response of the galactic disks. The models shown in Fig. 1 used "maximum disk" models with a central bulge which stabilized the disk against the growth of strong bar modes. In contrast, in models with no bulge, where the disk dominates the mass distribution within two disk scale lengths ("maximum disk" models) the perturbation easily drives a strong bar instability, which drives rapid inflow and strong starbursts while the galaxies are widely separated. Our results argue strongly that the structural properties of the progenitor galaxies are of extreme importance in determine the star-forming evolution of the merger [24,25]. Similar models by Barnes & Hernquist show that if the density of the bulge is reduced, it provides less stability and the results become more sensitive to orbital geometry and dynamical resonances [3].

How robust are these results given the range in modeling techniques and parameters? We can compare these MH96 simulations to similar galaxy merger models by other investigators using different methodologies. Bekki [4] used sticky particle hydrodynamical techniques to study the evolution of gas in merging galaxies, along with a slightly steeper $n = 2$ Schmidt law. For a similar 1:1 mass ratio merger, these models also show two distinct phases of inflow, associated first with the initial self-gravitating response of the disks, then later with the final merging. The calculated star formation rates are comparable to those of MH96, although the star formation rates of Bekki are slightly higher likely due to the use of a more responsive Schmidt law. Nonetheless, since in both methodologies star formation is tied to the buildup of dense gas, the similarity in results argues that sticky particle and SPH techniques model global gas dynamics in similar fashion.

A more sophisticated treatment of both the gas dynamics and the star formation law was done recently by Gerritsen & Icke [8,9]. In these SPH simulations, the full thermodynamic conditions in the gas were followed, allowing for shock heating and radiative cooling as the merger evolves. Star formation occurs only in gas which is found to satisfy the thermal Jeans instability, and feedback is incorporated using a "pressure particle" technique, in which a star-forming particle is given an overpressure which exerts a force on surrounding gas particles. The merger simulations of Gerritsen & Icke show the same two-phase inflow as those of MH96 and Bekki, but the detailed star-forming response is somewhat different. The use of the Jeans instability criterion limits the star forming response to some degree, as much of the dense gas is at sufficiently high temperatures to inhibit star formation. This effect is exacerbated by energy input from the starburst itself, which self-regulates to some extent the ability to achieve high star formation rates. In these models star formation rates are boosted by an order of magnitude during the final merging, but are unable to achieve the truly ultraluminous levels of $\gtrsim 100$ $M_\odot$ yr^{-1}.

In summary, the different modeling techniques give comparable results in terms of the global dynamics and triggering of gaseous inflows in merging galaxies. This is a comforting situation, as the first step in triggering starburst (and indeed AGN) activity seems to be the transport of gas into the nuclei of galaxies. The link between the dynamical properties of the host galaxy, the triggering of inflows, and – at least qualitatively – the triggering of luminous activity should be fairly robust. On the other hand, the small scale physical conditions in the gas and the detailed, quantitative estimates of star formation response in merging systems are still inexorably tied to the extreme uncertainties in the star formation physics and feedback process.

3   Challenges and Lessons Learned

While models of generic galaxy collisions do well at explaining the general dynamics and evolution of merging galaxies, more severe tests of the models are possible by comparing the model predictions to specific, well-studied examples of colliding galaxies. Here I describe four systems which provide different constraints or challenges to the numerical models.

3.1   The Antennae

The nearest and best-studied interacting system, the Antennae provides a high resolution look at the physical conditions inside merging galaxies. Dynamical models exist [1] describing the collisional evolution of the system, which place it several $\times 10^8$ years past the initial collision and with a merger due in a few $\times 10^8$ years more. In an interesting contrast to dynamical models, observations show that the region with the highest concentrations of molecular gas and star formation are not the nuclei, but rather the overlap region between the two galaxies [30,34]. Hydrodynamical modeling of the Antennae by Englmaier et al. [7] indicates that the overlap region is in fact an extended gaseous bridge connecting the two galaxies along our line of sight. However, these models are less successful at explaining the high gas content in the overlap region.

Does the lack of concentrated gas and star formation in the overlap in the models a sign of shortcomings in the modeling technique? Perhaps not. The choice of gas distribution in the progenitor disks will play an important role in the hydrodynamic evolution of the merger. Material in the bridge/overlap region comes from gas at a variety of initial radii in the disk, and Englmaier et ali. argue that using an exponential gas distribution (as opposed to the flat gas distribution used in their model) may result in more gas in the overlap. However, making the initial gas distribution more centrally concentrated could result in stronger gaseous inflows (since more gas is contained in the inner disk) and a larger quantity of nuclear gas, so that this solution may involve some fine tuning. Nonetheless, more realistic gas distributions have solved other shortcomings of the dynamical modeling (most notably, the gas/star offset in the tidal tail of NGC 3690 [15,22]), and may provide a solution here as well. If that turns out not to be the case, a more critical look is needed at the hydrodynamic techniques employed.

3.2 '   The Cartwheel

Early models of the Cartwheel [21] and other ring galaxies confirmed their collisional origin. When a smaller galaxy punches through a larger disk galaxy on a radial orbit, it sets up an outwardly propagating kinematic wave in the galaxy. The compression of gas along this wave results in enhanced star formation in the outer ring. More problematic, however, was the presence of the radial spokes in the Cartwheel, which can not be described by purely collisionless kinematics. Further hydrodynamic modeling [13,23] showed these spokes to be a consequence of dissipation and infall of gas from the outer ring, setting up radial streams of material from the outer ring (see Fig. 2). However, these models predicted that the inner ring should then be gas rich and actively star-forming, in marked contrast to the observed paucity of HI or active star formation in the inner ring [16,17].

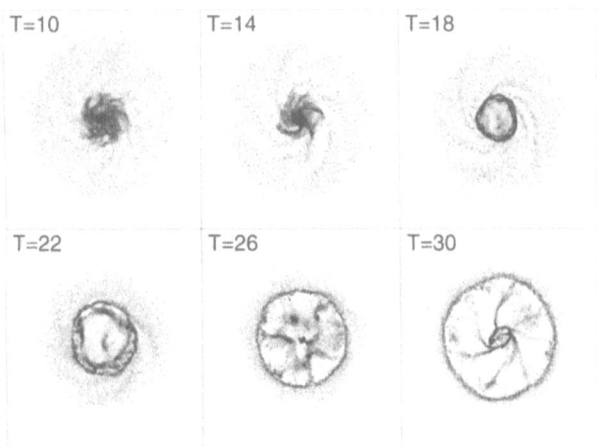

Fig. 2. Evolution of the gas in the Cartwheel model of [23].

Subsequent observations have partially explained this discrepancy. Rather than being rich in neutral III, the inner ring is dominated by molecular gas, with an inferred H_2 mass of $1.5 - 6 \times 10^9$ $M_\odot$ [18]. The preponderance of molecular gas in the inner ring may simply reflect initial conditions, as the molecular gas fraction in normal spiral galaxies rises in the central regions. However, if the central gas in the Cartwheel came from gas swept into the HI-rich outer ring as the models suggest, efficient conversion of the gas from neutral to molecular form must have taken place, perhaps due to collisional compression of the gas. Such a process has also been suggested to be occurring in NGC 7252 [14], which shows HI-rich tidal debris but a molecular rich central core.

If the inner ring of the Cartwheel is rich in molecular gas, why has star formation been suppressed? The answer may be that star formation in the inner region is in fact occurring, but is highly obscured. ISO mid infrared imaging of the Cartwheel has shown that the inner ring and nucleus provide 30% of the Cartwheel's luminosity at $15\mu m$ [6]. If this mid IR luminosity comes from dust heating by young stars, the star formation rates in the inner regions may be much higher than indicated by the $H\alpha$ data. However, it is also possible that shock heating from gas falling in from the spokes may provide sufficient energy to power the $15\mu m$ emission [32].

In the Cartwheel, several of the once-questioned qualitative predictions of the numerical models have been been vindicated by better datasets. The inner ring is gas-rich and shows morphological features suggestive of the predicted infall of gas needed to explain the radial spokes [32]. Star formation may also power the strong infrared emission detected from the inner ring. Nonetheless, there is still room for improvement for the quantitative predictions. The most serious shortcoming is that the models of MH94 predicted that the star formation in the inner ring should actually be stronger than that in the outer ring, which does not appear to be the case even in the mid-IR. This inconsistency may be due to the over-simplicity of the Schmidt law, and point towards a more detailed dependence of star formation on the kinematic and thermodynamic properties of the gas, through, for example, density threshold models for star formation.

3.3 Stephan's Quintet

A generic result of both modeling and observations of interacting systems is that the starburst activity is contained within the main body of the merging systems. While observations have shown that extended tidal debris often contains star forming clumps, these objects are typically low mass, with star formation rates of < 0.1 M$_\odot$ yr^{-1}. Numerical simulations show these systems to be spawned when gas is compressed along tidal caustics and driven into collapse locally [3], leading to the formation of low mass, self-gravitating structures in the expanding tidal tails.

Interestingly, multi-wavelength observations of the compact group Stephan's Quintet suggest a process driving starburst activity well outside of the main bodies of the merging galaxies which is completely different from that produced in the dynamical models. Hα and 15μm ISOCAM imaging show emission *between* the galaxies in the group [36]; assuming a normal IMF, the star formation rates associated with this emission are in the range of $0.66 - 0.81$ M$_\odot$ yr^{-1}, an order of magnitude greater than typically observed in tidal HII regions. This emission is spatially coincident with concentrated X-ray and radio continuum emission which is thought to arise from shock heating of the intragroup gas in Stephan's Quintet [28]. Based on these data, Xu et al. argue that a high speed collision between one of the galaxies (NGC 7318b) and the intragroup gas has driven starburst activity along the collisional front [36]. This interpretation is not unambiguous, however, since the highest regions of star formation along this purported shock front are spatially coincident with a tidal tail from NGC 7318a. Nonetheless, the combined multiwavelength data do support a collisional shock origin for the extended star formation. If this is the correct interpretation, improvements to the modeling techniques will be needed to handle the effects of hot intragroup gas and collisionally induced star formation.

3.4 NGC 6240 and Other Superwind Galaxies

Currently the most serious challenge to the dynamical models is the generation of large scale starburst winds. Many of the luminous starburst mergers show evidence for such winds in the form of high velocity ionized cones of gas or extended soft X-ray emission. For example, Fig. 3 shows HST Hα narrow band imaging of the merging galaxy NGC 6240, clearly showing a complicated superwind morphology.

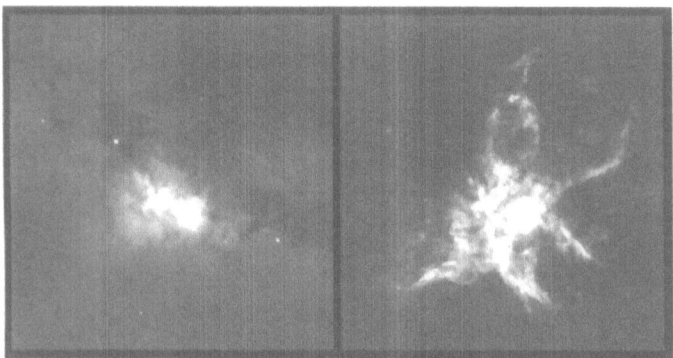

Fig. 3. HST image of NGC 6240. Left: F814W image showing the two nuclei and dust obscuration. Right: Narrow band Hα image. from [33].

Modeling outflows in the context of the dynamical N-body simulations remains problematic. Off all the phenomena exhibited by interacting galaxies, outflows depend most sensitively on most inadequate physical implementation of the model, the physics of star formation and feedback. The current techniques lack both the physics and the multi-phase spatial resolution necessary to reproduce the types of starburst winds seen in Fig. 3. Given the importance of these winds in creating hot gaseous halos around galaxies, distributing metals into the IGM, and feeding hot intra-cluster gas, more accurate modeling techniques are sorely needed to incorporate this process into the dynamical models.

4 E+A Galaxies: Problems with the Standard Picture?

With strong Balmer absorption lines and the lack of optical emission lines, E+A galaxies are believed to be systems in which star formation has recently been truncated. Spectral modeling of E+As by Zabludoff and collaborators indicate that the most extreme systems – those with the largest Balmer absorption equivalent widths and bluest colors – can only be explained by the truncation of a recent (< 1 Gyr), strong ($10\times$ increase) starburst. Originally identified in cluster environments, the recent realization that E+As are also abundant in the field [37] lends credence to the notion that field galaxy interactions may be responsible for the creation of E+A galaxies. This view leads to a picture where the extreme E+A galaxies may represent a transition from luminous starburst mergers to passively evolving field ellipticals.

Recently, we [38] have undertaken a systematic study of the properties of extreme E+A galaxies identified by Zabludoff et al. in the Las Campanas Red-shift Survey. The datasets obtained include *Hubble Space Telescope* imaging and HI observations using the VLA. Of the five systems imaged with HST, four are spheroid dominated systems, with three of these showing extended tidal features indicative of a recent merger. The fifth object, known as EA1, is the bluest E+A of the Zabludoff sample [37], and morphologically appears to be an ongoing merger (Fig. 4) much like many observed luminous starburst galaxies. Yet EA1 shows no sign of ongoing star formation in terms of optical [OII] emission [37]. Why has star formation ceased in this object, while it is undergoing the strong merger evolution which typically drives luminous starburst activity?

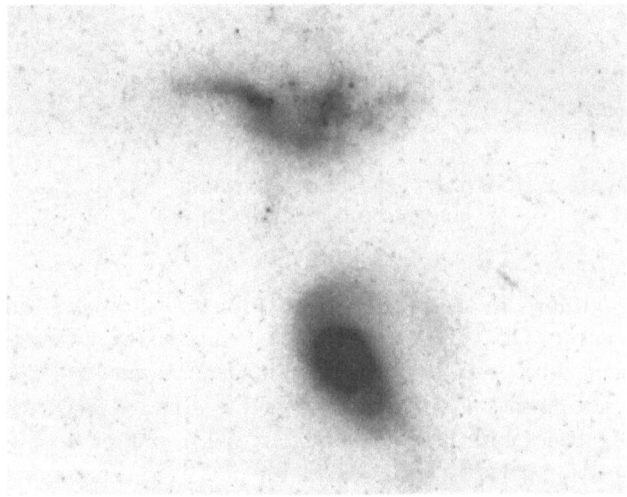

Fig. 4. HST F814W image of EA1, from [38]. EA1 is the upper system, showing a clear merger morphology. The role, if any, of the lower galaxy in triggering the events in EA1 is unclear.

VLA imaging of EA1 raises further questions. EA1 contains 3.5×10^9 $M_\odot$ of HI gas [5], so it clearly possesses fuel to drive starburst activity. Furthermore, the system is undetected in 1.3 GHz radio continuum, indicating that we are not simply witnessing a highly obscured starburst – indeed, the upper limit on the star formation rate from the radio continuum limits is $< 0.3 M_\odot$ yr^{-1} [5].

The combined properties of EA1 make it a unique object with no known counterpart in the local universe. Morphologically it appears to be a late stage merger, in the same dynamical phase as many of the extreme infrared starbursts, and it contains a significant amount of cold gas to drive star formation. Yet it shows a poststarburst spectrum with no sign of ongoing star formation. All nearby poststarburst or E+A-like systems involve either a *post*-merger system (and often still have residual star formation, like NGC 7252 [29]), or distant interactions which are well past the collisional phase (for example NGC 7714 [10]).

Given that dynamical models and morphological similarities to ULIRGs both argue that EA1 should be rapidly forming stars, what has quenched star formation in this system? It clearly *had* the capability of forming stars in the recent past (hence its poststarburst spectrum), but that star formation has been truncated. It may be that EA1 has exhausted its molecular gas reservoir and is unable to replenish it, or that the gas is now too diffuse to drive further star formation. Unfortunately, at this time we lack CO data for this system, and our C-array VLA data lacks the spatial resolution to map the density distribution of the HI gas.

A caveat to the confusing nature of EA1 is the fact that we may simply be seeing projection effects – EA1 could in principle be a widely separated system viewed along the line connecting the galaxies. While the morphology suggests the system is a late stage merger, we cannot rule out projection effects with the extant data. A second unknown is the role played by the companion galaxy to the south – perhaps it somehow triggered the previous starburst and/or stripped gas out of EA1. We continue to gather data on EA1 to test these scenarios. Nonetheless, at the current time, the properties of this peculiar E+A galaxy seem to provide challenges for our present understanding of the merger/starburst connection in galaxies.

References

1. J.E. Barnes: ApJ **331**, 699 (1988)
2. J.E. Barnes & P. Hut: Nature **324**, 446 (1986)
3. J.E. Barnes & L. Hernquist: ApJ **471**, 115 (1996)
4. K. Bekki: ApJ **502**, L133 (1998)
5. T.-C. Chang, J.H. van Gorkom, A.I. Zabludoff, D. Zaritsky, & J.C. Mihos: ApJ, in press (2001)
6. V. Charmandaris, O. Laurent, I.F. Mirabel, P. Gallais, M. Sauvage, L. Vigroux, C. Cesarsky, & P.N. Appleton: A&A **341**, 69 (1999)
7. P. Englmaier, D. Rigopoulou, & S. Mengel: in *Gas and Galaxy Evolution*, ed. by J.E. Hibbard, M.P. Rupen, & J.H. van Gorkom (ASP, San Francisco in press)
8. J.P.E. Gerritsen: Star Formation and the Interstellar Medium in Galaxy Simulations. PhD Thesis, University of Groningen (1997)
9. J.P.E. Gerritsen & V. Icke, in *Galaxy Interactions at Low and High Redshift*, ed. by J.E. Barnes & D.B. Sanders (Kluwer, Dordrecht 1999) pp. 213–216
10. R.M. González Delgado, M. García-Vargas, J. Goldader, C. Leitherer, & A. Pasquali: ApJ, **513**, 707 (1999)
11. L. Hernquist: ApJS **64**, 715 (1987)
12. L. Hernquist & N. Katz: ApJS **70**, 419 (1989)
13. L. Hernquist & M. Weil: MNRAS **261**, 804 (1993)
14. J.E. Hibbard, P. Guhathakurta, J.H. van Gorkom, & F. Schweizer: AJ **107**, 67 (1994)
15. J.E. Hibbard & M.S. Yun: AJ **118**, 162 (1999)
16. J.L. Higdon: ApJ **467**, 241 (1996)

17. J.L. Higdon & J.F. Wallin: ApJ **474**, 686 (1997)
18. C. Horellou, V. Charmandaris, F. Combes, P.N. Appleton, F. Casoli, & I.F. Mirabel: A&A **340**, L51 (1998)
19. S. Jogee, these proceedings
20. N. Katz: ApJ **391**, 502 (1992)
21. R. Lynds & A. Toomre: ApJ **209**, 382 (1976)
22. J.C. Mihos: ApJ, in press
23. J.C. Mihos & L. Hernquist: ApJ **437**, 611 (1994)
24. J.C. Mihos & L. Hernquist: ApJ **464**, 641 (1996)
25. J.C. Mihos, S.S. McGaugh, & W.J.G. de Blok: ApJ **477**, L79 (1997)
26. J.C. Mihos, D.O. Richstone, & G.D. Bothun: ApJ **400**, 153 (1992)
27. J.F. Navarro & S.D.M. White: MNRAS **265**, 271 (1993)
28. W. Pietsch, G. Trinchieri, H. Arp, & J.W. Sulentic: A&A **322**, 89 (1997)
29. F. Schweizer: ApJ **252**, 455 (1982)
30. S.A. Stanford, A.I. Sargent, D.B. Sanders, & N.Z. Scoville: ApJ, **349**, 492 (1990)
31. M. Steinmetz & E. Muller: MNRAS **276**, 549 (1995)
32. C. Struck, P.N. Appleton, K.D. Borne, & R.A. Lucas: AJ **112**, 1868 (1996)
33. R. van der Marel et al. : in preparation (2001)
34. L. Vigroux et al. : A&A **315**, L93 (1996)
35. Z. Wang, F. Schweizer, & N.Z. Scoville: ApJ **396**, 510 (1992)
36. C. Xu, J.W. Sulentic, & R. Tuffs: ApJ **512**, 178 (1999)
37. A.I. Zabludoff et al. : ApJ **466**, 104 (1996)
38. A.I. Zabludoff et al. : in preparation

Evidence for Large Stellar Disks in Elliptical Galaxies

Andreas Burkert and Thorsten Naab

Max-Planck-Institut für Astronomie, 69242 Heidelberg, Germany

Abstract. High-resolution numerical simulations of galaxy mergers are analysed. The global structure and isophotal shapes of the merger remnants are in good agreement with the observations. Whereas equal-mass mergers lead to anisotropic, boxy ellipticals, unequal-mass mergers result in disky and isotropic systems. The line-of-sight velocity distributions show small deviations from a Gaussian distribution. In all cases we find that the retrograde wings are steeper and that the prograde wings are broader than a Gaussian distribution. This is in contradiction with the observations which show broader retrograde and steeper prograde wings in all ellipticals. This fundamental difference between observation and theory can be explained if all ellipticals, even anistropic boxy ones, contain extended stellar disk components with luminosities of order 10% to 20% the total luminosity and scale radii of order the effective radii of the spheroids.

1 Introduction

Elliptical galaxies are believed to form by major mergers of spiral galaxies [15]. Numerous numerical simulations have indeed demonstrated that mergers lead to spheroidal stellar systems with surface brightness profiles that are in good agreement with the observations of elliptical galaxies (e.g. [1,9,10,2]). During the merging epoch the systems are far from dynamical equilibrium, resulting in bursts of star formation [13,4], the formation of massive star clusters and the infall of gas into the central regions and the formation and feeding of massive central black holes. When the merger remnants have settled into dynamical equilibrium, phase mixing and violent relaxation will erase the information about the initial conditions. However, as violent relaxation is incomplete, ellipticals should still show fine structure in their isophotal shape and velocity distribution which provides insight on their formation.

This conclusion is confirmed by numerical simulations. Heyl et al. [11] showed that the line of sight velocity distributions contain information about the initial disk orientations. Naab et al. [14] and Bendo & Barnes [6] demonstrated that the isophotal shapes of the remnants are determined primarily by the initial mass ratio of the merging galaxies.

In this paper we focus on signatures for gas infall and star formation during the formation of ellipticals. We demonstrate that all ellipticals must contain a second disk-like substructure that most likely formed by gas infall with subsequent star formation from gas-rich progenitors.

2   The Line-of-Sight Velocity Distribution of Merger Models

We have performed very high-resolution N-body simulations with more than 10^5 particles of spiral-spiral mergers with different mass ratios and orbital parameters. The spirals are constructed in dynamical equilibrium using the method described by [10]. Each galaxy consists of an exponential disk, a spherical, non-rotating bulge and a dark halo component. An analysis of the isophotal shapes and the global kinematical properties of the merger remnants is presented in [14]. We find a strong dependence on the initial mass ratio of the merging components. Whereas mergers with mass ratios $m_1 : m_2 \leq 2 : 1$ lead to boxy and anisotropic systems, unequal mass mergers with mass ratios $m_1 : m_2 > 2 : 1$ form disky and isotropic ellipticals.

Another interesting property is the line-of-sight velocity distribution and its deviations from a Gaussian. Following the procedure discussed by [5] we have placed slits with thickness 0.2 the effective radius along the apparent long axi of our projected merger remnants. The slit is subdivided into grid cells with length 0.15 the effective radius. All particles within a cell are binned in line-of-sight velocity. The resulting profile is parametrized using Gauss-Hermite functions [5,6] and the amplitude H_3 of the third-order basis function is determined by least squares fitting. In addition, in each grid cell the mean projected radial velocity v and the radial velocity dispersion σ of all particles is determined. In agreement with the observations, the deviations from Gaussians are small, of order a few percent

3   Comparison with Observations

Figure 1 (left panel) shows the local correlation between H_3 and v/σ as measured by [5] for their observed sample of galaxies. In all cases H_3 and v/σ have opposite signs, that is the prograde wings of the line-of-sight velocity distributions are always steeper than the retrograde wings. This result holds not only for ellipticals in low-density environments studied by [5] but also for ellipticals in the Coma cluster [12]. The right panel of Fig. 1 shows the result of a representative merger simulation, an unequal mass merger which leads to a disky, isotropic elliptical. The correlation between v/σ and H_3 is opposite to the observed one. The profiles have broad prograde wings and narrow retrograde wings. The same signature is found in all remnants, independent of whether they are disky or boxy, isotropic or anisotropic. The only exception are equal mass mergers of counter-rotating disks which lead to very anisotropic ellipticals with no signature of rotation.

4   Disk-Like Subcomponents in Elliptical Galaxies

One possible solution is the existence of a second disk component. In order to test this assumption we have placed stellar disks in the equatorial plane of

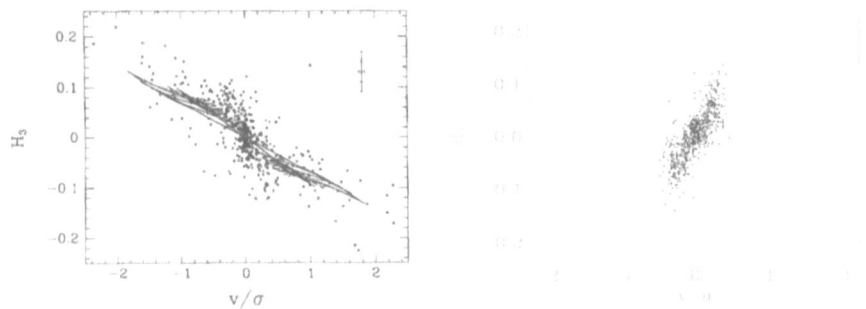

Fig. 1. *Left panel:* Observed local correlation between H_3 and v/σ [5]. Typical error bars are plotted in the upper right corner. The lines show model predictions for two-integral models [7,8]. *Right panel:* Correlation beween H_3 and v/σ for a typical merger remnant.

the merger remnants. The disks rotate in centrifugal equilibrium with the total gravitational potential. Adopting an exponential disk surface brightness profile, the only free parameters are the ratio of disk mass to spheroid mass and the scale length of the disk, normalized to the scale radius of the spheroid. The results are summarized in Fig. 2. Disks with small masses or radii do not change the line-of-sight velocity distribution of the stellar component, especially in the outer regions where v/σ is large. Disks with mass ratios and radii in the region shown by the filled dots in Fig. 2, on the other hand, lead to a significant change in the line profiles. The prograde wings become steeper than the retrograte ones also at an effective radius and beyond, in very good agreement with the observations. If the disks become too massive, the surface brightness profiles of the spheroids change from de Vaucouleurs profiles to exponential profiles which is again not in agreement with the observations. We therefore can conclude that disks with 10% to 20% the luminosity of the spheroids and scale lengths of order the effective radii of the spheroids can explain the observed correlation between H_3 and v/σ in ellipticals.

The origin of extended disk components in ellipticals is not understood up to now. Bekki [3] discussed the formation of nuclear disks in mergers which however have scale lengths that are small compared to the length scales of the spheroids. We recently started a new set of merger simulations of very gas-rich spiral galaxies. For unequal mass mergers, which lead to disky ellipticals, the gas settles indeed into extended disks if star-formation is suppressed during the early merging phase. However, in the case of equal-mass mergers, which produce boxy ellipticals, tidal torques lead to efficient angular momentum loss in the gaseous component, resulting in gas infall into the center and the formation and growth of central massive black holes. No extended gaseous disks are formed in this case. The origin of large stellar disks in boxy ellipticals is therefore still unclear.

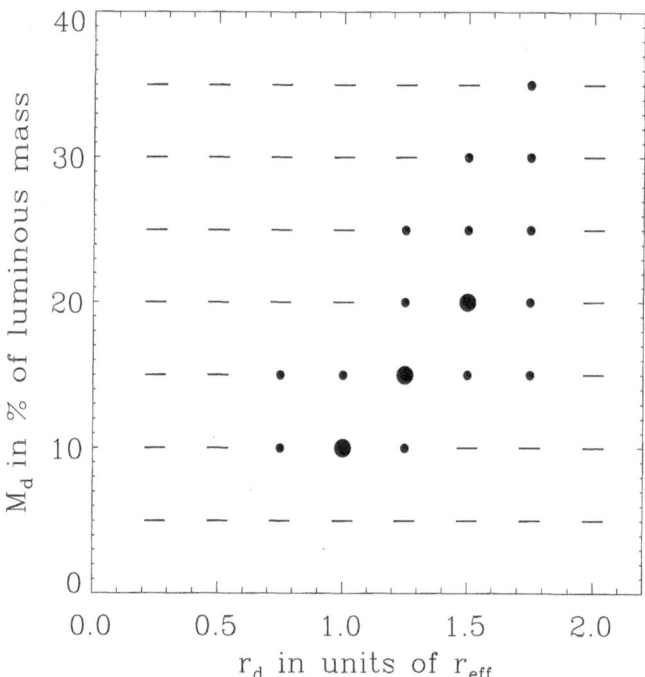

Fig. 2. Values of disk masses M_d versus disk scale lengths r_d that have been added to a merger remnant. Big dots represent solutions that provide an excellent fit to the observed H_3 versus v/σ relationship. Small dots show values that provide a reasonable fit. Minus signs show combinations that fail.

References

1. J.E. Barnes: ApJ **331**, 699 (1988)
2. K. Bekki: ApJ **502**, L133 (1998)
3. K. Bekki: ApJ **545**, 753 (2000)
4. K. Bekki: ApJ **546**, 189 (2001)
5. R. Bender, R.P. Saglia, O.E. Gerhard: MNRAS **269**, 785 (1994)
6. G.J. Bendo, J.E. Barnes: MNRAS **316**, 315 (2000)
7. W. Dehnen, O.E. Gerhard: MNRAS **261**, 311 (1993)
8. W. Dehnen, O.E. Gerhard: MNRAS **268**, 1019 (1994)
9. L. Hernquist: ApJ **400**, 460 (1992)
10. L. Hernquist: ApJ **409**, 548 (1993)
11. J.S. Heyl, L. Hernquist, D.N. Spergel: ApJ **463**, 69 (1996)
12. D. Mehlert, R.P. Saglia, R. Bender, G. Wegner: A&AS **141**, 449 (2000)
13. J.C. Mihos, L. Hernquist: ApJ **464**, 641 (1996)
14. T. Naab, A. Burkert, L. Hernquist: ApJ **523**, L133 (1999)
15. A. Toomre, J. Toomre: ApJ **178**, 623 (1972)

Starbursts in Ultraluminous Infrared Galaxies – Fueling and Properties

Paul P. van der Werf

Leiden Observatory, P.O. Box 9513, 2300 RA Leiden, The Netherlands

Abstract. The properties of starbursts in ultraluminous infrared galaxies are discussed, with particular emphasis on the fueling, the amount of extinction and the intrinsic properties of the nuclear starbursts. It is shown by the example of NGC 6240 that the H_2 vibrational lines can be used to measure the rate of gas inflow into the potential well, which is sufficient to fuel a nuclear starburst of the intensity required to account for the far-infrared emission. It is shown that in Arp 220 the faintness of all tracers of ionized gas can be accounted for by Lyman continuum absorption by dust within the ionized regions, combined with significant (but not extreme) extinction; there is no reason to invoke the presence of extreme extinction, an old starburst, or an additional non-stellar power source in Arp 220.

1 Introduction

Are ultraluminous infrared galaxies (ULIGs) powered by intense bursts of star formation or is an additional source of energy, such as an active galactic nucleus (AGN) required? In order to address this question, it is instructive to compare the near-infrared spectra of ULIGs and lower luminosity starburst galaxies. In starburst galaxies of low or moderate luminosity, the dominant emission line is the Brγ line, underlining the importance of massive young stars in the energetics of these objects. The brightest H_2 rovibrational line, the H_2 $v = 1{\rightarrow}0$ S(1) line, is typically fainter [7,8,10,11,21,20]. In contrast, in ULIGs the H_2 $v = 1{\rightarrow}0$ S(1) line is significantly brighter than the Brγ line, which is often not even detected [5,9]. This behaviour shows that ULIGs are not just scaled-up starburst galaxies. An extreme case of is presented by NGC 6240 (Fig. 1), where the H_2 $v = 1{\rightarrow}0$ S(1) line is 40 times brighter than Brγ. The low ratio of Brγ luminosity to FIR luminosity in ULIGs has been used to argue against star formation as the power source of ULIGs, for instance in the nearby ULIG Arp 220, where a starburst with the low Brγ luminosity observed can account for at most 10% of the bolometric luminosity of the galaxy [1,12]. However, spectroscopy of ULIGs at longer wavelengths with the Infrared Space Observatory (ISO) revealed bright emission lines from powerful starbursts, which are the dominant power source in most of the objects studied, but are obscured at shorter wavelengths [4]. Motivated by these results, in this paper the physical processes revealed by the K-band spectra of ULIGs are reexamined in the context of the starburst scenario. In Sect. 2 the origin of the H_2 line emission is discussed, while Sect. 3 addresses the faintness of Brγ.

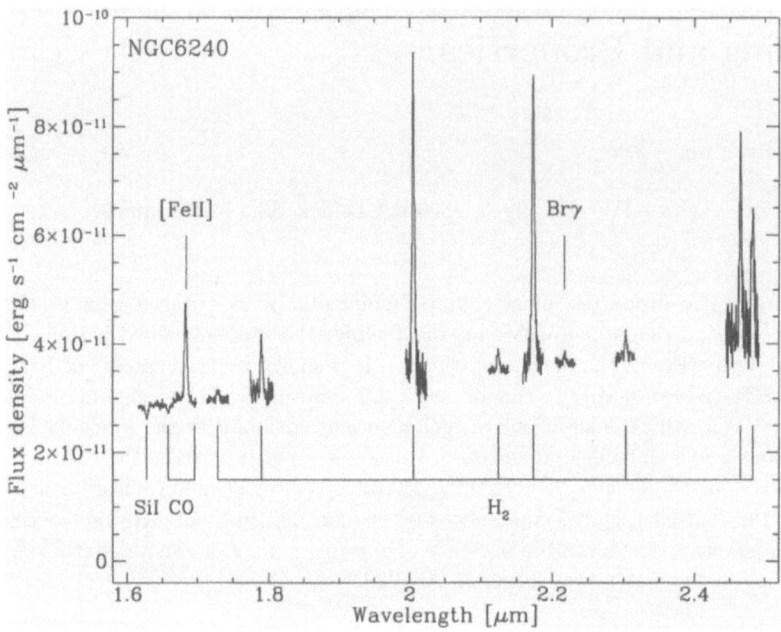

Fig. 1. Spectra of NGC 6240 in selected parts of the near-infrared H and K bands, integrated over a 4.''4 region [19].

2 Origin of the H_2 Emission in ULIGs

The most extreme vibrational H_2 emission is found in the nearby luminous merger NGC 6240: $7 \cdot 10^7$ $L_\odot$ is emitted in the H_2 $v = 1 \to 0$ S(1) line alone (for $H_0 = 75$ km s^{-1} Mpc^{-1} and with no correction for extinction). This line contains 0.012% of the bolometric luminosity of NGC 6240, which is considerably higher than any other galaxy [18]. Together the vibrational lines may account for 0.1% of the total bolometric luminosity.

Imaging of the H_2 $v = 1 \to 0$ S(1) emission from NGC 6240 has shown that the H_2 emission peaks *between* the two remnant nuclei of the merging system [18]. This morphology provides a unique constraint on the excititation mechanism, since it argues against any scenario where the excitation is dominated by the stellar component (e.g., UV-pumping, excitation by shocks or X-rays from supernova remnants). Instead, the favoured excitation mechanism is slow shocks in the nuclear gas component, which, as shown by high resolution interferometry in the CO $J = 2 \to 1$ line [15], also peaks between the nuclei of NGC 6240.

What is the role of these shocks? In the shocks mechanical energy is dissipated and radiated away, mostly in spectral lines (principally H_2, CO, H_2O and [O I] lines). This energy is radiated away at the expense of the orbital energy of the molecular clouds in the central potential well. Consequently, the dissipation of mechanical energy by the shocks will give rise to an infall of molecular gas to the

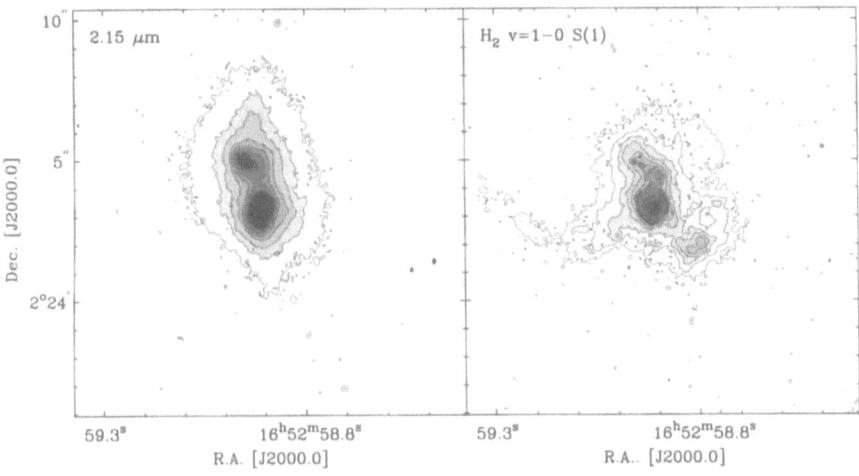

Fig. 2. High resolution imaging of NGC 6240 in the H_2 $v = 1{\to}0$ S(1) line and the 2.15 µm continuum with NICMOS/HST [19].

centre of the potential well. Therefore, *the H_2 vibrational lines measure the rate of infall of molecular gas* into the central potential well. This conclusion can be quantified by writing

$$L_{\rm rad} = L_{\rm dis}, \tag{1}$$

where $L_{\rm rad}$ is the total luminosity radiated by the shocks and $L_{\rm dis}$ the dissipation rate of mechanical energy, giving rise to a molecular gas infall rate $\dot{M}_{H_2}$ given by

$$L_{\rm dis} = \frac{1}{2}\dot{M}_{H_2}v^2, \tag{2}$$

where v is the circular orbital velocity at the position where the shock occurs.

Using a K-band extinction of $0^{\rm m}15$ [18], the total luminosity of H_2 vibrational lines from NGC 6240 becomes $7.2 \cdot 10^8$ $L_\odot$; inclusion of the purely rotational lines observed with ISO approximately doubles this number, so that $L_{\rm rad} = 1.5 \cdot 10^9$ $L_\odot$.

In order to use this number to estimate $\dot{M}_{H_2}$, it is necessary to establish more accurately the fraction of the H_2 emission that is due to infalling gas. Observations with NICMOS on the Hubble Space Telescope (HST) provide the required information (Fig. 2) [19]. The NICMOS image shows that the emission consists of a number of tails (presumably related to the superwind also observed in $H\alpha$ emission), and concentrations associated with the two nuclei, and a further concentration approximately (but not precisely) between the two nuclei. The relative brightness of the H_2 emission from the southern nucleus is deceptive, since this nucleus is much better centred in the filter that was used for these observations than the other emission components, in particular the northern nucleus. Taking this effect into account, it is found that 32% of the total H_2 flux is associated with the southern nucleus, 16% is associated with the northern nucleus, and

12% with the component between the two nuclei, the remaining 40% being associated with extended emission. Using inclination-corrected circular velocities of 270 and 360 km s^{-1} for the southern and northern nucleus respectively [16], and of 280 km s^{-1} for the central component [15], the mass infall rates derived using (1)–(2) are 80 M$_\odot$ yr^{-1} for the southern nucleus, 22 M$_\odot$ yr^{-1} for the northern nucleus and 28 M$_\odot$ yr^{-1} for the central component.

The derived molecular gas inflow rate to the two nuclei is remarkably close to the mass consumption rate by star formation of approximately 60 M$_\odot$ yr^{-1}, indicating that the H$_2$ emission from the nuclei directly measures the fueling of the starbursts in these regions. This analysis shows that the central regions of NGC 6240 are being fueled at a rate sufficient to maintain starburst activity at the level required to account for the FIR luminosity.

3   Dusty, Compact Starbursts in ULIGs

Accepting the starburst model, the faintness of Brγ and other recombination lines remains to be addressed. As pointed out in Sect. 1, it is evident that selective extinction towards the regions of most recent massive star formation plays a significant role in suppressing the Brγ emission. Can this effect be quantified?

The nearby ULIG Arp 220 has been studied in detail with the ISO satelite. Based on an upper limit on the ratio of the [S III] 18 and 33 μm lines, and a high Brα over Brγ flux ratio, an extinction $A_V = 50 \pm 10^m$, located purely in an absorbing foreground screen was proposed [14]. However, the supporting arguments have now weakened significantly. In the first place, a better understanding of the calibration of the [S III] 33 μm spectrum has made the upper limit on the [S III] 18 to 33 μm flux ratio less strict by approximately a factor of two [4]. Secondly, the Brα flux from ISO [14] is almost certainly an overestimate: the Brα line displays, on top of a highly structured baseline, a double-peaked structure, which is absent in any other line (including long-wavelength lines such as the well-detected [S III] 33 μm line). The velocity difference between the two peaks in Brα is approximately 600 km s^{-1} and therefore cannot be attributed to motion of the two nuclei of Arp 220, which have a radial velocity difference of approximately 200 km s^{-1} [12]. The same velocity difference of 200 km s^{-1} is found in long-slit Brγ spectra of the Arp 220 nuclei [6]. An integrated high-resolution Brγ spectrum (Fig. 3) shows no trace of a double-peaked structure, indicating that the high Brα flux found with ISO is most likely dominated by structure in the spectral baseline. The extinction derived from the Brα/Brγ ratio should thus be used as an upper limit. A further argument against an obscuring foreground screen with $A_V \approx 50^m$ is furnished by the derived Lyman continuum fluxes, which increase as shorter wavelength tracers are used [4], a behaviour suggesting that the extinction has been overestimated. Finally, a strong limit on the presence of an intense, highly obscured, but otherwise normal starburst follows from the upper limit to the free-free emission at millimetre wavelengths [12]. These results indicate that either a foreground screen with a lower visual extinction, or a model with mixed emission and absorption needs to be adopted.

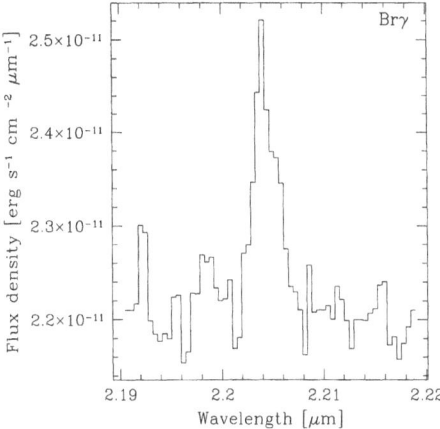

Fig. 3. Integrated spectrum of Brγ emission from Arp 220 [17].

This result does however *not* imply the presence of a strongly aged starburst, or an additional source of power in Arp 220, since the preceding analysis ignored the role of Lyman continuum absorption by dust within the ionized regions. If most of the ionizing radiation is absorbed by dust grains rather than hydrogen atoms, a dust-bounded (rather than hydrogen-bounded) nebula results, and all tracers of ionized gas (recombination lines, fine-structure lines, free-free emission) will be suppressed. If the H II regions in Arp 220 are principally dust-bounded, the observational properties of Arp 220 can be accounted for, even with only moderate extinction. Since the dust would also absorb far-ultraviolet radiation longwards of the Lyman limit, the formation of photon-dominated regions would also be suppressed, and the thus the same mechanism can account for the faintness of the 158 μm [C II] line in Arp 220 and other ULIGs [3].

Is the Arp 220 starburst dominated by dust-bounded H II regions? The *average* molecular gas density in the $\sim 10^{10}$ M$_\odot$ nuclear molecular complex in Arp 220 is $n_{H_2} \sim 2 \cdot 10^4$ cm^{-3} [12]. The strong emission from high dipole moment molecules such as CS, HCO$^+$ and HCN argues for even higher densities: $\sim 10^{10}$ M$_\odot$ of molecular gas (i.e., *all* of the gas in the nuclear complex) has a density $n_{H_2} \sim 10^5$ cm^{-3} [13]. At such densities the ionized nebulae created by hot stars are *compact* or *ultracompact* H II regions, where 50 to 99% of the Lyman continuum is absorbed by dust [22]. Observationally, hydrogen-bounded and dust-bounded H II regions can be distinguished by the quantity $R = L_{FIR}/L_{Br\gamma}$: for a wide range of parameters, $R < 3570$ implies that the nebula is hydrogen-bounded, while $R > 35700$ implies that the nebula is dust-bounded [2]. For Arp 220, the Brγ luminosity of $1.3 \cdot 10^6$ L$_\odot$ (from the spectrum in Fig. 3, with a distance of 77 Mpc) implies $R = 1.6 \cdot 10^5$ assuming an obscuring foreground screen with $A_V = 20^m$ (a model consistent with the results discussed above). Even with a foreground extinction of $A_V = 50^m$ (which is most likely an overestimate, as discussed above), a ratio $R = 1 \cdot 10^4$ would result, so that even in that case the absorption of Lyman continuum radiation by dust would play a significant role. The star formation takes place in (ultra)compact H II regions, where all of the usual tracers of ionized gas (recombination lines, fine-structure lines, free-free emission) are *quenched*, not extincted. While this result significantly complicates the interpretation of diagnostics of massive star formation in ULIGs, it is safe to conclude that the properties of Arp 220 can be accounted for by an intense, and significantly (but not extremely) obscured starburst. There is no reason to invoke the presence of extreme extinction, a strongly aged starburst, or an additional power source in Arp 220.

References

1. L. Armus, D.L. Shupe, K. Matthews, B.T. Soifer, G. Neugebauer: ApJ **440**, 200 (1995)
2. M. Bottorff, J. Lamothe, E. Momjian, E. Verner, D. Vinković, G. Ferland: PASP **110**, 1040 (1998)
3. J. Fischer et al.: 'An LWS spectroscopic survey of infrared bright galaxies'. In: *The universe as seen by ISO*, ed. P. Cox, M.F. Kessler (ESA SP-427, ESA Publications Division, Noordwijk, 1999) pp. 817–820
4. R. Genzel et al.: ApJ **498**, 579 (1998)
5. J.D. Goldader, R.D. Joseph, R. Doyon, D.B. Sanders: ApJ **444**, 97 (1995)
6. J.E. Larkin, L. Armus, R.A. Knop, K. Matthews, B.T. Soifer: ApJ **452**, 599 (1995)
7. A.F.M. Moorwood, E. Oliva: A&A **203**, 278 (1988)
8. A.F.M. Moorwood, E. Oliva: A&A **239**, 78 (1990)
9. T.W. Murphy, B.T. Soifer, K. Matthews, L. Armus, J.R. Kiger: AJ **121**, 97 (2001)
10. P.J. Puxley, T.G. Hawarden, C.M. Mountain: MNRAS **234**, 29P (1988)
11. P.J. Puxley, T.G. Hawarden, C.M. Mountain: ApJ **364**, 77 (1990); erratum ApJ **372**, 73 (1991)
12. N.Z. Scoville, M.S. Yun, P.M. Bryant: ApJ **484**, 702 (1997)
13. P.M. Solomon, S.J.E. Radford, D. Downes: ApJ **348**, L53 (1990)
14. E. Sturm et al.: A&A **315**, L133 (1996)
15. L.J. Tacconi, R. Genzel, M. Tecza, J.F. Gallimore, D. Downes, N.Z. Scoville: ApJ **524**, 732 (1999)
16. M. Tecza, R. Genzel, L.J. Tacconi, S. Anders, L.E. Tacconi-Garman, N. Thatte: ApJ **537**, 178 (2000)
17. P.P. van der Werf, F.P. Israel: in preparation (2001)
18. P.P. van der Werf, R. Genzel, A. Krabbe, M. Blietz, D. Lutz, S. Drapatz, M.J. Ward, D.A. Forbes: ApJ **405**, 522 (1993)
19. P.P. van der Werf, A.F.M. Moorwood, F.P. Israel: in preparation (2001)
20. L. Vanzi, G.H. Rieke: ApJ **479**, 694 (1997)
21. L. Vanzi, G.H. Rieke, C.L. Martin, J.C. Shields: ApJ **466**, 150 (1996)
22. D.O.S. Wood, E. Churchwell: ApJS **69**, 831 (1989)

The Stellar Populations of Nearby Radiogalaxies

Itziar Aretxaga[1], Elena Terlevich[1], Roberto J. Terlevich[2,1], Garret Cotter[3], and Angeles I. Díaz[4]

[1] INAOE, Aptdo. Postal 51 y 216, Puebla, Pue., Mexico
[2] IoA, Univ. of Cambridge, UK
[3] Cavendish Lab., Univ. of Cambridge, UK
[4] Dpto. Física Teórica, UAM, Spain

Abstract. We present results on a sample of 7 nearby luminous radiogalaxies ($z < 0.08$, $P_{178MHz} \gtrsim 10^{25}$ W Hz^{-1} Sr^{-1}), which mostly correspond to the FR II class. In two cases, Hydra A and 3C 285, the Balmer and $\lambda4000$ break indices constrain the spectral types and luminosity classes of the stars involved, revealing that the blue spectra are dominated by blue supergiant and/or giant stars. The ages derived for the last burst of star formation in Hydra A are between 7 and 40 Myr, and in 3C 285 about 10 Myr. The rest of the narrow-line radiogalaxies (four) have $\lambda4000$ break and metallic indices consistent with those of elliptical galaxies. The only broad-line radiogalaxy in our sample, 3C 382, has a strong featureless blue continuum and broad emission lines that dilute the underlying blue stellar spectra.

1 Introduction

In recent years direct evidence that star formation plays an important role in Active Galactic Nuclei (AGN) has been gathered [13,5,2,7] through the detection of stellar atmospheric features that indicate that most of the UV-NIR light coming from the nuclei of Seyfert 2 and LINERs is dominated by young starbursts. The evidence of these starbursts strongly supports some kind of Starburst-AGN connection. However, it is still to be demonstrated that starbursts play a key role in *all* kinds of AGN.

One of the most stringent tests *to assess if all AGN have associated enhanced nuclear star formation* is the case of lobe-dominated radio-sources, whose host galaxies have relatively red colours when compared to other AGN varieties. In this paper we address the stellar content associated with the active nuclei of a sample of FR II radiogalaxies, the most luminous class of radiogalaxies which possess the most powerful central engines and radio-jets [11]. The presence of extended collimated radio-jets, which fuel the extended radio structure over $\gtrsim 10^8$ yr, strongly suggests the existence of a supermassive accreting black hole in the nuclei of these radiogalaxies. This test addresses the question of whether AGN that involve conspicuous black holes and accretion processes also contain enhanced star formation.

2 Spectra and Analysis

Our sample of radiogalaxies was extracted from the 3CRR catalogue [8] with the only selection criteria being edge-brightened morphology, which defines the

FR II class of radiogalaxies and redshift $z < 0.08$. Six out of a complete sample of ten FR II radiogalaxies that fulfill these requirements were randomly chosen. In addition to this sub-sample of FR IIs, we observed the unusually luminous FR I radiogalaxy Hydra A (3C 218). This has a radio luminosity which is an order of magnitude above the typical FR I/FR II dividing luminosity.

Spectroscopic observations of this sample were performed using the double-arm spectrograph ISIS mounted in the 4.2m William Herschel Telescope in 1997 November 7–8 and 1998 February 19–20.

2.1 Line and Continuum Measurements

The CaT index was detected in all of the objects, although in three cases (3C 285, 3C 382 and 4C 73.08) it was totally or partially affected by residuals left by the atmospheric band corrections and the measurement of its strength was thus precluded. For the remaining cases, the strength was measured in the rest-frame of the galaxies against a pseudo-continuum [13]. The values corrected for velocity dispersions ($202 - 292$ km/s) range between 6.2 and 7.0 Å.

Stellar populations can be dated through the measurement of the $\lambda4000$ or Balmer breaks. In intermediate to old populations the discontinuity at $\lambda4000$ results from a combination of the accumulation of the Balmer lines towards the limit of the Balmer absorption continuum at $\lambda3646$ (the Balmer break) and the increase in stellar opacity caused by metal lines shortwards of $\lambda4000$.

Table 1 lists the values of the $\lambda4000$ break index, $\Delta4000$Å, measured in the spectra of the 6 narrow-line radiogalaxies and the elliptical galaxy in our sample. This excludes 3C 382, which has a spectrum dominated by a strong blue continuum and broad-emission lines, and shows very weak stellar atmospheric features and no break. We adopted a definition [4] which quantifies the ratio of the average flux-level of two broad bands, one covering the break (3750-3950) and one bluewards of the break (4050-4250).

Hydra A and 3C 285 have spectra which are much bluer than those of normal elliptical galaxies. In order to quantify better the strength of the break and the ages of the populations derived, we have performed a bulge subtraction using as

Table 1. Break indices.

object	original spectra $\Delta4000$Å mag	bulge-subtracted spectra $\Delta4000$Å mag	D	λ_1 Å
3C 98	2.1			
3C 192	1.9			
Hydra A	1.4	1.2	0.26–0.28	3746–3752
3C 285	1.6	1.0	0.48–0.53	3740–3746
4C 73.08	2.2			
DA 240	2.2			
NGC 4374	2.3			

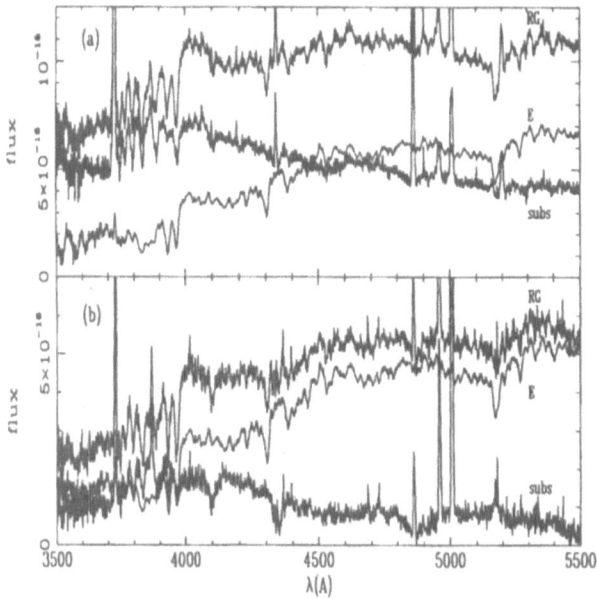

Fig. 1. Bulge-subtraction for (a) Hydra A and (b) 3C 285. We show the original spectra of the radiogalaxy (RG), the scaled template spectra of the bulge population (E), and the resulting bulge-subtracted spectra (subs).

template the spectrum of NCC 4374, scaled to eliminate the G-band absorption of the radiogalaxies. Since the velocity dispersion of the stars in NGC 4374 and in the radiogalaxies are comparable inside the spectral resolution of our data, no further corrections were needed. The G-band absorption is prominent in stars of spectral types later than F5 and it is especially strong in types K. Thus, by removing a scaled template of NGC 4374, we are isolating the most massive stars $(M \gtrsim 1\,M_\odot)$ in the composite stellar population of the radiogalaxies. Figure 1 shows the bulge subtractions obtained on these two radiogalaxies.

We measured on the bulge-subtracted spectra $\Delta 4000\text{Å}$ and also the Balmer break index as defined by the classical $D\lambda_1$ method of stellar classification [12]. The latter quantifies the Balmer discontinuity in terms of the logarithmic difference of the continuum levels (D) and the effective position of the break (λ_1). The method places a pseudo-continuum on top of the higher order terms of the Balmer series in order to measure the effective position of the discontinuity. The functional dependences on the effective temperature and gravity of the stars are sufficiently different for D and λ_1 to satisfy a two-dimensional classification.

The $D\lambda_1$ method could only be reliably applied in the cases of Hydra A and 3C 285. For the other radiogalaxies, the bulge-subtractions led to results that did not allow the identification of the absorption features and/or the break in an unambiguous way due to the resulting poor S/N.

3 Discussion

3.1 Comparison with Elliptical Galaxies and Population Synthesis Models

The radiogalaxies 3C 98, 3C 192, 4C 73.08 and DA 240 have indices of the order of 1.9 to 2.3, which overlap with those of normal E galaxies, $\Delta4000\text{Å}= 2.08 \pm$ 0.23. These values correspond to populations dominated by stars of *ages 1 to 10 Gyr old*, if one assumes coeval population synthesis models [9].

However, Hydra A and 3C 285 have indices in the range 1.4 to 1.6, typical of coeval populations which are 200 to 500 Myr old. Once the bulge population is subtracted, the $\Delta4000\text{Å}$ indices of Hydra A and 3C 285 decrease to 1.2 and 1.0 respectively, which are typical of systems younger than about 60 Myr.

Hamilton [4] found a sequence of increasing $\Delta4000\text{Å}$ from B0 to M5 stars, with values from 1 to 4 mag respectively. A comparison with the sequence he found leads us to conclude that the break in the bulge subtracted spectrum of Hydra A is dominated by B or earlier type stars while that of 3C 285 is dominated by A type stars. The index $\Delta4000\text{Å}$ does not clearly discriminate luminosity classes for stars with spectral types earlier than G0.

The equivalent width of the H10 absorption line in these two radiogalaxies give further support to the interpretation of the Balmer break as produced by a young stellar population. In Hydra A we find after bulge subtraction EW(H10)$\approx 3.9\,\text{Å}$, which, according to synthesis models [3] gives ages of 7 to 15 Myr for an instantaneous burst of star formation, and 40 to 60 Myr for a continuous star formation mode, in solar metallicity environments. In the case of 3C 285, EW(H10) $\approx 6\,\text{Å}$ would imply an age older than about 25 Myr for a single-population burst of solar metallicity.

3.2 The Blue Stellar Content

A better estimate of the spectral type and luminosity class of the stars that dominate the break in Hydra A and 3C 285 comes from the classification of Barbier and Chalonge. In Fig. 2 the solid squares connected by lines represent the maximum range of possible $D\lambda_1$ values measured in these radiogalaxies.

The Balmer break index is sensitive to the positioning of the pseudo-continuum on top of the higher order Balmer series lines, which in turn is sensitive to the merging of the wings of the lines, enhanced at large velocity dispersions. In order to assign spectral types and luminosity classes to the stars that dominate the break, therefore, it is not sufficient to compare the values we have obtained with those measured in stellar catalogues. The values measured for the radiogalaxies can be corrected for their intrinsic velocity dispersions; we have chosen instead to recalibrate the index using template stars of different spectral types

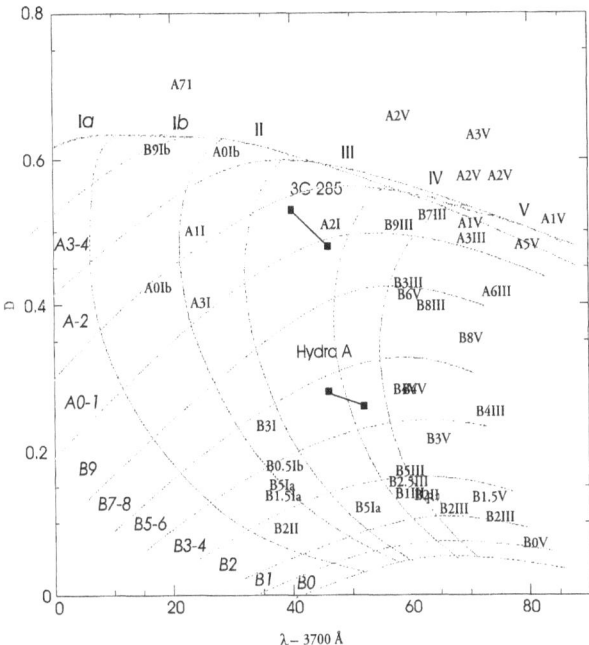

Fig. 2. Barbier and Chalonge index plane. The location of the range of values for 3C 285 and Hydra A is indicated by solid squares and connecting lines. We have also plotted the indices measured in the stellar library, that have been broadened to mimic the width of the Balmer lines in the radiogalaxies, represented by their corresponding spectral type and luminosity class (*e.g.* A2I). The grid of solid lines traces the original locus of unbroadened stars. The big symbols at the edges of the grid represent the correspondance into spectral classes and luminosity classes of the frame defined by the grid.

and luminosity classes convolved with gaussian functions, until they reproduce the width of the Balmer lines observed in the radiogalaxies (FWHM$\approx$ 12.5 Å). We used the B0 to A7 stars from a stellar library [6], which were observed with 4.5 Å resolution. The values of the $D\lambda_1$ indices measured in these broadened stars are represented in Fig. 2 by their respective classification. By comparison we also plot the grid traced by the locus of unbroadened stars [12]. The broadening of the lines shifts the original locus of supergiant stars from the $\lambda_1 \lesssim 3720$ Å range to the $3720 \lesssim \lambda_1 \lesssim 3740$ Å range, occupied by giant stars in the original (unbroadened) classification. Giant stars, in turn, shift to positions first occupied by dwarfs. Most dwarfs have Balmer line widths comparable to those of the radiogalaxies, and thus their locus in the diagram is mostly unchanged.

The value of the D index indicates that the recent burst in *Hydra A is dominated by B3 to B5 stars*, and the effective position of the Balmer break (λ_1) indicates that *these are giant or supergiant stars*, respectively. These stars have masses of 7 and 20 $M_\odot$. From the stellar evolutionary tracks of massive stars with standard mass-loss rate at $Z_\odot$ or $2 Z_\odot$ we infer that these stars must have *ages between 7 to 8 Myr (B3I) and 40 Myr (B5III)*. Note that the B4V stars in Fig. 2, near the location of Hydra A, cannot originate the break and at the same time follow the kinematics of the nucleus: the stars producing the Balmer absorption lines are dynamically decoupled from those producing the metallic lines [10]. Any dwarf star located in the stellar disk of Hydra A would show absorption lines that have been broadened beyond the 12.5 Å of FWHM we measure in this radiogalaxy, and its position would have been shifted further into larger values of λ_1.

The location in the $D\lambda_1$ plane of 3C 285 indicates that the break is produced by A2I stars. These are 15 $M_\odot$ stars. Again, *ages of 10 to 12 Myr* are found for the last burst of star formation in this radiogalaxy. The interpretation of the blue excess in terms of A type stars is further supported by the detection of the Ca II H line in the bulge-subtracted spectrum.

3.3   The Red Stellar Content

We have measured the CaT in a control sample of elliptical galaxies. We find that the range of CaT in elliptical galaxies, 5 to 7.5 Å, comprises the range of values of the radiogalaxies.

Mixed populations of young bursts which contain red supergiants, superposed on old populations can also yield values of the CaT[1] between 4 and 8 Å. Since metal rich giant stars have CaT values ranging from 6 to 9Å we regard our observations of the CaT index in radiogalaxies, as being compatible with ages 1 to 15 Gyr.

References

1. M.L. García-Vargas, M. Mollá, S. Bressan: AAS **130**, 513 (1998)
2. R.M. González Delgado et al.: ApJ **505**, 174 (1998)
3. R.M. González Delgado, C. Leitherer, T. Heckman: ApJS, in press
4. D. Hamilton: ApJ **297**, 371 (1985)
5. T.M. Heckman et al.: ApJ **482**, 114 (1997)
6. G.H. Jacoby, D.A. Hunter, C.A. Christian: ApJS **56**, 257 (1984)
7. D. Kunth, T. Contini: in IAU Symp. 193 *'Wolf-Rayet Phenomena in Massive Stars and Starburst Galaxies'*, eds. K.A. van der Hucht et al., (San Francisco, ASP, 1999) p. 725.
8. R.A. Laing, J.M. Riley, M.S. Longair: MNRAS **204**, 151 (1983)
9. M. Longhetti, A. Bressan, C. Chiosi, R. Rampazzo: AA **345**, 419 (1999)
10. J. Melnick, Gopal-Krishna, R.J. Terlevich: AA, **318**, 337 (1997)
11. S. Rawlings, R. Saunders: Nat. **349**, 138 (1991)
12. B. Strömgren: in 'Basic Astronomical Data, Stars and Stellar Systems', vol. 3, p. 123.
13. E. Terlevich, A.I. Díaz, R. Terlevich: MNRAS **242**, 285 (1990)

Part IV

Star Formation and the ISM

Starbursting in Spiral Arms and ULIRGs

Nick Scoville[1] and Maria Polletta[2]

[1] California Institute of Technology, Pasadena, CA 91125, USA
[2] Observatory of Geneva, Sauverny, Switzerland

Abstract. We present results from HST Hα and Pα imaging of HII regions in M51 which are used to define the properties of OB star clusters in the 'starburst' regions of spiral arms. These data indicate an *observed* Hα luminosity function which truncates on the high end at $L_{H\alpha} = 10^{39}$ erg sec^{-1}. This is far below the high end seen in ground-based imaging – we believe this is due to the fact that lower resolution ground-based imaging often blends multiple centers of OB star formation and the highest luminosity regions were really complexes rather single OB star clusters. On the other hand, based on the observed Pα/Hα ratios, we find the typical extinctions are high (2–3 mag) and when extinction corrections are applied to the observed Hα, the intrinsic Hα luminosities get up to $\sim 10^{40}$ erg sec^{-1}. This upper limit to the luminosity function corresponds to a cluster mass of $\sim$few$\times 10^3$ $M_\odot$ (for a salpeter IMF between 1 and 120 $M_\odot$; it is approximately at the point at which the IMF is first populated to ~ 100 $M_\odot$. We suggest that this limiting cluster mass may be understood physically if accretion (and thus cluster formation) in cloud cores is terminated when radiation pressure on the surrounding dust exceeds the self-gravity of the star cluster – this occurs when the highest mass stars are formed. In ULIRGs, the augmented star formation is probably due to the collision of massive clouds since, often, large numbers of bright clusters may be found in the overlap regions of the colliding galaxies . In the extreme object Arp 220, high resolution NICMOS and mm-wave interferometric imaging indicates double nuclei separated by ~ 350 pc, each of which is embedded in a massive gas and dust disk. Although the geometry in Arp 220 is clearly different than a spherical cloud core forming a single oB star cluster, the star formation may also be self-regulated at $\sim 10^3$ $M_\odot$ yr^{-1} by the radiation pressure arising from the starburst. The disks will swell vertically and self-regulate their star formation.

1 Introduction

OB star formation in galactic spiral arms is probably enhanced due to the concentration of giant molecular clouds resulting from orbit crowding in the spiral density wave. In the luminous infrared galaxies, nuclear starbursts are fueled by extraordinarily large masses of gas and dust concentrated at radii of a few hundred pc by viscous accretion and the torques associated with galactic merging. In these two very different environments, dissipative gas dynamics play a critical role in setting the initial conditions for the local starbursts. In both instances, the star formation efficiency (as measured by the local luminosity to ISM-mass ratio) is several$\times 10^2$ $L_\odot$ per $M_\odot$. These facts suggest a degree of commonality between spiral arm star formation and nuclear starbursts despite the much greater extent (and therefore integrated output) of the latter. Here we describe our recent high resolution study of HII region properties in M51 and review existing results for Arp 220.

2   M51 HII Regions

M51 has been the focus of numerous ground based Hα studies [7,11,22,12,13,9,21] and these studies have contributed much of what is currently known regarding OB associations in other spiral galaxies. We have recently completed a comprehensive study of M51 comprised of HST (WFPC2 and NICMOS) imaging of the Hα and Pα emission lines [19]. The $0.1 - 0.2''$ resolution available with HST corresponds to $4.6 - 9.3$ pc. These sizes correspond to those of individual resolved Galactic HII regions (eg. the Orion Nebula). Related mm-CO interferometry has been presented in [1]. The former probes the stellar disk, the dust, and the OB star formation while the latter probes the dense, molecular ISM which is the birthsite of OB star clusters.

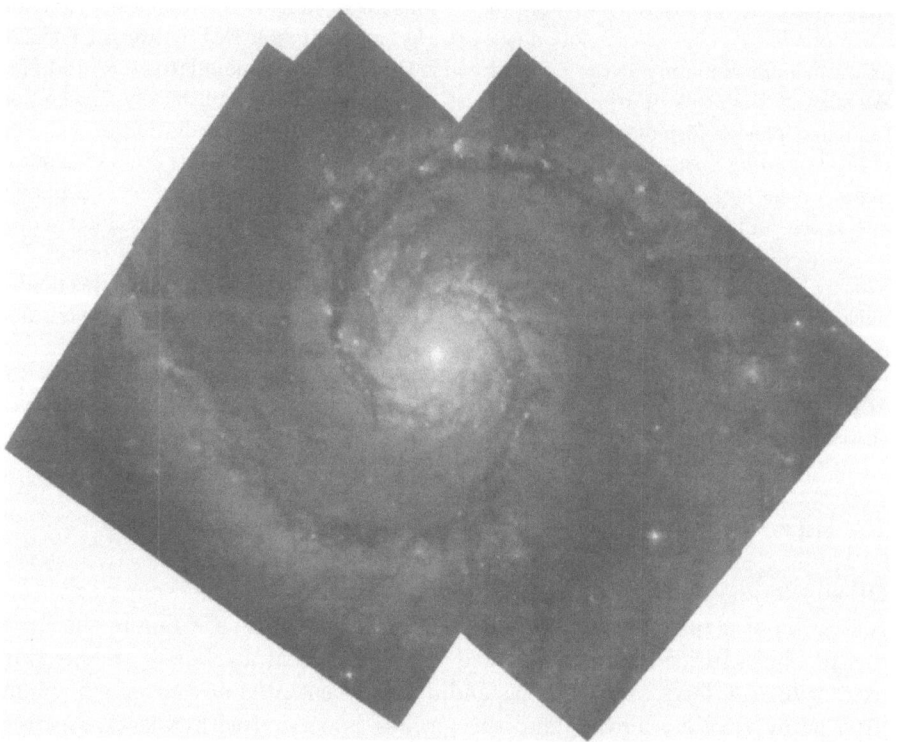

Fig. 1. The HST Hα and optical continuum are shown for the central $281 \times 223''$ of M51. North is at top.

In Fig. 1, the Hα in the central $281\times223''$ of M51. Hα emission extends out $10''$ from nucleus at PA $\sim$ -15° and bright, discrete Hα emission regions outline the spiral arms. The locations of bright Hα emission are closely associated with the dark dust lanes, but relative to the dust, the Hα is often displaced to the outside or leading edge of the arms. Although generally Pα shows the same emission regions, many of the arm HII regions have considerable reddening.

As described in [19], we developed automated algorithms for defining the boundaries of the HII regions and measuring their properties. A sample of 1746 HII regions with at least 3 pixels above 5 σ were measured from the Hα image and over a third of these were well detected in Pα. The 'apparent' Hα luminosity function (uncorrected for extinction) is shown in Fig. 2 together with those derived by [12] and [9]. All luminosity functions exhibit roughly similar shapes and a power-law fit to the apparent luminosity function yields α = -1.21$\pm$0.01, compared to -0.55 and -0.27 derived for the samples of [12] and [9] over the range $\sim 4 \times 10^{37} - 5\times10^{38}$ erg s^{-1}.

Most significantly though is the fact that the HST luminosity function has its highest luminosity at about 1/5'th of that in the ground-based luminosity functions. This is difference is due to the higher resolution of the HST imaging resolving many of the HII regions into multiple regions. However, the lower 'apparent' luminosities obtained here are to a large extent compensated by the extinction corrections we derive. The mean extinction derived from the Pα/Hα ratios for the discrete Hα emission regions is $A_{H\alpha}$ = 0.798$\times$ $<A_V>$ = 2.55 mag. If this is applied on-average to all HII regions, then the observed Hα luminosity functions are increased by a factor of 9. The mean sizes are 28 — 33 pc in arm, interarm and nuclear regions. Both the shapes of the distributions and the mean sizes are the same for all three areas. The majority of the HII regions have size $\leq$ 40 pc and thus may, in fact, be due to a single or a few OB star cluster(s). In the full sample of M51 regions (Fig. 2), the most luminous is at an 'apparent' $L_{H\alpha}$ = 10^{39} erg sec^{-1}; if this region has a modest extinction, it's luminosity would be a few times that of W49. (The 30Dor cluster in the LMC is a few times more luminous than W49 and the diameter of its stellar cluster is $\sim$ 20 pc. The most luminous Hα region in M51 requires a total Lyman continuum emission rate of a few$\times10^{51}$ sec^{-1} to maintain its ionization. This is similar to the maximum found by [8] for Galactic regions.

3 Limiting OB Star Cluster Mass

The existence of an upper 'cutoff' luminosity (Q $\sim$ 10^{51} sec^{-1}) could be due to a physical limitation which limits the formation of higher mass clusters. The observed cutoff corresponds to a cluster of total mass few$\times10^3 M_\odot$ for a Salpeter IMF with stars between 1 and 120 $M_\odot$. This mass cluster also corresponds approximately to the cluster mass at which the main sequence is first populated up to the upper cutoff of 120 $M_\odot$. The parent molecular clouds are much more massive and one must ask: why don't the OB star clusters generally build up to much greater mass?

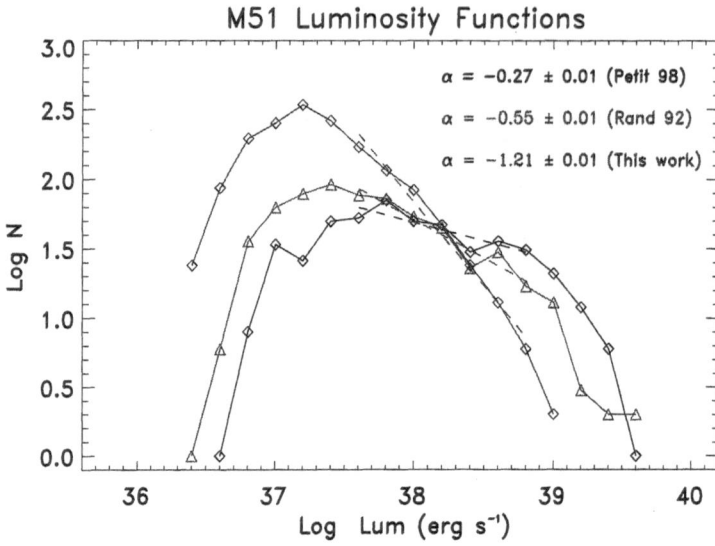

Fig. 2. The observed Hα luminosity function for all HII regions defined and measured in the WFPC2 images (not corrected for extinction) is compared with the luminosity functions derived from ground-based imaging by [12] and [9]. The fact that the HST luminosity function appears to shift to lower luminosity is due largely to the ability to separate individual HII regions which become blended in ground-based images (see text). Typical extinctions of $A_{H\alpha} = 0.798$ $A_V = 2.5$ mag will shift the luminosity functions a factor of 9 higher.

   To understand this apparent mass limit for the clusters, we have considered the buildup of a star cluster in a molecular cloud core. Initially, the cluster will contain only a few low-mass stars and the gas dynamics will be entirely determined by the self-gravity of the cloud core and cluster. Neglecting rotational or magnetic stresses, this is clearly the *most favorable* situation for accretion and maximal growth of the cluster. However, as the cluster becomes more massive and populates the upper main sequence, the higher luminosity-to-mass ratio of high-mass stars will result in increased radiation pressure on the surrounding dust — eventually terminating further accretion to the cloud core. It is easy to show that for a spherical cloud core with a standard abundance of dust to gas, the formation of a massive cluster in a molecular cloud core is likely to be terminated at the point when the luminosity-to-mass ratio of the cluster reaches ~ 1000 $L_\odot / M_\odot$ when radiation pressure begin to dominate the self-gravity of the cluster [19]. Radiation pressure effectively terminates further gas and dust accretion to the central core. However, we also find that a radiatively-compressed shell will then propagate outwards at a few km/s, possibly triggering a second wave of star formation out to a few pc radius.

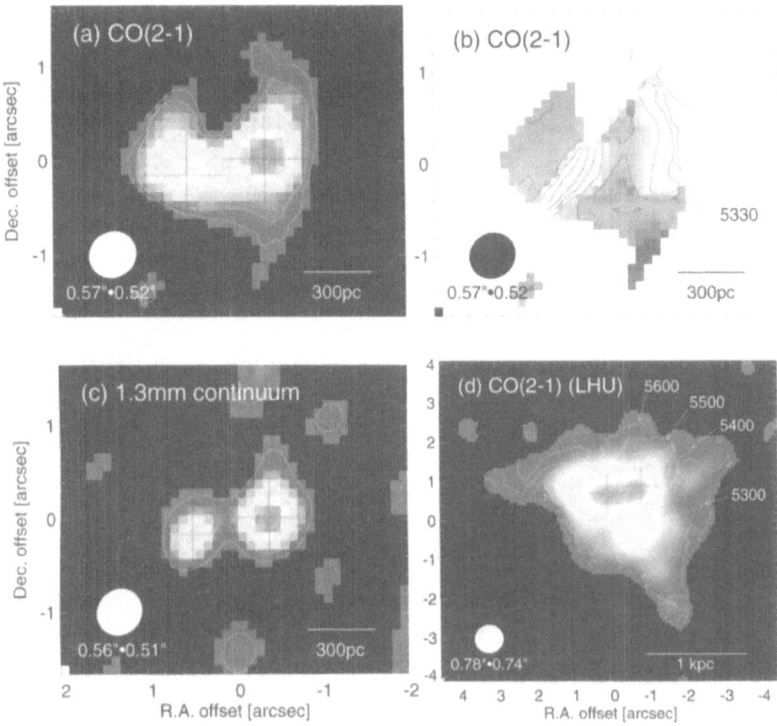

Fig. 3. The merging nuclei of Arp 220 are shown in 0.5″resolution imaging of the CO(2–1) and dust continuum emission. These data clearly resolve the two nuclei and reveal for the first time counter-rotating disks in each nucleus. The panels show: a) continuum-subtracted CO(2–1) (using only high resolution data), b) the CO mean velocities, c) the 1.3 mm dust continnum, and d) the total CO emission including both low and high resolution interferometry [14]. Crosses indicate the 1.3 mm continuum positions of the nuclei.

4   Arp 220 – A 'Prototypical' ULIG

Mm-wave imaging provides a unique capability to probe the starbursts in dusty ULIRG nuclei. More than 20 luminous ($\geq 10^{11}L_\odot$) infrared galaxies have now been imaged, primarily at OVRO and IRAM [16,17,4,3,20]. Virtually all display massive concentrations of molecular gas in the central few kpc.

Arp 220, at 77 Mpc, is one of the nearest and the best known ultra-luminous merging system ($L_{8-1000\mu m} = 1.5 \times 10^{12}$ $L_\odot$). Visual wavelength images reveal two faint tidal tails, indicating a recent tidal interaction [6], and high resolution ground-based radio and near-infrared imaging show a double nucleus [2,5]. The radio nuclei are separated by 0.″98 at P.A. ~90° [2], corresponding to 350 pc. To power the energy output seen in the infrared by young stars requires a star formation rate of ~10^2 M$_\odot$ yr^{-1}. Arp 220 has been the subject of a number of OVRO and IRAM interferometer studies imaging in the 2.6 mm CO line

[16], 3 mm HCN [10], and 1.3 mm CO [17,4,14]. The CO (2–1) line emission, mapped at $1''$ resolution, showed two peaks separated by, $0.9''$, and an inclined disk of molecular gas [17,4]. These peaks correspond well with the double nuclei seen in near-infrared and radio continuum images. The $0.5''$ resolution CO and 1.3 mm continuum maps obtained recently by [14] using OVRO are displayed in Fig. 1. These reveal *counter-rotating* disks of gas in each of the nuclei. The kinematic data clearly require very high mass concentrations in each nucleus, consistent with their being individual galactic nuclei. The fact that they are counter-rotating is consistent with the concept that more complete merging may be associated with counter-rotating precursor galaxies in which there can be greater angular momentum cancellation. The masses in each nucleus are apparently dominated by the molecular gas – a common finding of the the ULIG galaxy studies [3].

The total molecular gas content for Arp 220 is $9 \times 10^9 M_\odot$ based on the CO (2–1) emission and a CO-to-H_2 conversion ratio which is 0.45 times the Galactic value [17]. This enormous mass (approximately two times that of the total Galactic ISM) is contained entirely within $R < 1.5$ kpc and approximately $5 \times 10^9 M_\odot$ is apparently concentrated in a thin disk in the nuclear region at radii < 250 pc. The gas probably collected in the center of the merging system as a result of torques associated with the encounter and the high dissipation in the gas. The inferred mean extinction through the disk in the center of Arp 220 is $A_V = 1000$–2000 mag [17]. In the ULIRG disks, the star formation may also be regulated by radiation pressure since a high star formation rate leads to higher luminosity and greater radiation pressure support of the disks in the vertical direction (and thus a reduced stasr formation rate). This is much the same situation as the OB star clusters forming in a molecular cloud core in as much as the ULIRG disks are probably self-gravitating and the radiation is once again due to the young, high mass stars.

5   Induced Starburst Activity

In normal spiral galaxies like the Milky Way, most of the star formation proceeds at a steady, low efficiency rate. This may be seen from the fact that estimates for the current star formation rate in the Galaxy are typically $\sim$ few $M_\odot$ yr^{-1} while the supply of star forming molecular gas is $\sim 3 \times 10^9$ $M_\odot$, implying a gas–star cycling time of 10^9 yrs. Yet, the majority of the molecular gas is at average density 250 H_2 cm^{-3} with a free-fall collapse time of 10^7 yrs. Therefore, on the timescale of gravitational free-fall, the formation of stars occurs with an efficiency of only 1%. It is generally believed that the clouds are supported against collapse by a combination of turbulent motions and magnetic fields. The magnitude of the turbulent or magnetic support can be judged by the fact that the molecular emission linewidths of individual Galactic GMCs are typically 10 times the thermal speed of sound in the gas (i.e. M = 10). We conclude that in order *to change the internal state of a GMC and elevate the rate of collapse to form stars, the disturbance must be highly supersonic and in an external medium of density comparable to that of the molecular gas (unless the Mach number M >>10).*

What better way to *'arouse or wake up'* a supersonically supported cloud than to collide it supersonically with another cloud – the ram pressure resulting from the collision will automatically exceed the internal magneto-turbulent pressure! In principle, the cloud-cloud collision rate will depend quadratically on the local volume density of the clouds, their cross sections, and on the cloud-cloud velocity dispersion. In the spiral arms of a galaxy like the Milky Way, the number-density of clouds is enhanced due to orbit crowding in the arms and the OB star formation rate is possibly enhanced due to cloud-cloud collisions [15]. In interacting and merging galaxies, the cloud collision rate will be enhanced: first in the individual disks due to tidally-elevated velocity dispersion (specifically, the normally circular, 'non-intersecting' orbits are disrupted); later due to the passage of the disks through each other; and lastly due to the concentration of the gas at smaller radii (i.e. smaller volume) in the merger nucleus. The first two effects are clearly evidenced in the NICMOS ULIRG sample [18] – for example many luminous star forming regions can be seen in the tidal tails (eg. IRAS 22491-18) and in the galaxy overlap region (eg. NGC 6090 and VV114), where massive concentrations of molecular gas are also seen [3,24]. It is interesting to note that evidence of greatly enhanced star formation due to molecular cloud collisions is also found in the NGC 4038/39 system (the 'Antennae') since the galaxy overlap region exhibits multiple CO emission velocities along the line of sight precisely coinciding with the location of very strong 15μm emission seen with ISO [23].

6   Concluding Remarks

Interactions and merging plays a fundamental role in the evolution of galaxies – producing the most luminous starburst galaxies and very likely luminous AGN. The dynamical effects of the interactions are most dramatic in the ISM since it is dissipative and has high filling factors. The torques and increased velocity dispersions due to the galactic encounters lead to rapid transport of the dense molecular ISM to the nuclear regions and high rates of cloud-cloud collisions. The shock-induced cloud compression might trigger formation of super starburst clusters in the regions of physical overlap of the galactic disks and in tidal bridges/tails. In many of the ultra-luminous IR galaxies, over 10^9 $M_\odot$ of H_2 gas is found at radii $<< 500$ pc (comparable with the Galactic ISM mass, but within an area 50 times smaller). Within the central region, it now appears this gas dissipatively settles into a thin disk. These ultra-massive nuclear gas disks are presumably the sites of the nuclear starburst activity which may be regulated by a balance of self-gravity and radiation pressure support.

References

1. S. Aalto, S. Hüttemeister, N.Z. Scoville, P. Thaddeus: ApJ **522**, 165 (1999)
2. W.A. Baan, A.D. Haschick: ApJ **454**, 745 (1995)
3. P.M. Bryant, N.Z. Scoville: AJ **117**, 2632 (1999)
4. D. Downes, P.M. Solomon: ApJ **507**, 615 (1998)
5. J.R. Graham et al.,: ApJ **354**, L5 (1990)
6. R.D. Joseph, G.S. Wright: MNRAS **214**, 87 (1985)
7. R.C. Kennicutt, B.K. Edgar, P.W. Hodge: ApJ **337**, 761 (1989)
8. C.F. McKee, J.P. Williams: ApJ **476**, 144 (1997)
9. H. Petit, C.T. Hua, D. Bersier, G. Courtes: A&A **309**, 446 (1996)
10. S.J. Radford, et al.: in Proc. of IAU Sym. 146, 'Dynamics of Galaxies and Their Molecular Cloud Distributions', eds. F. Combes & F. Casoli, (Dordrecht, Kluwer, 1991), p.303
11. R.J. Rand, S.R. Kulkarni: ApJ **349**, L43 (1990)
12. R.J. Rand: AJ **103**, 815 (1992)
13. M. Rozas, J.E. Beckman, J.H. Knapen: A&A **307**, 735 (1996)
14. K. Sakamoto, N.Z. Scoville, M.S. Yun, M. Crosas, R. Genzel, et al.: ApJ **514**, 68 (1999)
15. N.Z. Scoville, D.B. Sanders, D.P. Clemens: ApJ **310**, L77 (1986)
16. N.Z. Scoville, A.I. Sargent, D.B. Sanders: ApJ **366**, L5 (1991)
17. N.Z. Scoville, M.S. Yun, P.M. Bryant: ApJ **484**, 702 (1997)
18. N.Z. Scoville, A.S. Evans, R. Thompson, G. Rieke, D. Hines, F. Low, N. Dinshaw, J. Surace, L. Armus: AJ **119**, 991 (2000)
19. N.Z. Scoville, et al., (2001), in preparation
20. L.J. Tacconi, R. Genzel, M. Tecza, J.F. Gallimore, D. Downes, N.Z. Scoville: ApJ **524**, 732 (1999)
21. D. Thilker, R. Braun, R. Walterbos: AJ **120**, 3070 (2000)
22. J.M. van der Hulst, R.C. Kennicutt, P.C. Crane, A.H. Rots, A&A **195**, 38 (1988)
23. C.D. Wilson, N.Z. Scoville, S.D. Madden, V. Charmandaris: ApJ **542**, 120 (2000)
24. M.S. Yun, N.Z. Scoville, R.A. Knop: ApJ **430**, L109 (1994)

Extreme Starbursts and Molecular Clouds in Galaxies

Philip M. Solomon

Physics and Astronomy, SUNY, Stony Brook, NY 11794, USA

Abstract. The extraordinary starbursts found in ultraluminous IR galaxies occur in molecular gas concentrated in compact very massive "clouds" which we call "Extreme Starbursts". They have one thousand times the mass but are only a few times larger than GMCs. High-mass star formation in sufficiently dense and massive structures does not disrupt further star formation; it is a runaway process. Star formation remains embedded in the molecular gas and there is little or virtually no optical–UV radiation. In the early universe extreme starbursts may be more frequent and they may be the mode of star formation in high redshift submillimeter sources.

1 Introduction

All star formation takes place in molecular clouds and most star formation takes place in Giant Molecular Clouds (GMCs), the most massive objects in the Galaxy [15] and the dominant form of the molecular interstellar medium. The distribution of GMCs in the Milky Way molecular ring [11], as traced by CO emission, is very different from that of atomic hydrogen and similar to that of H II regions, demonstrating that it is not the ISM distribution as a whole that governs star formation rates but the distribution of molecular gas. An understanding of star formation rates and starbursts in galaxies requires an understanding of the physical conditions in GMCs and their relation to galactic dynamics. Due to the limited resolution of current millimeter wave arrays the Milky Way is the only galaxy in which a large number of GMCs have been identified and mapped and we must turn to Milky Way GMCs to examine star formation efficiencies on the scale of individual clouds.

The high-mass star formation rate within and on the edge of GMCs can be estimated from the far infrared luminosity emitted by dust heated by embedded OB stars. The association of molecular clouds and FIR emission from OB star formation regions in a section of the Milky Way is shown in Fig. 1. The mass of molecular gas can be determined from the CO luminosity or, more accurately for individual clouds, from the virial theorem utilizing the CO kinematics. The ratio of the FIR luminosity to the CO luminosity, $L_{\mathrm{FIR}}/L_{\mathrm{CO}}$, or to the cloud mass, $L_{\mathrm{FIR}}/M_{\mathrm{VT}}$, is a measure of the rate of star formation per solar mass of the cloud and is an indicator of star formation efficiency. The star formation rate is $\dot{M}_* \sim (2.5 \times 10^{-10})\, L_{\mathrm{IR}}\ [\mathrm{M_\odot/yr}]$ and the gas depletion time is $\tau = (M_{cloud}/\dot{M}_*)$ years [4]. A FIR and CO survey of 60 GMCs [7] shows that for active OB star

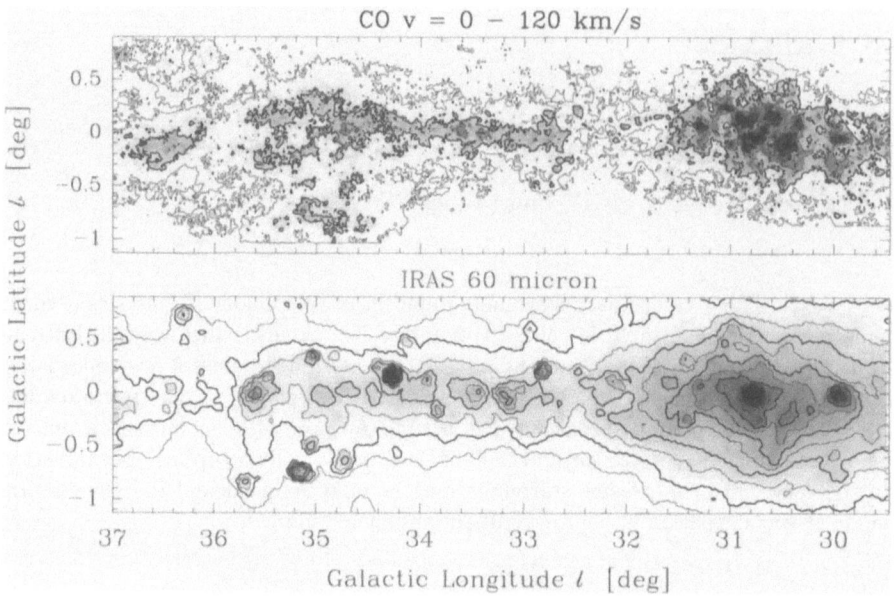

Fig. 1. Molecular clouds and FIR emission from a section of the inner Milky Way. The CO (1–0) emission in this view includes all velocities along the line of site blending many clouds. All of the strong FIR emission is associated with individual clouds and embedded H II regions. The CO is from data obtained at FCRAO (T. Mooney) with 50 arc second resolution.

forming clouds L_{FIR}/L_{CO} is typically ~ 15 and typical molecular gas depletion times are $\approx 10^9$ years. Even in the most active individual clouds (M17A and W49), $L_{FIR}/L_{CO} = 100$ and the gas depletion time of 2×10^8 years is two orders of magnitude greater than the cloud dynamical timescale. By this measure star formation is clearly not very efficient in ordinary GMCs.

The efficiency of star formation in galaxies must be measured relative to the potential for star formation determined by the available mass of molecular gas. Normal spiral galaxies, even those with moderately high IR luminosities, have a range of L_{FIR}/L_{CO} similar to GMCs, indicating star formation efficiencies similar to those of the Milky Way. In normal spirals, the rate of star formation is proportional to the mass of molecular gas. In contrast, far IR starbursts, luminous infrared galaxies (LIGS) and ultraluminous infrared galaxies (ULIGs) have L_{FIR}/L_{CO} ratios 3 to 20 times higher than normal spirals. Most of these are closely interacting and merging systems [9]. CO observations of 30 IRAS identified ultraluminous galaxies out to z = 0.3 show that while all without exception have high CO luminosities, they also all have abnormally high star formation efficiencies [14]. While their CO luminosities are at the very high end of normal galaxies, their FIR luminosities are more than an order of magnitude above normal spirals. This situation is summarized in Fig. 2, which shows that the star formation efficiency of ULIGs as indicated by L_{FIR}/L_{CO} is not only

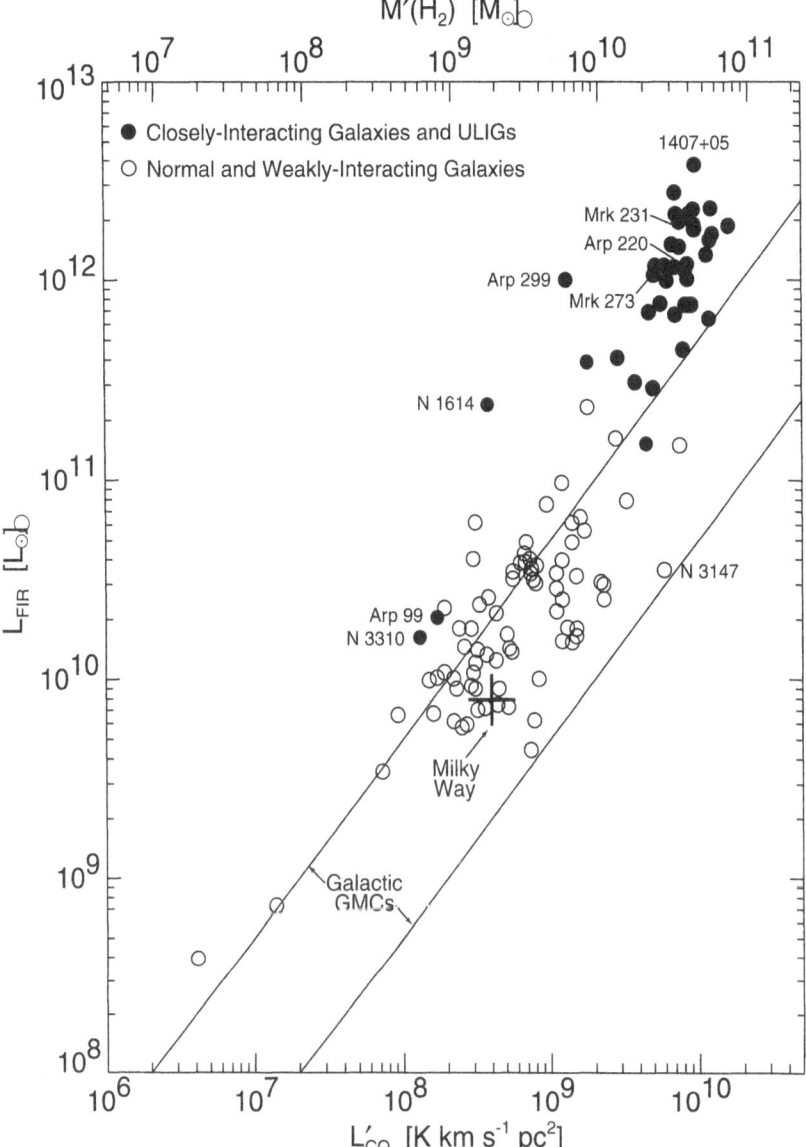

Fig. 2. FIR luminosity vs. CO (1–0) luminosity (lower scale) and molecular gas mass (top scale). (Adapted from [12] and [14]). Solid circles indicate closely-interacting galaxies, mergers and ultraluminous galaxies; open circles are isolated and weakly-interacting galaxies. The solid lines labeled GMCs bracket the L_{FIR}/L_{CO} for Galactic giant molecular clouds with active OB star formation. The top axis is the molecular, H_2, mass assuming a Milky Way CO to H_2 conversion factor. This is an overestimate for ULIGs which have an even higher ratio of $L_{FIR}/M(H_2)$. Ultraluminous galaxies have a higher $L_{FIR}/M(H_2)$ than any Galactic GMC.

higher than in normal spirals, but is higher than that of any individual GMC. CO luminosity and molecular mass are not very good indicators of high-mass star formation rates. L_{CO} is not actually very well correlated with L_{FIR}. When infrared luminous galaxies are included, L_{FIR}/L_{CO} ranges over a factor of 100. This large dispersion is further complicated by the interpretation of CO emission from ULIGs in terms of H_2 mass.

The large nuclear concentration of molecular gas in ultraluminous galaxies has been mapped in the millimeter lines of CO by several groups during the past decade. Scaling to the CO signal strengths from Milky Way molecular clouds, however, soon led to a paradox for many of the sources — the estimated gas mass was equal to or larger than the dynamical mass indicated by the linewidths and source size. For Arp 220, for example, nearly all of the mass in the central few hundred parsecs was in the form of molecular gas [10]. To resolve this dilemma, we [2,14] showed that in the extreme environment in the central few hundred pc of ultraluminous galaxies, much of the CO luminosity must come from an inter-cloud medium that fills the whole volume, rather than from clouds bound by self gravity, and therefore the CO luminosity traces the geometric mean of the gas mass and the dynamical mass, rather than just the gas mass. This has the effect of lowering the molecular mass for a given CO luminosity and increasing the star formation efficiency.

What are the physical differences between the molecular gas in ULIGs and normal spirals which account for the extraordinary infrared starbursts? To answer this question, we have carried out comprehensive observations of the global properties of the molecular gas in ULIGs and also of the detailed morphology and kinematics.

2 Dense Molecular Gas: HCN, a Molecular Starburst Indicator

Until recently, CO has been the only molecular tracer systematically observed in a wide sample of galaxies, particularly ULIGs. CO traces molecular hydrogen, H_2 at densities ≥ 300 cm^{-3}, typical of densities in GMCs, but far below the densities in GMC cloud cores, the actual sites of star formation. To measure the mass of *dense* molecular gas it is essential to use a molecular transition with a much higher density threshold. A particularly useful molecule in this respect is HCN, which has a moderate abundance but a high dipole moment, and therefore a short lifetime and requires a large H_2 density $n(H_2) \geq 3 \times 10^4$ cm^{-3} for significant excitation and emission. Initial measurements of the HCN (1–0) line from 5 ULIGs [13] showed extraordinarily strong emission. Arp 220, Arp 193, Mrk 231, and NGC 6240 all showed HCN line luminosities greater than the CO luminosity of the Milky Way, and about 30 times higher than the HCN luminosity of normal spirals. The unexpected huge HCN luminosity and the high ratio of L_{CO}/L_{HCN} indicate that a large fraction of the total molecular gas in ULIGs is at a high density similar to that in star forming cloud cores, rather than the envelopes, of GMCs.

Although the HCN sample contained only 10 galaxies, the results indicated that the ratio of far infrared to HCN luminosity is similar in ultraluminous galaxies and normal spirals, including the Milky Way, which suggests that the star formation rate per solar mass of *dense* gas is independent of the infrared luminosity. Subsequent HCN measurements (Fig. 3) of the global HCN luminosity in 60 galaxies covering 3 orders of magnitude in IR luminosity [5,6] confirm this result, demonstrating that the $L_{FIR} - L_{HCN}$ correlation is substantially better than that for $L_{FIR} - L_{CO}$, and show that L_{FIR}/L_{HCN} is almost independent of FIR luminosity (There is a very small effect amounting to a factor of 1.7 for the most luminous galaxies). This indicates that the star formation rate is proportional to the dense molecular gas content of a galaxy, and is also evidence that the power source in ULIGs is primarily star formation, not AGNs. Normal spiral galaxies clearly powered by star formation have the same ratio of far infrared luminosity to HCN luminosity as ultraluminous galaxies. The dense molecular gas in normal spirals and ultraluminous galaxies is the star forming material being processed into stars with equal efficiency in all galaxies. HCN luminosity is a star formation indicator.

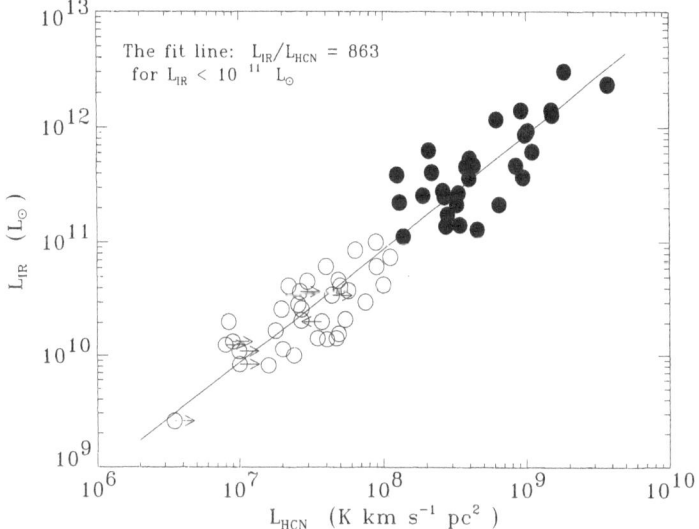

Fig. 3. FIR luminosity vs. HCN (1–0) line luminosity for 65 galaxies (from [5] and [6]). Solid circles represent luminous and ultraluminous galaxies with $L_{FIR} \geq 10^{11}$ $L_{\odot}$. Open circles are all galaxies with $L_{FIR} < 10^{11}$ $L_{\odot}$. Ultraluminous and normal galaxies fall along the same line with a constant ratio of L_{FIR}/L_{HCN}. HCN emission traces dense molecular gas with $n(H_2) \geq 3 \times 10^4$ cm^{-3}.

3 Rotating Nuclear Rings and Extreme Starbursts in Ultraluminous Infrared Galaxies

High spatial resolution CO maps of ULIGs (eg [10,16,3]) all show evidence of central molecular concentrations and rotation. A survey with the IRAM Array of the kinematics of 10 ULIGs included 5 galaxies mapped at the unprecedented resolution of 0.5 arc seconds [3]. Analysis of the CO intensity distribution and kinematics utilizing models of CO excitation and radiative transfer yields a detailed kinematic picture dominated by rotating rings or discs on a scale of a few hundred parsecs (eg Mrk 231, Mrk 273, Arp 220, Arp 193) up to 2 kpc (for VII Zw31), with compact *Extreme Starburst Regions* embedded in the rings or disks. An example of the CO (2–1) data is shown in Fig. 4 for Arp 220, the nearest ULIG, and therefore the object with the highest spatial detail. Most of the CO emission originates from a disk rotating at 330 km/s with a characteristic radius of 340 pc (1 arc second); the strongest CO emission and most of the 1.3 mm dust continuum emission, indicative of the luminosity source, originates from the two smaller sources Arp 220 East and West coincident with the extended nonthermal radio continuum peaks, usually interpreted as the "nuclei" of the merging galaxies. Each of these objects shows its own internal kinematics; the East source shares a kinematic axis with the main disk and the West source has a kinematic axis rotated by 110° from the East source and main disk axis. (see Fig. 21 in [3]). A map with similar resolution obtained at OVRO [8] yields similar results, but with the East and West kinematic axes offset by 180° interpreted as counter rotating nuclei. A full interpretation of the relation between these two sources and the larger disk will have to wait for even higher resolution data.

An analysis of the molecular mass, total dynamical mass and mass of young stars required to account for the extreme starburst in the East and West sources [3] shows that most of the mass must be in molecular gas and young stars, not in old stars from a pre-existing nuclear bulge. The characteristic radii of these extreme starburst regions are only 70 and 110 pc, with total dynamical masses determined from a virial theorem analysis of rotation of about 3 and 1.5×10^9 $M_\odot$, yet each of these regions is producing between 3 and 5×10^{11} $L_\odot$. Two other Extreme Starburst Regions have been identified in Mrk 273 and Arp 193. The high-resolution CO (2–1) maps show a remarkable molecular-line source in the Mrk 273 nuclear disk — a bright, $0.35'' \times < 0.2''$ CO core, that resembles the West nucleus of Arp 220. This is the most luminous extreme starburst region identified in the sample of 10 galaxies. It has an infrared luminosity of about 6×10^{11} $L_\odot$, generated from a current molecular mass of 1×10^9 $M_\odot$ in a region with a radius of only 120 pc (see Table 1). The extended nonthermal continuum emission [1] coincident with a high mass of dust and gas leaves little doubt that this region is powered by star formation. To put this in perspective, the entire molecular core has a radius about 5 times that of a very IR luminous Milky Way GMC (for example W51), but with about 3,000 times the molecular mass and $\approx 10^5$ times the IR luminosity from OB stars.

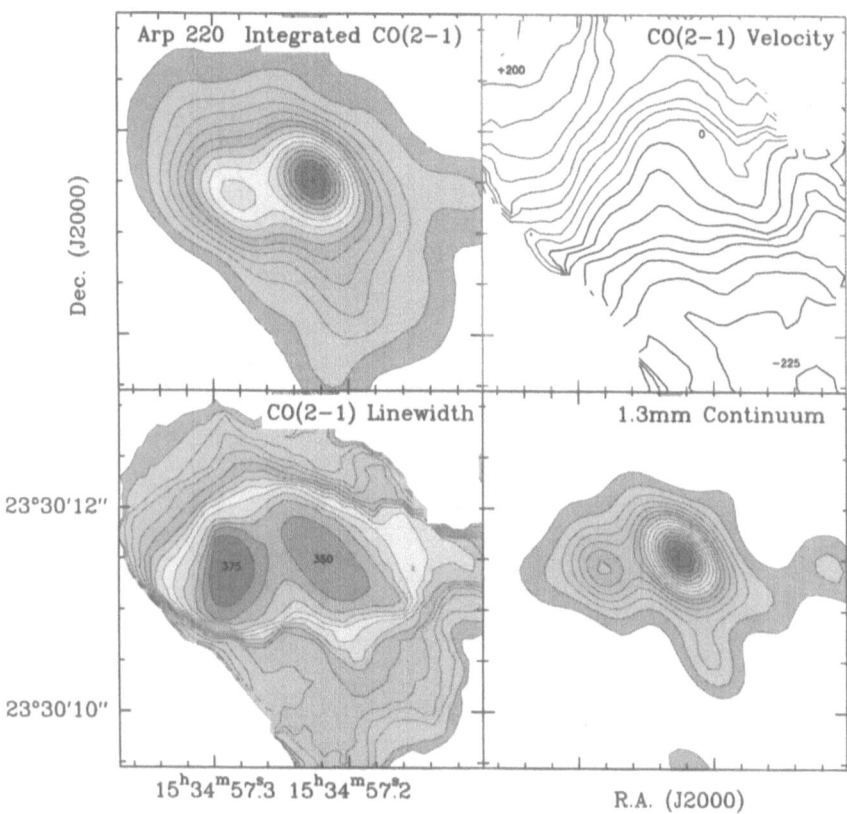

Fig. 4. *Arp 220 :* CO (2 1) integrated intensity, velocity, linewidth (FWHM), and the 1.3 mm continuum. CO integration limits: $(-320, +300 \, \mathrm{km \, s^{-1}})$. Beam $= 0''.7 \times 0''.5$. One arc second corresponds to 350 pc. The two sources of extreme starbursts can be seen in the integrated CO intensity and the 1.3 mm dust continuum where they completely dominate the emission. The strong deviations from circular rotation in the disk, indicated by the sharp twists in the velocity contours, are due to the West source which is rotating at an angle shifted by $110°$ from the main disk (from [3]).

Table 1 (adapted from Table 12 of [3]) lists the properties of 4 Extreme Starburst Regions identified in the 3 closest galaxies in the sample. They are the most prodigious star formation events in the local universe, each representing about 100 times as many OB stars as 30 Doradus. We are observing these objects near the peak of their star formation, when about half of the gas has been turned into stars. The duration of the starburst, limited by the molecular gas supply and the current star formation rate is about 5 to 10 $\times 10^{6}$ yrs. This short lifetime is consistent with the even shorter dynamical timescale of these compact (≈ 100 pc) starbursts. These are not only extreme starbursts, they are extremely efficient

Table 1. Properties of Extreme Starburst Regions.

	Arp 193 SE core	Mrk 273 core	Arp 220 west	Arp 220 east
Reference radius:				
R(pc)	150	120	68	110
Gas mass:				
$M_{gas}(< R)$ $(10^9\,M_\odot)$	0.6	1.0	0.6	1.1
Mean gas density:				
$< N(H_2) >$ (cm^{-3})	2×10^3	5×10^3	2×10^4	8×10^3
Total mass:				
$M_{tot}(< R)$ $(10^9\,M_\odot)$	1.4	2.6	1.4	3.2
Estimated mass in new stars:				
$M_{new\star}(< R)$ $(10^9\,M_\odot)$	0.8	1.6	0.8	2.1
Luminosity:				
$L_{FIR}(< R)$ $(10^{12}\,L_\odot)$	0.2	0.6	0.3-0.5	0.2-0.4
Luminosity to mass ratio:				
$L_{FIR}/M_{new\star}$ $(L_\odot/\,M_\odot)$	300	360	380	100

starbursts. The high L/M for these starbursts also suggests that they may require a high-mass IMF.

The high average density in the Extreme Starburst Regions solves the puzzle of the origin of the extraordinary HCN emission associated with ultraluminous galaxies discussed above. In Arp 220, the HCN lines have the same velocities as the East and West "nuclei". The East and West "extreme starbursts" alone account for most of the HCN emission. These dense, compact sources have a hydrogen column density of $0.6 \times 10^{25}\,cm^{-2}$ and mean density of 20,000 cm^{-3}, enough to thermalize the lower rotational levels of HCN by a combination of collisions and radiative trapping even if the average density is the local density. Within the Arp 220 East and West sources, the HCN emission may thus have the same intrinsic brightness temperature as the CO (2–1) emission. Using the observed sizes and linewidths, the HCN luminosity of these two regions alone is $L_{HCN} \approx 7 \times 10^8$ K km s^{-1} pc^2, which is 3/4 of the observed total [13]. These two regions thus emit only 1/4 of the CO luminosity but most of the HCN luminosity. It is likely that high density, extreme starburst regions exist in almost all ultraluminous galaxies and are the source of most of the star formation and most of the HCN emission. The origin of the HCN emission in the high density gas of the Extreme Starburst Regions directly relates the HCN emission to star formation.

4   Summary

The extraordinary starbursts found in ultraluminous IR galaxies occur in molecular gas concentrated in compact very massive "clouds" which we call "Extreme

Starbursts". They have one thousand times the mass of GMCs, but are only a few times larger. The entire structure, containing about a billion solar masses of molecular gas, has an average H_2 density characteristic of molecular cloud cores which represent only a few percent of the molecular mass in ordinary GMCs. It appears that high-mass star formation in sufficiently dense and massive structures does not disrupt further star formation. Since the star formation remains embedded in the molecular gas, there is little or virtually no optical–UV radiation escaping, and extreme star formation can be traced only in the far infrared. If the density and mass are sufficiently great, star formation is a runaway process. In the local universe this occurs primarily in galaxy mergers. In the early universe extreme starbursts may be more frequent and they may be the mode of star formation in the population of high redshift submillimeter sources.

References

1. J.J. Condon, Z.P. Huang, Q.F. Yin, T.X. Thuan: Ap. J. **378**, 65 (1991)
2. D. Downes, P.M. Solomon, S.J.E. Radford: Ap. J. **414**, L13 (1993)
3. D. Downes, P.M. Solomon: Ap. J. **507**, 615 (1998)
4. J.S. Gallagher, D.A. Hunter: 'Measuring Star Formation Rates in Blue Galaxies'. In: *Star Formation in Galaxies*, ed. by C. Lonsdale (US GPO, Washington 1987) pp. 167–177
5. Y. Gao: Dense Molecular Gas in Galaxies and the Evolution of Luminous Infrared Galaxies. Ph.D. Thesis, SUNY, Stony Brook (1996)
6. Y. Gao, P.M. Solomon: to be published in Ap. J. (2001)
7. T. Mooney, P.M. Solomon: Ap. J. **334**, L51 (1988)
8. K. Sakamoto et al.: Ap. J. **514**, 68 (1999)
9. D.B. Sanders et al.: Ap. J. **325**, 74 (1988)
10. N.Z. Scoville, A.I. Sargent, D.B. Sanders, B.T. Soifer: Ap. J. **366**, L5 (1991)
11. N.Z. Scoville, P.M. Solomon: Ap. J. **199**, L105 (1975)
12. P.M. Solomon, L.J Sage: Ap. J. **334**, 613 (1987)
13. P.M. Solomon, D. Downes, S.J.E. Radford: Ap. J. **387**, L55 (1992)
14. P.M. Solomon, D. Downes, S.J.E. Radford, J.W. Barrett: Ap. J. **478**, 144 (1997)
15. P.M. Solomon, D.B. Sanders: 'Giant Molecular Clouds as the Dominant Component of Interstellar Matter in the Galaxy'., In:*Molecular Clouds in the Galaxy*, ed by P. Solomon, M. Edmunds (Pergamon Press, Oxford 1980) pp. 41–74
16. M.S. Yun, N.Z. Scoville: Ap. J. **451**, L45 (1995)

Starbursts: Triggers and Evolution

Shardha Jogee

Division of Physics, Mathematics, and Astronomy, MS 105-24, California Institute of
Technology, Pasadena, CA 91125, USA

Abstract. Why do the circumnuclear (inner 1–2 kpc) regions of spirals show vastly different star formation rates (SFR) even if they have a comparable molecular gas content? Why do some develop starbursts which are intense short-lived ($t \ll 1$ Gyr) episodes of star formation characterized by a high star formation rate per unit mass of molecular gas (SFR/M_{H2}), which I refer to as star formation efficiency (SFE). I address these questions using high resolution ($2''$ or 100–200 pc) CO ($J=1{\rightarrow}0$) observations from the Owens Valley Radio Observatory, optical and NIR images, along with published radio continuum (RC) and Brγ data. The sample of eleven galaxies includes the brightest nearby starbursts comparable to M82 and control non-starbursts. More detailed results are in [8] and [10].

1 External Disturbances and Large-Scale Stellar Bars

The sample galaxies have developed large molecular gas reservoirs of several $\times 10^8$ to several $\times 10^9$ $M_\odot$ in the inner kpc radius, assuming a standard CO-to-H_2 conversion factor. As shown in Fig. 1, the circumnuclear SFR per unit mass of molecular gas spans more than an order of magnitude for a given circumnuclear molecular gas content.

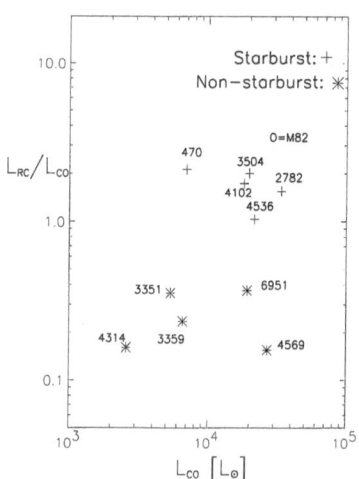

Fig. 1. The sample galaxies are shown. L_{RC} is the RC luminosity at 1.5 GHz [1] and L_{CO} is the single dish CO luminosity [14], both measured in the central $45''$.

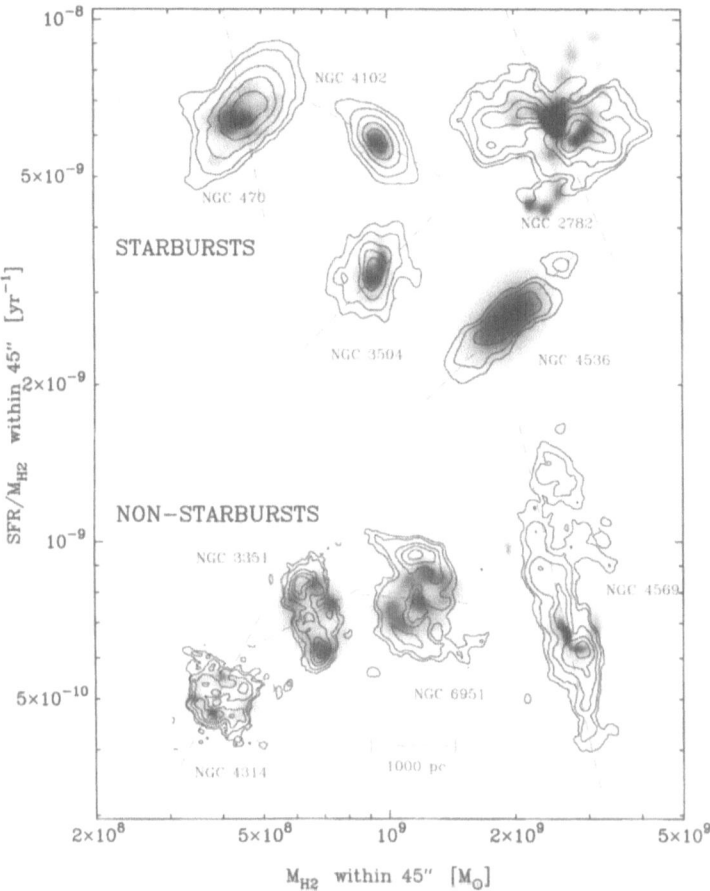

Fig. 2. In the SFR/M_{H2} vs. M_{H2} plane, the CO intensity (contours) is overlaid on the star formation (greyscale), as traced by RC in NGC 4102, NGC 2782, and NGC 6951, and by Hα in the others. The dotted line is the P.A. of the large-scale stellar bar/oval. The synthesized CO beam is typically 100–200 pc.

A spontaneously or tidally induced m = 2 instability such as a large-scale stellar bar and minor mergers/interactions can help to drive gas towards the inner kpc [6]. In optical and NIR images, all our sample galaxies show a large-scale stellar bar/oval whose position angle is marked on Fig. 2. The stellar bar may have been recently tidally triggered in NGC 3359 which has a steep abundance gradient along the bar [11], and in NGC 4569 which has a warped disk, an asymmetric bar and disturbed CO properties [8]. While the galaxies in our sample are not major mergers, all of them except for NGC 6951, show evidence for recent tidal interactions or mergers with mass ratios ranging from minor (1:10) to intermediate (1:4). NGC 2782 and NGC 470, which have the largest mass-ratio interactions, are starbursts.

The presence of a large-scale bar in starbursts and non-starbursts alike suggests that the starburst lifetime is short with respect to the timescale for bar destruction. Thus, within a given barred potential, star formation can change from an inefficient pre-burst phase in the early stages of bar-driven inflow, to a circumnuclear starburst, to a post-starburst after star formation rapidly consumes gas near the center in a few $\times 10^8$ years. The bar itself may be weakened or dissolve over timescales > 1 Gyr due to the development of a high central mass concentration (e.g., [5]; [3])

2   Morphology of the Circumnuclear Molecular Gas

What is the circumnuclear CO morphology and how does it relate to the properties of the barred potential? The molecular gas shows a wide variety of morphologies (Fig. 2) ranging from relatively axisymmetric annuli or disks (starbursts NGC 4102, NGC 3504, NGC 4536, and non-starbursts NGC 4314), elongated double-peaked and spiral morphologies (starburst NGC 2782 and non-starbursts NGC 3351 and NGC 6951) to extended distributions elongated along the large-scale bar (non-starburst NGC 4569). In NGC 4569, the gas extends out to a large (2 kpc) radius, at a similar P.A. as the large-scale stellar bar (Fig. 2), and shows complex non-circular motions. The optical, NIR, and CO properties of NGC 4569 suggest it is in the early stages of bar-driven/tidally-driven inflow of gas towards the inner kpc. In the other galaxies, the gas distribution is less extended, and in many systems it is concentrated inside the outer inner Lindblad resonance (ILR) of the large-scale bar. As shown in Fig. 3, both starbursts and non-starbursts host ILRs. In the sample, the bar pattern speed $\Omega_\mathrm{p} > 40$–115 km s^{-1} kpc^{-1}, the radius of the outer ILR is typically > 500 pc, and the radius of the inner IILR < 300 pc. Note that in NGC 2782 and NGC 470 which are claimed to host nuclear stellar bars [9,4], there is a strong misalignment ($\geq 40°$) between the CO distribution and both the major axis and minor axis of the large-scale stellar bar/oval.

3   Circumnuclear Star Formation Morphology and Efficiency

The starbursts and non-starbursts have circumnuclear SFR of 3–11 and 0.1–$2 \, M_\odot$ yr^{-1}, respectively, from RC and Brγ data. The SFE is therefore not a simple function of molecular gas content. The CO and star formation morphology are shown in Fig. 2. For a given CO-to-H_2 conversion factor, the starbursts have a larger peak gas surface density $\Sigma_{\mathrm{gas-m}}$ in the inner 500 pc radius than non-starbursts with a similar circumnuclear gas content (Fig. 4a). In the starbursts, both $\Sigma_{\mathrm{gas-m}}$ and Σ_{SFR} increase towards the inner 500 pc radius (Fig. 4a–b). Over the region of intense SF in several starbursts, $\Sigma_{\mathrm{gas-m}}$ remains close to the critical density (Σ_{crit}) for the onset of gravitational instabilities [13], despite an order of magnitude variation in Σ_{crit} (Fig. 5b and e). In the non-starbursts, there

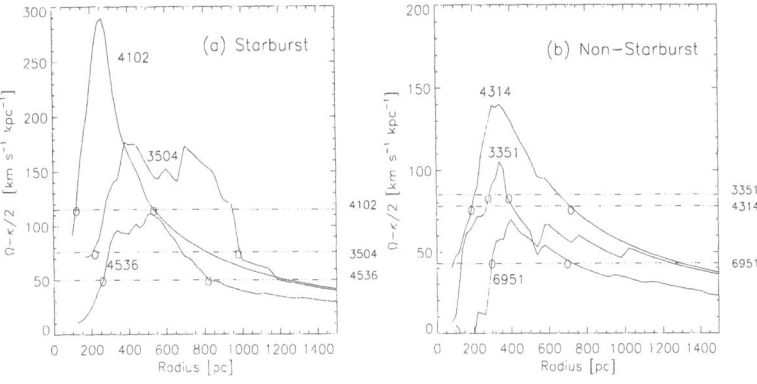

Fig. 3. $(\Omega - \kappa/2)$ is plotted against radius. The bar pattern speed Ω_p is drawn as horizontal lines and estimated by assuming that the corotation resonance is near the end of the bar. Under the epicycle theory for a weak bar, the intersection of $(\Omega - \kappa/2)$ with Ω_p defines the locations of the ILRs.

are gas-rich regions with no appreciable star formation, for instance, inside the ring of HII regions in NGC 3351 and NGC 4314, at the CO peaks in NGC 6951, and in the extended gas in NGC 4569 (Fig. 2). The gas surface density, although high, is still sub-critical in regions of inhibited star formation, as illustrated for NGC 4314 in Fig. 5e–f. In NGC 4569, the large local shear in the extended gas with large non-circular kinematics along the large-scale stellar bar may inhibit star formation. I suggest that circumnuclear starbursts produce a high SFE by developing supercritical surface densities in a large fraction of the gas close to the center, while sub-critical densities and large local shear may limit the SFE of non-starbursts.

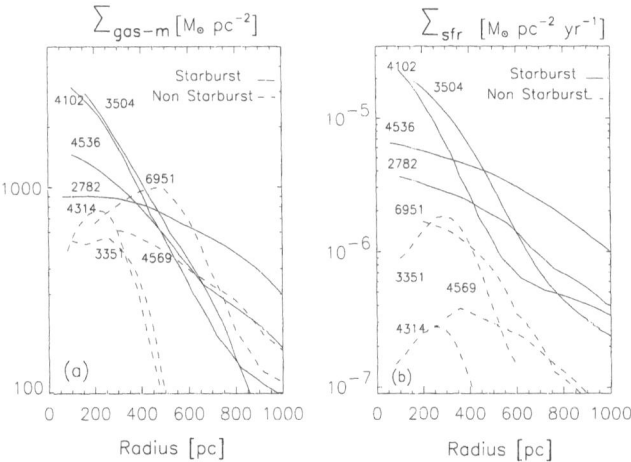

Fig. 4. (a), (b) show the azimuthally averaged molecular gas surface density (Σ_{gas-m}) and SFR per unit area (Σ_{SFR}). The extinction-corrected Σ_{SFR} profiles are convolved to a similar resolution of 100–200 pc for all the galaxies.

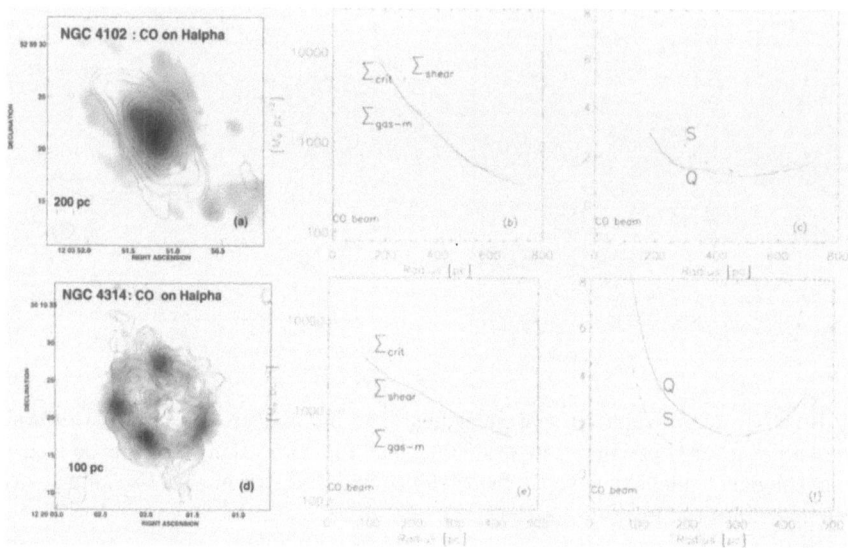

Fig. 5. (a, d) the CO distribution (contours) on the Hα (greyscale). **(b, e)** $\Sigma_{\mathrm{gas-m}}$, Σ_{crit}, and Σ_{shear}. **(c, f)** the Toomre Q and shear S parameters [8]. Quantities are plotted starting at a radius equal to the CO beam size ($\sim 2''$). In the non-starburst NGC 4314, Q reaches its lowest value (1–2) in the ring of HII regions between r = 250–400 pc while at lower radii where there are no HII regions, Q increases to 6, indicating sub-critical gas densities. In the starburst NGC 4102, Q remains $\sim$1–2 between a radius of 250–700 pc, over the region of intense star formation, although Σ_{crit} varies by roughly an order of magnitude.

4 The Extreme Molecular Environment in the Inner Kpc

Table 1 illustrates how molecular gas in the inner kpc and the outer disk differ markedly. This has important implications for the circumnuclear region. First, the high molecular gas density (several 100–$1000\,\mathrm{M_\odot}\ \mathrm{pc^{-2}}$) and mass fraction (10–30%) will lead to enhanced self-gravity and clumpiness of the gas (e.g., [12]). The two-fluid disk of gravitationally coupled gas and stars will be more unstable to gravitational instabilities than a purely stellar disk (e.g., [7]). Second, in the presence of a large epicyclic frequency (several 100–$1000\,\mathrm{km\ s^{-1}\ kpc^{-1}}$) and velocity dispersion (10–$40\,\mathrm{km\ s^{-1}}$), gravitational instabilities can overcome Coriolis and pressure forces only at very high gas densities (few 100–$1000\,\mathrm{M_\odot\ pc^{-2}}$). However, once triggered, they now grow on a timescale (t_{GI}) as short as a few Myrs, comparable to the lifetime of an OB star. These conditions can enhance the fraction of gas converted into stars before a molecular cloud is disrupted by massive stars. Third, a high pressure, high turbulence ISM may favor more massive clusters (e.g., [2]) and it is relevant that many sample galaxies show super star clusters in the inner kpc. Fourth, a comparison of Columns 3 and 4 in

Table 1. Molecular Gas Properties in the Circumnuclear Region.

Quantities	Outer Disk of Sa–Sc	Inner r = 500 pc of sample galaxies	Inner r = 500 pc of Arp 220
(1) $M_{gas,m}$ ($M_\odot$)	$\leq$ few $\times 10^9$	few $\times (10^8-10^9)$	3×10^9
(2) M_{gas}/M_{dyn} (%)	< 5	10–30	40–80
(3) SFR ($M_\odot$ yr^{-1})	—	0.1–11	> 100
(4) Σ_{gas-m} ($M_\odot$ pc^{-2})	1–100	500–3500	4×10^4
(5) σ (km s^{-1})	6–10	10–40	90
(6) κ (km s^{-1} kpc^{-1})	< 100	800–3000	> 1000
(7) Σ_{crit} ($M_\odot$ pc^{-2})	< 10	500–1500	2200
(8) t_{GI} (Myr)	> 10	0.5–1.5	0.5
(9) λ_J (pc)	few $\times$ 100–1000	100–300	90

The rows are : (1) molecular gas mass (2) ratio of molecular gas mass to dynamical mass; (3) star formation rate; (4) molecular gas surface density; (5) gas velocity
dispersion; (6) critical Toomre density for the onset of gravitational instabilities (7)
epicyclic frequency; (8) growth timescale of the most unstable wavelength (Q/κ) (9)
Jeans length

Table 1 suggests that the prototypical ultra luminous infrared galaxy (ULIRG)
Arp 220 may be a scaled-up version of the starbursts in our sample. ULIRGs
may be starbursts which have built an extreme molecular environment (density
and linewidths) in the inner few 100 pc of a deep stellar potential well through
major mergers or interactions.

References

1. J.J. Condon, G. Helou, D.B. Sanders, D. B., T.B. Soifer: ApJS **73**, 359 (1990)
2. B.G. Elmegreen: ApJ **411**, 170 (1993)
3. D. Friedli, W. Benz: A&A **268**, 65 (1993)
4. D. Friedli, H. Wozniak, M. Rieke, P. Bratschi: A&AS **118**, 461 (1996)
5. H. Hasan, C. Norman: ApJ **361**, 69 (1990)
6. L. Hernquist, J.C. Mihos: ApJ **448**, 41 (1995)
7. C.J. Jog: MNRAS **278**, 209 (1996)
8. S. Jogee: Ph.D. Thesis, Yale University (1999)
9. S. Jogee, J.D.P. Kenney, B.J. Smith: ApJ **526**, 665 (1999)
10. S. Jogee, J.D.P. Kenney, & N. Z. Scoville: in preparation (2001)
11. P. Martin, J. Roy: ApJ **445**, 161 (1995)
12. I. Shlosman, J. Frank, M.C. Begelman: Nature **338**, 45 (1989)
13. A. Toomre: ApJ **139**, 1217 (1964)
14. J.S. Young, S. Xie, L. Tacconi, P. Knezek, P. Viscuso, L. Tacconi-Garman, N.
    Scoville, S. Schneider, et al.: ApJS **98**, 219 (1995)

The Genesis of Super Star Clusters
as Self-Gravitating HII Regions

Jonathan C. Tan[1] and Christopher F. McKee[2]

[1] Dept. of Astronomy, UC Berkeley, Berkeley, CA 94720, USA
[2] Depts. of Physics and of Astronomy, UC Berkeley, Berkeley, CA 94720, USA

Abstract. We examine the effects of ionization, radiation pressure and main sequence winds from massive stars on self-gravitating, clumpy molecular clouds, thereby modeling the formation and pre-supernova feedback of massive star clusters. We find the process of "turbulent mass loading" is effective in confining HII regions. Extrapolating typical Galactic high-mass star forming regions to greater initial gas cloud masses and assuming steady star formation rates, we determine the timescales for cloud disruption. We find that a dense ($n_c \simeq 2 \times 10^5$ cm^{-3}) cloud with initial mass $M_c \simeq 4 \times 10^5$ M$_\odot$ is able to form $\sim 2 \times 10^5$ M$_\odot$ of stars (50% efficiency) before feedback disperses the gas after ~ 3 Myr. This mass and age are typical of young, optically visible super star clusters (SSCs). The high efficiency permits the creation of a bound stellar system.

1 Introduction

Most Galactic stars are born in highly clustered regions [12,5], where the disruptive effects of massive stars are paramount. The multitude of dusty high redshift sources and the intensity of the far infrared background they produce also imply that a major fraction ($\sim 1/2$) of total cosmic star formation has occurred in starbursts, replete with massive stars [20]. A significant fraction of star formation in local starbursts occurs via the creation of super star clusters (SSCs) [22,24], each with hundreds to thousands of OB stars crammed into a few parsecs. At least some SSCs are gravitationally bound [8,9] and their masses ($\sim 10^5 - 10^6$ M$_\odot$ [19,15]) and sizes suggest we may have found globular clusters in their infancy.

Massive stars violently disrupt their surroundings with ionizing and non-ionizing photons, protostellar, main sequence and post-main sequence winds, and supernovae. Adding to this complexity is the extremely dense, clumpy and turbulent nature of the gas in which high-mass stars are born. We present a simplified model of pre-supernova feedback, to examine how the efficient star formation required to produce bound clusters may occur in the presence of vigorous energy injection from many massive stars.

2 Initial Conditions – a Clumpy Molecular Cloud

We consider a spherical cloud of radius, R_c, mass M_c, and mean H density n_c. The cloud consists of dense clumps embedded in a uniform inter-clump medium. The clumps are distributed uniformly within a central core of radius 0.2 R_c, and with a r^{-1} distribution outside, which mimics observed molecular cloud profiles. We choose mean clump mass and volume fractions of $f_m = 0.8$ and $f_V = 0.03$ respectively. Our clumps, of uniform density n_{cl}, have a mass spectrum $d\mathcal{N}_{cl}/d\ln m_{cl} \propto m_{cl}^{-0.6}$ between upper, $m_{cl,u}$, and lower, $m_{cl,l}$, limits. Clump masses span a range of 10^3, with $m_{cl,u} = 0.025 M_c$, so no one clump dominates the cloud. The clump velocities are set so the cloud is in virial equilibrium.

Plume et al. [18] determined sizes and masses for 25 regions of Galactic high-mass star formation. The mean properties of this sample were $R_c \sim 0.5$ pc, $M_c \sim 3800\,M_\odot$ (virial mass) and thus $n_c \simeq 2 \times 10^5\,\mathrm{cm}^{-3}$ ($\simeq 3.2 \times 10^4\,M_\odot\,\mathrm{pc}^{-3}$). Note that these regions are only a small fraction of the host Giant Molecular Cloud. Our adopted clump mass and volume fractions imply clump densities of $n_{cl} = 6.5 \times 10^6\,\mathrm{cm}^{-3}$ and an inter-clump density of $5.0 \times 10^4\,\mathrm{cm}^{-3}$. These properties form the basis of our model **A**. Determining the gas properties of the precursors to SSCs is more difficult as current observations only probe scales down to $\sim 30 - 100$ pc [7,23], while the star clusters have a typical radius ~ 4 pc [22]. For simplicity we consider models with 10 (**B**) and 100 (**C**) times the mass of **A**, but with the same mean density. With the same values of f_V and f_m the clump and interclump densities are also the same. For each cloud we shall consider a fixed star formation rate, ϕ_{50}, which converts 50% of the initial cloud into stars over 3 Myr – i.e., the approximate time before the first supernova is expected to occur.[1] The properties of our initial gas clouds are listed in Table 1.

Table 1. Model Parameters.

Model	$M_c\,/(M_\odot)$	$R_c\,/(\mathrm{pc})$	$\sigma_{1D}\,/(\mathrm{km\,s}^{-1})$	$\phi_{50}\,/(M_\odot\,\mathrm{yr}^{-1})$
A	4×10^3	0.5	2.8	5.8×10^{-4}
B	4×10^4	1.1	5.9	6.6×10^{-3}
C	4×10^5	2.3	12.9	6.7×10^{-2}

3 Feedback Processes

To model the feedback processes, we need the total number of hydrogen ionizing ($\lambda < 912$) photons emitted per second (S), the bolometric luminosity (L_{bol}) and the mass flux ($\dot{M}_w$) and velocity (v_w) of the stellar wind from the forming star cluster. We derive these quantities from the STARBURST99 model [13] at solar metallicity. As we are interested only in the first few Myr of stellar evolution, we approximate the values of these quantities to be constant for a given stellar mass. Motivated by observations of the stellar population of the R136 cluster in 30 Doradus [4], we consider a stellar initial mass function (IMF) represented by

---

[1] This rate excludes the formation of an initial star cluster of $\sim 250\,M_\odot$ necessary to produce an ionizing luminosity of 10^{49} ionizing photons per second (see below).

a broken power law, $d\mathcal{N}_*/d\ln m_* \propto (m_*/m_{*,0})^{-\alpha}$, with $\alpha = 1.35$ (Salpeter) for $m_* > m_{*,0} = 3\,M_\odot$ and $\alpha = 0.8$ for $m_* < m_{*,0}$. We choose lower and upper limits of stellar mass of $0.1\,M_\odot$ and $100\,M_\odot$, respectively. This increases the mass per unit of feedback (assuming negligible contribution from stars below $m_{*,0}$) by a factor of 1.35 compared to the standard STARBURST99 IMF ($\alpha = 1.35$ between $1\,M_\odot$ and $100\,M_\odot$). We find

$$S_{49} = 3.94 \times 10^3 M_{*,6}, \tag{1}$$

$$L_{\mathrm{bol},5} = 1.10 \times 10^4 M_{*,6}, \tag{2}$$

$$\dot{M}_{w,-6} = 3.09 \times 10^3 M_{*,6}, \tag{3}$$

$$\dot{M}_{w,-6} v_{w,2000} = 3.98 \times 10^3 M_{*,6}, \tag{4}$$

where $S_{49} = S/10^{49}$ photons s^{-1}, $L_{\mathrm{bol},5} = L_{\mathrm{bol}}/10^5\,L_\odot$, $\dot{M}_{w,-6} = \dot{M}_w/10^{-6}\,M_\odot$ yr^{-1}, $v_{w,2000} = v_w/2000$ km s^{-1} and $M_{*,6} = M_*/10^6\,M_\odot$. The number of $m_* > 8\,M_\odot$ stars (i.e. number of core collapse supernovae) is $\mathcal{N}_{\mathrm{SN}} = 1.4 \times 10^4 M_{*,6}$.

3.1   Ionization

Assuming spherical symmetry with the massive stars forming at the center of the cloud, we calculate the ionizing flux received at a given distance R from the star cluster by accounting for attenuation by clump shadowing, H recombinations and dust absorption, so that

$$\frac{dS}{dR} = -S\bar{A}_{\mathrm{cl}}\mathcal{N}_{\mathrm{cl}} - 4\pi R^2 \alpha_2 n_i n_e - \sigma_d n_i S, \tag{5}$$

where $\bar{A}_{\mathrm{cl}}$ is the mean cross sectional area of clumps at R, $\mathcal{N}_{\mathrm{cl}}$ is the space number density of clumps, α_2 is the recombination coefficient, n_i is the hydrogen nuclei number density of the ionized gas and $\sigma_d = 0.5 \times 10^{-21}\,\mathrm{cm}^2$ [1] is the dust absorption cross-section per H nucleus. We assume He is singly ionized and $n_{\mathrm{He}} = 0.1 n_i$. An HII region rapidly forms with typical size $R_{\mathrm{St}} \simeq 3.05 \times 10^{-2} S_{49}^{1/3} n_{i,5}^{-2/3}$ pc, for dust free, uniform density gas. The short sound crossing time justifies our assumption of uniform n_i. Thermal balance maintains a constant ionized gas temperature, $T_i \simeq 10^4$ K. Over-pressurized compared to the surrounding neutral medium, the HII region tends to expand.[2] However, when neutral clumps become exposed to ionizing photons, they implode and inject mass into the HII region [2]. A compressed neutral globule remains, which continues to evaporate more slowly [3]. We employ models for magnetically supported clumps so that the pressure of the ionized gas streaming from the surface of an imploded clump at distance R is

$$\frac{p_c}{k} \simeq 8.7 \times 10^7 n_{\mathrm{cl},7}^{1/21} \left(\frac{S_{49}}{R_{\mathrm{pc}}^2}\right)^{4/7} m_{\mathrm{cl}}^{-4/21}\ \mathrm{K\ cm}^{-3}$$

$$= 9.9 \times 10^9 n_{\mathrm{cl},7}^{1/21} \left(\frac{M_{*,6}}{R_{\mathrm{pc}}^2}\right)^{4/7} m_{\mathrm{cl}}^{-4/21}\ \mathrm{K\ cm}^{-3}, \tag{6}$$

---

[2] For the propagation of ionization fronts into the clump and interclump material, we assume the neutral gas has a temperature of 80 K.

where m_{cl} is the clump mass in solar mass units and n_{cl} is the initial clump density. Clumps may only photoevaporate if this pressure is greater than the wind ram pressure (below). Evaporating clumps are gradually ejected from the HII region by the rocket effect. However, clumps orbiting in the cloud potential continue to enter the HII region and inject fresh gas. The HII region is limited by a recombination front as the ionized gas flows out at approximately its sound speed. We have examined the details of this mechanism of "turbulent mass loading" and its implications for Galactic ultracompact HII regions elsewhere [21].

3.2 Winds

Stellar winds carve out a hot, low density cavity at the center of the HII region. The wind's thermal pressure is suppressed by cooling caused by mass injected by photoevaporating clumps [17]. Instead, we calculate the edge of the wind cavity by balancing the thermal pressure of the HII region with the wind ram pressure,

$$\frac{p_{w,ram}}{k} = 7.6 \times 10^5 \frac{\dot{M}_{-6} v_{w,2000}}{R_{pc}^2} \text{ K cm}^{-3} \simeq 3.0 \times 10^9 \frac{M_{*,6}}{R_{pc}^2} \text{ K cm}^{-3}. \quad (7)$$

In addition, this pressure contributes to the clump dynamics and may quench photoevaporation, particularly at small distances from very massive clusters. We treat photoevaporation mass injection into the wind via a supersonic mass-loaded wind model, e.g. [6]. In cases of extreme mass loading the ionizing photons may be trapped in the wind.

3.3 Radiation Pressure

Our fiducial clumps absorb most of the momentum flux of the star cluster's radiation. The pressure is

$$\frac{p_{rad}}{k} = 8.7 \times 10^9 \frac{M_{*,6}}{R_{pc}^2} \text{ K cm}^{-3}. \quad (8)$$

4 Numerical Method for Modeling SSC Formation

With an N-body code, we follow the dynamics, masses and sizes of a collection of several thousand clumps, in the fixed potential of the initial cloud and subject to the feedback processes described above. At the same time we model the coupled evolution of a spherically symmetric wind cavity and HII region. While we do not account for clump collisions, we note that their initial mass spectrum is probably created by a steady-state balance of collisional fragmentation and agglomeration. Our model does not yet include a physical mechanism for star formation and so we investigate the response of our fiducial gas clouds to different imposed steady star formation rates, such that 50% of the cloud would be converted to stars after 3 Myr. Stars are created from gas in the innermost clumps at these rates and artificially added to the central cluster. We start our models by instantly forming $\sim 250\,M_\odot$ of stars, equivalent to an ionizing luminosity of $S_{49} = 1$. We examine the time taken for the HII region to reach 90% of the initial cloud radius, at which point star formation is assumed to cease. Photoionization gradually destroys our model clouds, **A**, **B** and **C**, as shown in Fig. 1.

5   Discussion

The turbulent, clumpy and self-gravitating nature of our clouds impedes their destruction and allows star formation to proceed to higher efficiencies. For example, an HII region in a uniform, quiescent cloud with the same mean density and initial star cluster as model **A**, would destroy the cloud in $\sim 3 \times 10^5$ years, even with no additional star formation. However, with most of the cloud mass in dense clumps with velocities set by virial equilibrium, the process of "turbulent mass loading" confines the ionized gas for 1.5 – 2 Myr, even though by this time, five times the initial stellar mass has formed and feedback is correspondingly greater (model **A**). Reasonable models for clouds forming SSCs survive up to 3 Myr (models **B** and **C**). The fraction of the HII region occupied by the wind cavity increases with star cluster mass.

After a few Myr our neglect of Wolf-Rayet winds and supernovae becomes important. Furthermore, our use of a fixed gravitational potential, our assump-

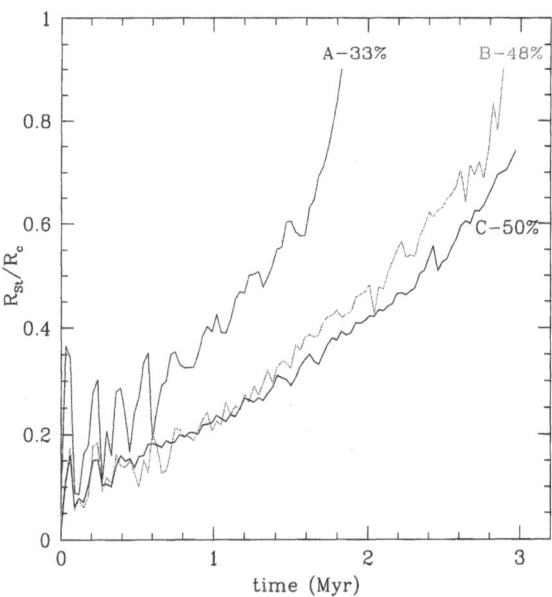

Fig. 1. Star cluster formation and cloud destruction at constant star formation rates, ϕ_{50}. **A**: $M_c = 4000\,M_\odot$ (Galactic case); **B**: $M_c = 4 \times 10^4\,M_\odot$; **C**: $M_c = 4 \times 10^5\,M_\odot$. Star formation efficiencies are quoted when $R_{St} = 0.9R_c$ or after 3 Myr.

tion that all massive stars are at the cloud center and the absence of protostellar winds in our model cause us to overestimate the cloud destruction time. Nevertheless, our results are qualitatively consistent with observed ages of the youngest optically visible SSCs, which are $\sim 5\,\mathrm{Myr}$ [22]. Younger clusters are still embedded in dense gas [7,15]. R136 in 30 Doradus, which is considered a small SSC, has recently dispersed its gas and is $\sim 1 - 2$ Myr old [14].

With steady star formation rates over the cloud lifetime, our results imply that star formation efficiency increases with initial cloud mass. The high efficiencies apparent in our SSC models allow for the creation of bound clusters [11], even in the presence of vigorous feedback. Since the mass loss is gradual, efficiencies as low as $\sim 30\%$ may result in loosely-bound clusters. Our Galactic model (**A**) is close to this limit, though the inclusion of additional feedback processes, such as protostellar winds, will reduce the efficiency.

We plan to extend our models to include additional feedback processes, such as protostellar winds, Wolf-Rayet winds and supernovae, and a physical star formation mechanism (e.g., photoionization regulated [16]) in the neutral clumps. We shall compare our models to observations of extra-galactic compact HII regions (e.g., [10]). We hope to predict minimum cloud masses and densities required to form bound stellar systems and probe in more detail the differences between Galactic and super star clusters.

References

1. J.A. Baldwin, et al.: ApJ, **374**, 580 (1991)
2. F. Bertoldi: ApJ, **346**, 735 (1989)
3. F. Bertoldi, C.F. McKee: ApJ, **354**, 529 (1990)
4. B. Brandl, et al.: In: *From Darkness to Light*, ed. by T. Montmerle, Ph. André (ASP conf. ser., 2000)
5. J.M. Carpenter: AJ, in press, (2000)
6. J.E. Dyson, R.J.R. Williams, M.P. Redman: MNRAS, **277**, 700 (1995)
7. A.M. Gilbert, et al.: ApJL, **533**, L57 (2000)
8. L.C. Ho, A.V. Filippenko: ApJL, **466**, L83 (1996)
9. L.C. Ho, A.V. Filippenko: ApJ, **472**, 600 (1996)
10. H.A. Kobulnicky, K.E. Johnson: ApJ, **527**, 154 (1999)
11. C.J. Lada, M. Margulis, D. Dearborn: ApJ **285**, 141 (1984)
12. E.A. Lada, K.M. Strom, P.C. Myers: 1993, In: *Protostars and Planets III*, ed. by E.H. Levy, J.I. Lunine (Univ. Arizona Press, Tucson 1993) p245
13. C. Leitherer, et al.: ApJS, **123**, 3 (1999)
14. P. Massey, D.A. Hunter: ApJ, **493**, 180 (1998)
15. S. Mengel, et al.: preprint, astro-ph/0010238 (2000)
16. C.F. McKee: ApJ, **345**, 782 (1989)
17. C.F. McKee, D. Van Buren, B. Lazareff: ApJ, **278**, L115 (1984)
18. R. Plume, et al.: ApJ, **476**, 730 (1997)
19. A. Sternberg: ApJ, **506**, 721 (1998)
20. J.C. Tan, J. Silk, C. Balland: ApJ, **522**, 579 (1999)
21. J.C. Tan, C.F. McKee: in preparation (2001)
22. B.C. Whitmore, et al.: AJ, **118**, 1551 (1999)
23. C.D. Wilson, N. Scoville, S.C. Madden, V. Charmandaris: ApJ, **542**, 120 (2000)
24. Q. Zhang, S.M. Fall: ApJL, **527**, L81 (1999)

Part V

Global Properties

Stellar Features in Integrated Starburst Spectra as Stellar Population Diagnostics

Daniel Schaerer

Laboratoire d'Astrophysique, Observatoire Midi-Pyrénées, 14, av. E. Belin, 31400 Toulouse, France (schaerer@ast.obs-mip.fr)

Abstract. We review the main stellar features observed in starburst spectra from the UV to the near-IR and their use as fundamental tools to determine the properties of stellar populations from integrated spectra. The origin and dependence of the features on stellar properties are discussed, and we summarise existing modeling techniques used for quantitative analysis. Recent results from studies based on UV, optical and near-IR observations of starbursts and active galaxies are summarised. Finally, we briefly discuss combined starburst + photoionization models including also observations from nebular emission lines. The present review is complementary to the recent summary by Schaerer [57] discussing more extensively nebular analysis of starbursts and related objects.

1 Introduction

The analysis of distinct spectral features in integrated spectra is at the base of numerous investigations on the stellar content of distant galaxies. Indeed, in addition to the overall continuum spectral shape, stellar absorption (and rarely also emission) lines carry crucial information on the presence of stars of various spectral types and luminosity class, and allow one thus in principle to "decompose" the integrated galaxy spectrum in its stellar constituents, and to determine their fundamental properties such as ages, IMF, the star formation history etc.

In objects such as active galaxies, where non stellar emission processes are thought to contribute to the emitted light, the study of possible stellar features allows one to constrain the relative stellar contribution, and thus to determine the efficiency of various emission processes (e.g. stellar versus non-stellar).

These basic properties illustrate the interest of spectroscopic studies of stellar features in starbursts and other galaxies.

The aim of the present review is to discuss the main stellar features observed in starburst spectra over the entire spectral range where such features are detectable, i.e. from the UV over the optical to the near-IR. At longer wavelength dust emission dominates and the stellar continuum and associated lines are not detectable anymore.

An "inventory" of the strongest stellar features is given for each spectral domain and recent results in the respective fields are summarised. In the last Section, we also briefly discuss the use of combined stellar and nebular emission line (hereafter EL) analysis for the study of stellar populations in the optical and IR.

The current review is complementary to a recent review [57] discussing new developments in multi-wavelength modeling tools, the current status of ionising fluxes from massive stars, and their importance for EL analysis of starbursts and related objects.

2   UV Features

The UV spectral range ($\sim$ 1000 – 3000 Å) of starbursts is rich in stellar lines originating in early type stars (mostly OB, also Wolf-Rayet), it contains few or weak nebular lines, and rather numerous interstellar (IS) absorption lines.

A rough inventory of the strongest *stellar* lines observed mostly in the $\sim$ 1200 – 1800 Å range follows (cf. the detailed work of [10]):

- Well known stellar wind lines (P-Cygni or EL) from O and Wolf-Rayet (WR) stars) are Si IV 1400, C IV 1550, N V 1240, He II 1640, and N IV1720. The following synthesis models include at least partly these lines and discuss their behaviour: [63,43,12,13,36–38,52].
- Other wind lines blue ward of Lyα include O VI+Lyβ+C II 1010-1060, discussed and modeled by [21], and potentially lines of C, N, P, S, and Ar in the range recently observed with FUSE (see [70,15]).
- The strongest photospheric lines from OB stars are Si II 1265, 1485, Si III 1295-1300, 1417, C II 1334, 1335, C III 1247, 1427, S V 1501 (see [10]).
- Other stellar features include "depressions" due to numerous Fe lines (at $\sim$ 1400, 1600, 1940), and also Fe II 2570-2615 and Mg II 2780-2825 features at longer wavelengths (e.g. [52,68])

It is important to note that many of these lines can also be formed in the interstellar medium (cf. [27,53]). A careful separation of the stellar and interstellar component is necessary in many cases (see [10]). The wavelength range between $\sim$ 1800 and 3000 Å remains still little explored. A similarly detailed understanding of this spectral range is of importance for studies of galaxies in the $z \sim$ 1–2 redshift range (see e.g. [5]).

Given the dependence of the various lines on stellar luminosity, age, and also metallicity (cf. below), e.g. well studied for the stellar wind lines (e.g. [78,35]), the features can be used to constrain the parameters of the integrated population, such as age, SF history, and IMF, by means of evolutionary synthesis techniques (e.g. [43,36]). The most up-to-date model suited to such analysis is *Starburst99* [37,10]).

These techniques have been extensively applied to the interpretation of UV spectra of nearby starbursts (mostly HST spectra), especially by Leitherer, Heckman, González Delgado and collaborators (some references given below) and by Mas-Hesse & Kunth [43,44]). Summarised in one sentence the main result of these studies is that all the objects contain young bursts ($\lesssim$ 10–20 Myr) characterized by instantaneous burst or continuous star formation, the distinction being often difficult to draw, which are populated with a rather normal Salpeter-like IMF with stars up to $M_{up} \sim$ 60 – 100 $M_\odot$. In a recent study [77] (cf. these proceedings) have examined the stellar populations in the field of NGC 5253 and find a possible indication for a steeper IMF, although other explanations (e.g. age effects) are possible.

The similarity of the spectra of many high redshift galaxies (e.g. Lyman break galaxies) with the local starbursts is now well recognized and offers many exciting possibilities. For example, from the beautiful spectrum of the lensed z $\sim$ 2.7 galaxy 1512-cb58 of [49] these authors and de Mello et al. [10] derive a constant star formation rate, an IMF slope between Salpeter (2.35) and $\sim$ 2.8, and find indications for a subsolar metallicity, in agreement with EL measurements from [71].

Obviously it is of great interest to derive/estimate the metallicity (Z) from stellar UV lines. Since, for example, the strength of stellar wind lines depends on Z this is in principle possible. This can e.g. be done using the correlation of the equivalent width of Si IV with metallicity found by [29]. However, a priori, such a correlation should only be valid in a statistical sense, since the line strength also depends on age (cf. [36]). This should be less of a difficulty if the full wind line profiles can be analyzed. The inclusion of spectral libraries of metal-poor stars in evolutionary synthesis models has just been completed (e.g. [38], see also [26]). Alternative possibilities to derive the chemical composition include the use of IS absorption lines (cf. [49]), or the use of weak stellar features such as various Fe blends known to vary with Z (cf. [25]).

3   Stellar Features in the Optical

The inventory of the strongest stellar lines in the optical is as follows:

- Broad emission lines from Wolf-Rayet stars of various subtypes (WN, WC) are detected in some young starbursts: He II 4686 bump, C IV 5808, C III 5696, possibly also N III 4512, Si III 4565 (see e.g. [56,24]). Synthesis models treating these lines include [6,62].
- H and He absorption lines from OBA stars have been discussed and modeled by [11,46,21].
- The Ca II triplet $\sim$8498, 8542, 8662 has often been studied (e.g. [72,73,45,16]). Its origin in both late type giants and supergiants complicates a priori the analysis.
- Other metallic features and molecular bands originating in stars with F types and later are Ca II H+K 39XX, CH G band 4284-4318, Mg I+MgH 5156-5196, Na I 5880-5914, various TiO bands $\geq$ 6200. These are found e.g. in the template spectra of Bica & Alloin ([2] and subsequent papers) for clusters and in the starburst spectra of Storchi-Bergmann et al. [68].

A complete overview of all starburst studies exploiting these features is not possible here. I shall instead briefly summarise recent results on starburst (and possibly also AGN) studies using Wolf-Rayet features and metallic lines. The contribution of González Delgado (these proceedings) illustrates the use of H and He absorption lines.

3.1   WR Features as a Probe of the Most Massive Stars

Since WR are the descendents of the most massive stars, detections of their features provide the best indication of the presence of massive stars (Minitial $\gtrsim$ 25–60 M$_\odot$) and allow the constraint of the upper end of the IMF. A catalogue of all known galaxies with WR detections has been compiled by Schaerer et al. ([59]; on the Web at `http://webast.ast.obs-mip.fr/people/schaerer/`).

Starburst–AGN Connection Whereas early studies were mostly focused on detections of the Ca II triplet (e.g. [72,73]), the finding of stellar features including the so-called WR-bump, UV lines and the Ca II triplet in the Seyfert2 galaxy Mrk 477 [28] has considerably revived this subject. Since then other possible WR detections indicating important massive star populations have been made in Mrk 1210 (Sey2, [69]), Mrk 463E and Mrk 1 (Sey2, [22]), TF 1736+1122 (Sey 2, [76]), and in three PG QSO [41]. Given the strong nebular contamination due to He II λ4686, other unambiguous massive star features are required to fully clearly establish the presence of massive stars in these objects. Such attempts, aiming to detect the WR lines of C IV and/or C III which are not affected by nebular contamination, have been undertaken by [31] with 2-D spectroscopy in several Sey2.

From the analysis of WR and H+He absorption lines in their sample of 20 Sey2 galaxies, ([22]; cf. these proceedings) find that the blue and near-UV light of half of their objects is dominated by young and/or intermediate age stars. A similar result was found by [69] on a smaller sample.

"Normal" Starbursts (So-Called WR Galaxies) Studies of WR galaxies (mostly BCD, Irr, spirals) are summarised in the reviews of [55,56]. Including the detections of spectral signatures from both WN and WC stars in a fair number of objects covering a large metallicity range, the following overall conclusions emerge from the studies of [59] and [24]. Except possibly at the lowest metallicities a good agreement is found between the observations and the evolutionary synthesis models of [62]. From this comparison one finds clear indications for short bursts ($\Delta t \leq$ 2-4 Myr) in objects with subsolar metallicity, an IMF compatible with Salpeter, and a large upper mass cut-off of the IMF, in agreement with several earlier studies. In addition, the observed WC/WN star ratios provide new constraints for mass loss and mixing scenarios in stellar evolution models [59].

We have recently undertaken a first study of metal-rich starbursts (metallicities up to $\sim$ 2–3 times solar) with the aim of constraining the upper end of the IMF in such environments. From the strengths of the observed WR features we derive a conservative *lower limit* of $M_{\rm up} \gtrsim$ 30–40 M$_\odot$[60]. New observations are being obtained to improve the accuracy of this result. Direct studies of the stellar content are of prime importance, also to verify the reliability of indirect studies based on nebular line analysis (cf. below).

3.2 Population Synthesis Studies Using Metallic Lines

Early starbursts studies using metallic lines have mostly concentrated on the Calcium triplet (see references above). More complete analysis of starbursts and AGN spectra using numerous optical stellar features have recently been presented by two groups. Both approaches are based on "classical" population synthesis using either observed stellar templates [64,3] or cluster templates [50,51].

The former authors analyse 5000 – 8000 Å spectra of 12 starburst, Seyfert and LINERs, and use a synthesis technique yielding a mathematically unique solution to determine the relative contributions of different stellar populations. Regarding the importance of super metal-rich stars their results differ from other work (e.g. [8,22].

Raimann et al. [50,51] have analysed average spectra of H II galaxies, starbursts and Sey2 taken from the Terlevich et al. [74] catalogue. Their main result regarding H II galaxies is the finding of significant old ($\leq$ 500 Myr) underlying populations, which has an impact on the line diagnostics based on equivalent widths. Their conclusions are supported by the study of selected BCD's by [44] and by comparisons of predicted and observed EL trends in several large samples of H II galaxies [66].

4 Stellar Features in the Near–IR

This spectral range is in most cases dominated by features from late type stars (G, K, M) corresponding to red supergiants (RSG), asymptotic giant branch stars (AGB), or red giants. Typical ages for the appearance of these stars are $\gtrsim$ 10^7, 10^8, and 10^9 yr respectively. The strongest features in the K and K band are

- atomic transitions of Si I 1.59 μm, Na I 2.21, Fe I 2.23, 2.24, Ca I 2.26, Mg I 2.28
- molecular features, such as CO (6,3) 1.6, CO (2,0) 2.29 and many OH lines (all wavelengths given in μm here).

The following papers (incomplete selection) describe their dependence on stellar type and luminosity class and/or provide spectral libraries: [30,32,47,1,14]. Recently, some authors (e.g Gilbert & Graham, these proceedings) have begun to use also theoretical spectra of cool stars for population studies.

Rather than providing a detailed review of the many studies undertaken in this area, I will briefly recall some difficulties affecting near-IR studies of stellar populations. While obviously the traditional method of population synthesis, decomposing the integrated spectrum in various template constituents, can be equally applied to any wavelength range, the near-IR properties predicted by evolutionary synthesis models are unfavourably affected by uncertainties in post main-sequence stellar evolution and the modeling of cool stars.

Indeed, it is found that all current non-rotating stellar evolution models predict an incorrect variation of the relative red/blue supergiant lifetimes with metallicity Z [34][1]. In addition the predicted $T_{\rm eff}$ of RSG may be too high [48]. This implies, e.g. that the CO features predicted by evolutionary synthesis models based on these tracks are weaker than observed at subsolar Z [48]. Also, the predicted strong metallicity dependence of colors like V-K (see [6]) is therefore incorrect. Improvements are expected from new stellar models allowing for more realistic mixing scenarios including rotation (cf. [42]) and better understanding of mass loss in these phases.

Uncertain mass loss scenarii render the prediction of AGB stars, whose influence at long wavelength is non negligible for populations with ages $\gtrsim 1$ Gyr (e.g. [4]), rather difficult, as best illustrated by [19]. For regions with small masses stochastical effects [33] and fluctuations of the IMF from the finite number of stars [7] also lead to an expected dispersion. While these effects are obviously of general nature, they are of particular importance for predictions involving the dominant contribution from stars with very short lived phases (e.g. AGB, WR).

5   Combined Stellar and Nebular Analysis of Starbursts

In various situations (e.g. starbursts with strong optical EL; IR observations – $\lambda \gtrsim 4$–10 μm where the continuum emission is dominated by dust and stellar features are thus completely absent) analysis of nebular emission lines are of interest to constrain the stellar population.

However, given the very nature of nebular physics, the EL are not only sensitive to the ionising spectrum carrying information of the stellar populations, but depend also strongly on the nebular geometry and chemical composition. The dependence on these additional parameters (essentially the so-called "ionisation parameter" U and composition) render EL studies of stellar populations more complex and thus require the use of sufficient observational constraints.

Recent developments and the current state-of-the-art of stellar ionising fluxes forming the input to photoionisation models have been reviewed by [57] and shall not be repeated here. In the following we briefly summarise the main recent studies undertaken in the optical and IR (cf. [57] for more details.)

5.1   Optical Studies

Recent tailored starburst and photoionisation models are presented in the studies of [17]: NGC 7714, [39]: NGC 2363, [40]: NGC 5461, [66]: I Zw 18, and [23]: NGC 604. Although somewhat different in each study, the general approach is summarised in [17].

Overall one finds that both the stellar and nebular lines give consistent results regarding the main properties of the stellar population, such as age, IMF etc. At a more detailed level, however, several of these studies encounter significant difficulties (e.g. the temperature sensitive ratio [O III] $\lambda4363/5007$ is

---

[1] Depending on the adopted set, the predicted RSG/BSG may well be correct for a certain metallicity (e.g. Geneva models ok for solar metallicity). The predicted metallicity variation does, however, not follow the observed trends.

underpredicted), which indicates that some physical processes are missing in the photoionisation models [66,39,67]. In short, although most observables can be reproduced by the combined starburst and photoionisation models — and the tool can thus be used to derive SB properties from the EL — one has to conclude that for accurate studies relying on nebular lines from H II regions (and presumably also more complex objects) some additional physical process(es) (possibly shocks, conductive heating at X-ray interfaces etc.) must be taken into account (cf. [57]).

5.2 Starburst + Photoionisation Models in the IR Domain

Analysis of IR observations (mostly from SWS and LWS on ISO) of starbursts based on combined SB + photoionisation models are just beginning to appear in the literature. In this context it is useful to keep some intrinsic difficulties in mind. Given the nature of objects and the large apertures involved, the integrated spectrum generally includes a large variety of regions. This fact, together with the complex geometries involved, render *a priori* the construction of photoionisation models difficult.

Simple models were constructed for case studies of Arp 299 and M82 by [9,54] to interpret their LWS (40-200 μm) spectra. Colbert et al. [9] find that the observed EL spectrum of M82 is compatible with an instantaneous burst at ages $\sim$ 3–5 Myr, a Salpeter IMF, and a high upper mass cut-off. Surprisingly, inspection of models with similar ingredients (cf. [65]), shows that the shorter wavelength data (see [18]) is clearly incompatible with the Colbert et al. [9] model, which predicts too hard a spectrum. In view of the few line ratios originating from the H II gas and the large number of free parameters the photoionisation model is underconstrained. A larger wavelength coverage or other constraints are required.

Förster-Schreiber [14] has described the geometry of clusters and gas clouds in M82 by a single "effective" ionisation parameter. This value has been adopted as typical for a sample of 27 starbursts in the SB + photoionisation models of [75]. Instead of modeling a simple stellar population their models are based on an ensemble of H II regions following an observed luminosity function, which overall leads to a reduction, albeit small, of the hardness of the ionising spectrum. From the ISO/SWS [Ne III]/[Ne II] line ratios they conclude that the observations are compatible with a high upper mass cut-off ($M_{up} \sim 50$–100 M$_\odot$). To reproduce the relatively low average [Ne III]/[Ne II] ratio, short timescales of SF are required. More detailed studies including additional observational constraints would be very useful to confirm this result.

A different approach has been taken by [61], who modeled two well studied objects (NGC 5253, II Zw 40) with a fairly well known massive star population and existing UV-optical-IR observations. While their model successfully reproduces the stellar features and the observed ionisation structure of H, He, and O (as revealed from the optical and IR lines), the predicted IR fine structure line ratios of [Ne III]/[Ne II], [Ar III]/[Ar II], and [S IV]/[S III] show too high an excitation. The origin of this discrepancy (atomic data? separate emission components?

other?) is still unknown. In any case this attempt to describe two relatively "simple" objects illustrates the current limitations and shows that further progress is needed for a proper understanding and use of the IR fine structure lines as reliable diagnostics. Improvement is expected from multi-wavelength analysis of simpler objects (e.g. Galactic and LMC H II regions, PN) and other ongoing work. Such studies should be crucial to reliably extend the diagnostic tools to the IR to fully exploit the enormous observational capabilities provided by recent and upcoming facilities in probing the properties of massive star formation from the local Universe to high reshift.

Acknowledgements: I thank the organisers for this very interesting and stimulating workshop and for financial support. Part of this work is also supported by the INTAS grant 97-0033.

References

1. B. Ali,et al.: AJ **110**, 2415 (1995)
2. E. Bica, D. Alloin: 1986, A&A **162**, 21 (1986)
3. C. Boisson, et al.: A&A **357**, 850 (2000)
4. G. A. Bruzual, S. Charlot: ApJ **405**, 538 (1993)
5. L. Campusano, et al.: A&A submitted
6. M. Cerviño, J. M. Mas-Hesse: A&A **284**, 749 (1994)
7. M. Cerviño, et al.: A&A **360**, L5 (2000)
8. R. Cid Fernandes Jr., et al.: MNRAS **297**, 579 (1998)
9. J. W. Colbert, et al.: ApJ **511**, 521 (1999)
10. D. F. de Mello, C. Leitherer, T. M. Heckman: ApJ **530**, 251 (2000)(DLH00)
11. A. I. Díaz: MNRAS **231**, 57 (1988)
12. M. N. Fanelli, R. W. O'Connell, T. X. Thuan: ApJ **321**, 768 (1987)
13. M. N. Fanelli, et al.: ApJS **82**, 197 (1992)
14. N. Förster-Schreiber: PhD thesis, Ludwig-Maximilian Universität, München (1998)
15. A. Fullerton, et al.: ApJ **538**, L43 (2000)
16. M. L. García-Vargas, et al.: A&AS **130**, 513 (1998)
17. M. L. García-Vargas, et al.: ApJ, **478**, 112 (1997)
18. R. Genzel, et al.: ApJ **498**, 579 (1998)
19. L. Giradi, G. Bertelli: MNRAS **300**, 533 (1998)
20. R. M. González Delgado: ApJ **489**, 601 (1997)
21. R. M. González Delgado: ApJS **125**, 489 (1999)
22. R. M. González Delgado, et al.: ApJ in press,(astro-ph/0008417) (2000)
23. R. M. González Delgado, E. Pérez: MNRAS in press, (astro-ph/0003067)(2000)
24. N. Guseva, Y. I. Izotov, T. X. Thuan: ApJ **531**, 776 (2000)
25. S. M. Haser, et al.: A&A **330**, 285 (1998)
26. S. R. Heap: In: *Evolution of Galaxies. I. Observational clues* ed. by J.M. Vilchez, G. Stasinska, Astrophysics and Space Science, in press (2000)
27. T. M. Heckman, C. Leitherer: AJ **114**, 69 (1997)
28. T. M. Heckman, et al.: ApJ **482**, 114 (1997)
29. T. M. Heckman, et al.: ApJ **503**, 646 (1998)
30. S. G. Kleinmann, D. N. B. Hall :ApJS **62**, 501 (1986)
31. D. Kunth, T. Contini: IAU Symp. **193**, 725 (1999)
32. A. Lancon, B. Rocca-Volmerange : A&AS **96**, 593 (1992)

33. A. Lancon, M. Mouhcine: In: *Massive Stellar Clusters* ed. by A. Lancon, C.M. Boily, ASP Conf. Series **211**, 34 (2000)
34. N. Langer, A. Maeder: A&A **295**, 685 (1995)
35. C. Leitherer, H. Lamers: ApJ **373**, 89 (1991)
36. C. Leitherer, et al.: ApJS **99**, 173 (1995)
37. C. Leitherer, et al.: ApJS **123**, 3 (1999) (Starburst99)
38. C. Leitherer, et al.: ApJ in press (astro-ph/0012358) (2000)
39. V. Luridiana, et al.: ApJ **527**, 110 (1999)
40. V. Luridiana, et al.: ApJ submitted (2000)
41. S. Lipari, et al.: ApJ in press (astro-ph/0007316) (2000)
42. A. Maeder, G. Meynet: ARAA **38**, 143 (2000b)
43. J. M. Mas-Hesse, D. Kunth: A&AS **88**, 399 (1991)
44. J. M. Mas-Hesse, D. Kunth: A&A **349**, 765 (1999)
45. Y. D. Mayya: ApJ **482**, L149 (1998)
46. K. Olofsson: A&AS **111**, 57 (1995)
47. L. Origlia, et al.: A&A **280**, 536 (1993)
48. L. Origlia, et al.: ApJ **514**, 96 (1999)
49. M. Pettini, et al.: ApJ **528**, 96 (2000)
50. D. Raimann, et al.: MNRAS **314**, 295 (2000a)
51. D. Raimann, et al.: MNRAS **316**, 559 (2000b)
52. C. Robert, et al.: in preparation (2001)
53. M. S. Sahu: AJ **116**, 1205 (1998)
54. S. Satyapal, et al.: in preparation (2001)
55. D. Schaerer: IAU Symp. **193**, 539 (1999a)
56. D. Schaerer: In: *Spectrophotometric Dating of Stars and Galaxies* ed. by I. Hubeny, S.R. Heap, R.H. Cornett, ASP Conf. Series **192**, p. 49 (astro-ph/9907164) (1999b)
57. D. Schaerer: In: *Stars, Gas and Dust in Galaxies: Exploring the Links* ed. by D. Alloin, G. Galaz, K. Olsen, ASP Conf. Series, in press (astro-ph/0007307) (2000)
58. D. Schaerer, T. Contini, D. Kunth: A&A **341**, 399 (1999a)
59. D. Schaerer, T. Contini, M. Pindao: A&AS **136**, 35 (1999b)
60. D. Schaerer, N. Guseva, Y. I. Izotov, T. X. Thuan: A&A **362**, 53 (2000)
61. D. Schaerer, G. Stasińska: A&A **345**, L17 (1999)
62. D. Schaerer, W. D. Vacca: ApJ **497**, 618 (1998)
63. K. Sekiguchi, K. S. Anderson: AJ **94**, 644 (1987)
64. M. Serote Roos, et al.: MNRAS **301**, 1 (1998)
65. G. Stasińska, C. Leitherer: ApJS **107**, 661 (1996)
66. G. Stasińska, D. Schaerer: A&A **351**, 72 (1999)
67. G. Stasińska, et al.: A&A submitted (2000)
68. T. Storchi-Bergmann, et al.: ApJS **98**, 103 (1995)
69. T. Storchi-Bergmann, et al.: ApJ **501**, 94 (1998)
70. G. Taresch, et al.: A&A **320**, 500 (1998)
71. H. I. Teplitz: ApJ **542**, 1 (2000)
72. E. Terlevich, et al.: MNRAS **242**, 48 (1990a)
73. E. Terlevich, et al.: MNRAS **242**, 271 (1990b)
74. R. Terlevich, et al.: A&AS **91**, 285 (1991)
75. M. D. Thornley, et al.: ApJ **539**, 641 (2000)
76. H. D. Tran, et al.: ApJ **516**, 85 (1999)
77. C. Tremonti, et al.: ApJ submitted (2000)
78. N. R. Walborn, et al.: In *IAU Atlas of O type spectra from 1200 to 1900 Å* (NASA Ref. Publ. 1155, 1985)

The Stellar Population of Nearby Starbursts

Rosa M. González Delgado[1] and Claus Leitherer[2]

[1] Instituto de Astrofísica de Andalucía (CSIC), Apdo. 3004, 18080 Granada, Spain
[2] Space Telescope Science Institute, 3700 San Martin Drive, Baltimore, MD 21218, USA

Abstract. This contribution presents the analysis of the ultraviolet and optical spectra of two typical nearby starburst galaxies. The goal is to investigate the IMF in metal-rich starbursts and the stellar population age. We found that the UV light in the inner 50 pc of NGC 3049 is provided by very young (a few Myr old) clusters, in which massive ($M \geq 50\,M_\odot$) stars formed. In contrast, no single age stellar population is able to explain the UV-Optical flux emitted from the central 300 pc of NGC 7714. Young (O) and intermediate (B and A) mass stars are present in this nuclear starburst. The duration of the starburst in NGC 7714 seems to be related to the interaction timescale with its companion galaxy NGC 7715.

1 Introduction

Starbursts are objects that have size of a few to 10^3 pc in which the total energetics are dominated by star formation and associated phenomena. Their stellar population is young ($\leq$ a few 10^8 yr), and in many of them massive stars (O and Wolf-Rayet) are the dominant stellar population. Important questions related to the starburst phenomena are to know how the star formation proceeds in these objects (burst vs continuous star formation) and how is the initial mass function (IMF). These questions can be answered by studying the stellar content. In most of the starbursts the stellar population is unresolved because they are at distances larger than ≥ 1 Mpc. Therefore, their stellar content can be infered through studies of the integrated stellar population. This contribution presents the analysis of the integrated light at ultraviolet and optical wavelengths of nearby starbursts to investigate their IMF and their stellar population age.

2 The UV Light: The Young Stellar Population

Starbursts show strong emission at UV and far-UV wavelengths, if they do not suffer too much dust obscuration. The continuum emission depends on the content of massive stars. At these wavelengths the spectrum shows strong absorption features. The most important are P-Cygni lines (such as CIVλ1550, SiIV λ1400 and NV λ1240) that are formed in the winds of massive stars. Stellar winds are driven by radiation pressure; thus, massive stars transform their radiative momentum into kinetic energy via absorption in metal lines. The shape of these lines reflects the stellar mass-loss rate, which is related to the stellar luminosity and thus to the stellar mass. Therefore, the shape of the line profiles are related to the content of massive stars, thus, to the IMF and age of the starburst [8]. As a consequence, changes in the line profiles are related with changes of the IMF and age of the starburst. Here, we present the results of the application of this technique to the metal-rich, barred starburst NGC 3049. This starburst was chosen to investigate the universality of the IMF, in particular the formation of very massive stars in metal-rich environments. In fact, studies on IR-luminous starburst galaxies [3] and metal rich HII regions [1] suggest a deficiency in the formation of massive stars, probably produced by the effect of the high metallicity.

2.1   The Starburst of NGC 3049

NGC 3049 is a Wolf-Rayet [13,10] barred early type spiral galaxy (SB(rs)ab) at a distance of 20 Mpc. HST images with STIS/MAMA (FUV) show that recent star formation is taking place along the bar. However, a significant fraction of the UV emission is produced within the central 2 arcsec (Fig. 1). The STIS/MAMA

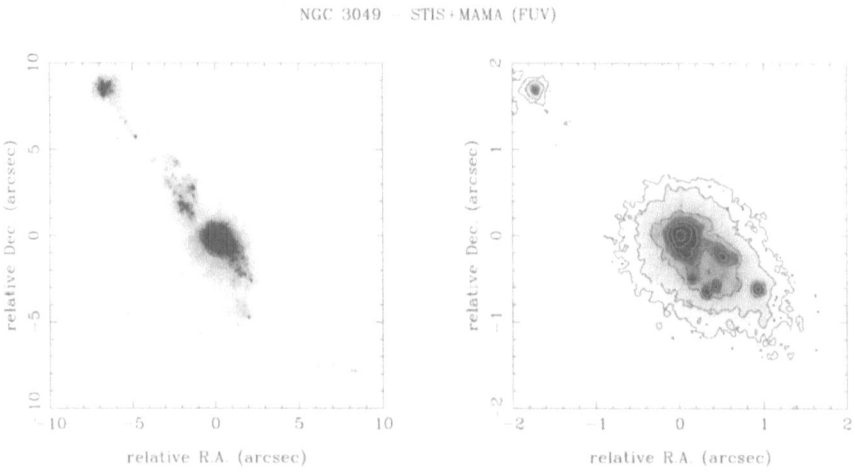

Fig. 1. STIS/MAMA (FUV) image of the starburst barred galaxy NGC 3049. The pixel size is 0.025 arcsec/pixel, 1 arcsec corresponds to 100 pc.

UV spectrum of the inner 50 pc shows a strong P-Cygni and HeII $\lambda1640$ wind lines (Fig. 2). The analysis of the wind lines leads to the following results: a) The stars formed following a Salpeter IMF, being the upper mass cut-off (Mup) larger than 50 $M_\odot$. b) They formed in an instantaneous burst 3–4 Myr ago. c) The metallicity of the starburst is oversolar, in agreement with the O/H abundance estimaded from the ionized gas.

Further constrains to the young stellar population of this starburst come from the spectral energy distribution (SED). In fact, the optical colors of the emission

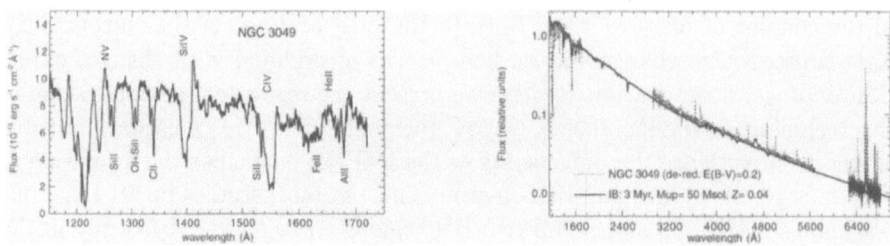

Fig. 2. *Left:* STIS/MAMA (G140L) spectrum of NGC 3049 obtained with a slit of width 0.5 arcsec (corresponding to 50 pc). The most important wind (NV, SiIV, CIV and HeII) and resonance interstellar lines (SiII, OI, CII, FeII and AlII) are labelled. *Right:* SED built with STIS (MAMA and CCD) observations is compared with an instantaneous burst 3 Myr old.

from the central 50 pc are compatible with an instantaneous burst formed 3–4 Myr ago and with Mup= 50 M$_\odot$ (Fig. 2), in agreement with the results obtained from the analysis of the wind lines. However, the SED emitted by the central 1×2 kpc indicates that an older underlying population contributes significantly to the optical colors [9]. These results suggest that the stellar population in starbursts is complex and several stellar populations that are not coeval can contribute to the optical integrated light on larger spatial scales.

3   The Optical Light:
The Intermediate Age Stellar Population

At optical wavelengths, the spectrum of starbursts is mainly dominated by nebular lines originating from the ionized gas. The emission-line spectrum depends on the radiation field from the ionizing cluster, and the abundance and geometry of the gas; thus, nebular lines are, as UV wind lines, a useful diagnostic to constrain the young stellar population (e.g. [2,6]).

However, the optical continuum of starbursts shows the higher order terms of the H Balmer and HeI lines (e.g. λ4471, λ4026, λ3819) in absorption or with absorption wings. These stellar lines can be more readily detected in absorption than the lower terms of the Balmer series, as Hα, because the strength of the gaseous Balmer lines in emission decreases rapidly with decreasing wavelength, whereas the equivalent width of the stellar absorption lines is constant with the wavelength [5]. These lines form in the photosphere of early type stars; thus, they are very sensitive to the presence of B and A stars. The strength of these lines and their profile are useful diagnostics to date starburst and post-starburst galaxies up to 1 Gyr (Fig. 3); thus, they can be used to constrain the intermediate age stellar population in starbursts. Here, we present evidence of the presence of an intermediate age stellar population in addition to the ionizing stellar cluster in the nuclear starburst of NGC 7714.

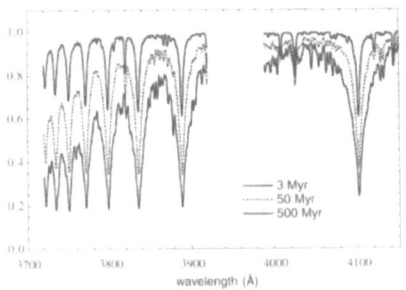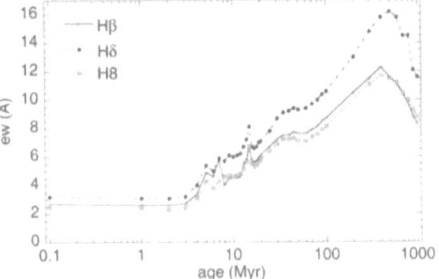

Fig. 3. *Left:* Synthetic spectra predicted for an instantaneous burst formed following a Salpeter IMF between 1–80 $M_\odot$ at ages 3, 50 and 500 Myr. *Right:* Equivalent widths of the Balmer lines Hβ, Hδ and H8.

3.1   The Nuclear Starburst of NGC 7714

NGC 7714 is a spiral SBb galaxy at a distance of 37 Mpc; it is interacting with the irregular galaxy NGC 7715. It is known as a prototypical nuclear starburst because it is a strong UV and FIR source [14]. The HST+WFPC2 image (Fig. 4) shows that the nuclear starburst in NGC 7714 has a very complex morphology, 10–20 clusters contribute to the nuclear emission. The HST UV spectrum of the central 1.7 arcsec (corresponding to 300 pc) shows very strong UV wind lines, and it is well fitted by a 4–5 Myr old instantaneous burst (Fig. 5). The emission-line ratios also suggest a very young stellar population as responsible for the ionization of the gas. The bolometric luminosity of these ionizing clusters is (0.5–1)$\times 10^{10}$ $L_\odot$, and the supernova rate estimated is about one event per century (see

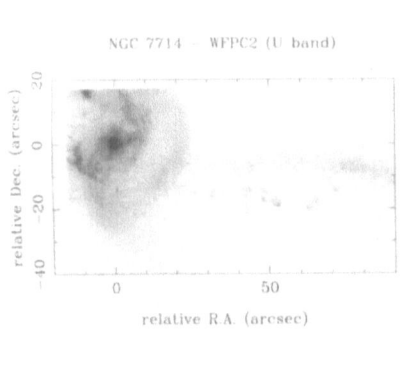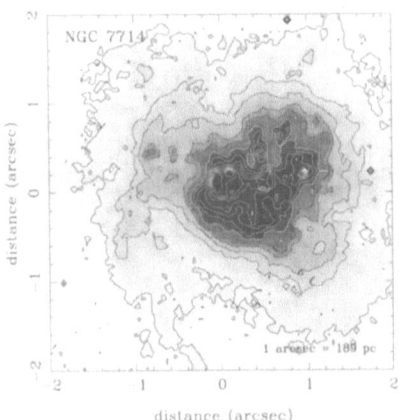

Fig. 4. HST+WFPC2 (U band) image of the starburst NGC 7714 *Left:* Whole galaxy. The companion galaxy is barely detected to the right of the image. A chain of HII regions connects the two galaxies. *Right:* Nuclear starburst of NGC 7714. It shows a complex morphology. Several clusters contribute to the emission. The brightest one is offset with respect to the apparent geometric center of the galaxy by 0.3 arcsec.

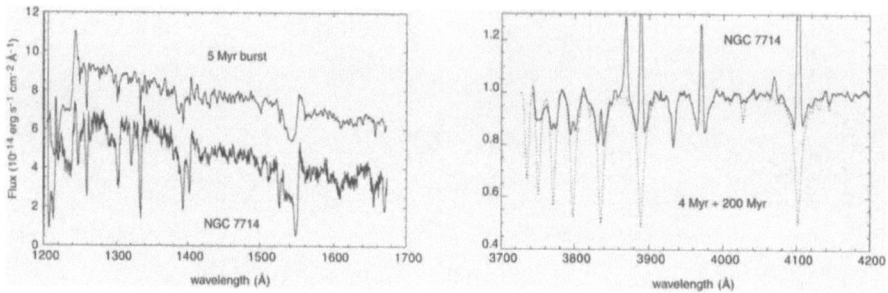

Fig. 5. *Left:* HST+GHRS spectrum of the central 1.7×1.7 arcsec of NGC 7714 and the synthetic spectrum of an instantaneous burst 5 Myr old. *Right:* Near-ultraviolet spectrum (full line) and the synthetic spectrum (dotted line) obtained by the combination of two instantaneous bursts 4 and 200 Myr old.

[4] for more details). However, the X-ray luminosity [12] and the non-thermal radio continuum [14] indicate a larger rate. This suggests the existence of an older stellar population that produces a larger supernova rate. Evidence of an intermediate age stellar population in the nucleus of NGC 7714 comes from the near-ultraviolet spectrum. The young ionizing clusters provide only 30% of the total flux detected at 4000 Å. An old stellar population contributes very little to these wavelengths, as suggested by the strong dilution of the G band and CaII H and K with respect to the values in elliptical galaxies. However, the strongest evidence of intermediate age stars in the nucleus comes from the detection of the high order Balmer series in absorption. The strength of the lines indicates an age of severals 10^8 yr (Fig. 5). This timescale of several 10^8 yr is similar to estimates of the interaction timescale with the companion galaxy NGC 7715 [11]. In fact, the optical spectrum of NGC 7715 (its color and the strength of the Balmer lines) indicate that the dominant stellar population in the nucleus of NGC 7715 has an age of ~300 Myr (Fig. 6). These timescales suggest that the starbursts in these galaxies may be triggered by the interaction between NGC 7714 and its companion. However, the star formation history in the nucleus of NGC 7714 seems to be more complex. The SED (from UV to near-infrared) requires the

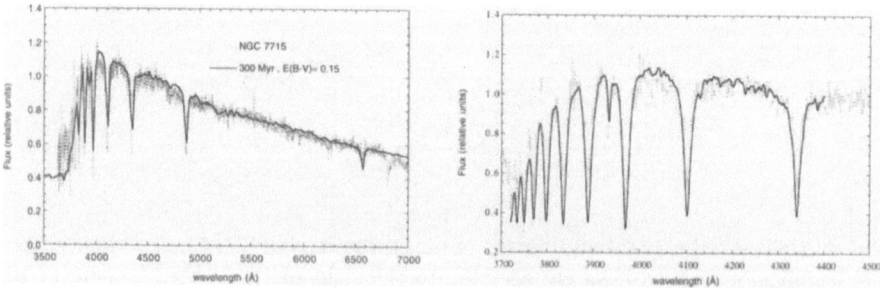

Fig. 6. Optical spectrum (dotted line) of the post-starburst galaxy NGC 7715. The synthetic spectrum (full line) is an instantaneous burst 300 Myr old.

existence of an intermediate age population of a few tens Myr that is highly obscured, that does not contribute to the UV light but dominates the near infrared flux [7].

4 Summary and Conclusions

- Stellar signatures at UV (wind resonance stellar lines) and optical (the higher order Balmer series and HeI lines in absorption) wavelengths combined with colors are good diagnostics to constrain the stellar population in starburst galaxies and the IMF.
- The UV, optical and near-infrared spectra of starbursts on spatial scales of several tens of parsecs can be well described by a single age stellar population, but not on scales of several 100 pc.
- Intermediate (10–few 100 Myr old) age stars contribute significantly to the optical and near-infrared continuum in starburst galaxies on spatial scales of several 100 parsecs.

Acknowledgments: We are grateful to our collaborators (Elizabeth Barton, Jeff Goldader, Tim Heckman, Ariane Lançon, and Grazyna Stasinska) for their contribution to several parts of this work. Rosa González Delgado is very grateful to the organizers for their financial support.

References

1. F. Bresolin, R.C. Kennicutt, D.R. Garnett: ApJ **510**, 104 (1999)
2. M.L. García-Vargas, R.M. González-Delgado, E. Pérez, D. Alloin, A. Díaz, E. Ter-levich: ApJ **478**, 112 (1997)
3. J.D. Goldader, R.D. Joseph, R. Doyon, D.B. Sanders: ApJ **474**, 104 (1997)
4. R.M. González Delgado, M.L. García-Vargas, J. Goldader, C. Leitherer, A. Pasquali: ApJ **513**, 707 (1999)
5. R.M. González Delgado, C. Leitherer, T. Heckman: ApJS **125**, 489 (1999)
6. R.M. González Delgado, E. Pérez: MNRAS **317**, 64 (2000)
7. A. Lançon, J.D. Goldader, C. Leitherer, R.M. González Delgado: ApJ, in press (2001)
8. C. Leitherer, C. Robert, T.M. Heckman: ApJS **99**, 173 (1995)
9. J.M. Mas-Hesse, D. Kunth: A&A **349**, 765 (1999)
10. D. Schaerer, T. Conti, D. Kunth: A&A **341**, 399 (1999)
11. B.J. Smith, J. Struck, R. Pogge: ApJ **483**, 754 (1997)
12. I.R. Stevens, D.K. Strickland: MNRAS **294**, 523 (1998)
13. W.D. Vacca, P.S. Conti: ApJ **401**, 543 (1992)
14. D.W. Weedman, F.R. Feldman, V.A. Balzano, L.W. Ramsey, R.A. Sramek, C.-C., Wuu: ApJ **248**, 105 (1981)

Star Formation Histories of Starbursts

Natascha M. Förster Schreiber[1], Michele D. Thornley[2], Dieter Lutz[3], Reinhard Genzel[3], Henrik W.W. Spoon[4], Amiel Sternberg[5], and Dietmar Kunze[3]

[1] CEA/DSM/DAPNIA/SAp, C.E. Saclay, Orme-des-Merisiers, Bât. 709, 91191 Gif-sur-Yvette CEDEX, France
[2] NRAO, 520 Edgemont Road., Charlottesville, VA 22903, USA
[3] MPE, Postfach 1312, 85741 Garching, Germany
[4] ESO, Karl-Schwarzschild-Strasse 2, 85748 Garching, Germany
[5] School of Physics and Astronomy, Tel Aviv University, Ramat Aviv, Tel Aviv 69978, Israel

Abstract. We present results of a mid-infrared *ISO* spectroscopic survey of 27 starburst galaxies, and of near-infrared integral field spectroscopy and mid-infrared *ISO* spectroscopy of the starburst galaxy M 82. Together with the application of starburst models, the data are consistent with the formation of massive stars ($\gtrsim 50 - 100$ M$_\odot$) in starburst environments, and support short decay timescales of starburst episodes ($\sim 10^6 - 10^7$ yr) indicating important negative feedback effects of starburst activity.

1 Introduction

In recent years, it has become clear that starbursts play an important role in galaxy formation and evolution at all epochs in the Universe. However, despite extensive studies, a detailed and quantitative understanding of the starburst phenomenon is still lacking. Crucial issues remain open such as, for instance, the stellar initial mass function in starbursts and the evolution of starburst activity. Progress has been hindered by severe dust obscuration often hampering studies at UV/optical wavelengths and by the scarcity of spatially resolved data. Studies at near- and mid-infrared wavelengths provide alternatives for obscured systems as these ranges are rich in starburst signatures and extinction effects are much less important than in the UV/optical regimes.

In this context, we have carried out a mid-infrared spectroscopic survey of starburst galaxies obtained with the Short Wavelength Spectrometer (SWS, [3]) on board *ISO* [10] [1], and a detailed investigation of M 82 based on near-infrared integral field spectroscopy from the MPE 3D instrument [22] and mid-infrared spectroscopy from the SWS. We interpreted the data with starburst models combining stellar evolutionary synthesis and photoionization modeling with the codes STARS and CLOUDY ([18], [5]) which employ the Geneva tracks, and recent non-LTE unified stellar atmospheres [14]. Starbursts are modeled as a col-

---

[1] *ISO* is an ESA project with instruments funded by ESA Member States (especially the PI countries: France, Germany, the Netherlands, and the United Kingdom) and with the participation of ISAS and NASA. The SWS is a joint project of SRON and MPE.

lection of evolving star clusters photoionizing the surrounding interstellar gas. Full accounts of our studies are given in [21], [7], and [8]; we discuss here their implications on the issues of massive star formation and evolution of starburst activity.

2   Mid-Infrared Fine-Structure Line Survey

We have gathered atomic fine-structure line fluxes using the SWS ($R \equiv \lambda/\Delta\lambda \sim 1000$) for 27 dusty starburst galaxies, including well-studied infrared-bright starbursts (15 objects) and additional sources selected on the basis of their 60 μm flux densities ($S_{60\,\mu\mathrm{m}} > 20\,\mathrm{Jy}$) and their visibility by *ISO* in the last months of the mission. The sample consists of solar metallicity systems except for NGC 5253 and II Zw 40. We focussed on the [Ne II] 12.8 μm and [Ne III] 15.6 μm lines as tracers of the massive star population. Using existing results from the literature, we also collected estimates of the intrinsic Lyman continuum luminosity L_{Lyc} and of the infrared luminosity L_{IR} assuming it is equal to the bolometric luminosity (L_{bol}) of the OB stars reprocessed by interstellar dust. We modeled the [Ne III] 15.6 μm/[Ne II] 12.8 μm and $L_{\mathrm{bol}}/L_{\mathrm{Lyc}}$ ratios as a function of burst age and timescale (t_{b} and t_{sc}) for a star formation rate $\propto \exp(-t_{\mathrm{b}}/t_{\mathrm{sc}})$, and of IMF upper mass cutoff (m_{up}) assuming a Salpeter slope [16] and a lower mass cutoff of 1 $M_\odot$. We adopted the nebular parameters we determined for M 82 ($n_{\mathrm{e}} = 300\,\mathrm{cm}^{-3}$, $\log U = -2.3\,\mathrm{dex}$, solar metallicity) and which we found appropriate for other sample galaxies for which constraints were available from the SWS or in the literature.

Figure 1 shows the neon ratios for the sample galaxies, as well as local templates of high-mass star-forming regions from similar observations. The majority of the galaxies exhibit low neon ratios, with only a few solar metallicity objects characterized by high ratios (NGC 5253 and II Zw 40 have the highest ratios, consistent with harder stellar spectral energy distributions and a hotter main sequence at low metallicities). The modeling results (Fig. 2) indicate that most of these starbursts are currently deficient in very massive stars, either because they never formed due to a truncation of the IMF at high masses or because they have disappeared as a result of aging effects. Suppressed formation of very massive stars in starbursts is however difficult to reconcile with direct evidence for the presence of $\gtrsim 50 - 100\,M_\odot$ stars in Galactic and near-extragalactic active star-forming regions such as the Galactic Center, NGC 3603, and the R 136 cluster in 30 Dor (e.g. [13]; [6], [4], [12]). Therefore, we favoured the scenario in which very massive stars form in starbursts and aging effects are responsible for the observed range in neon ratios. Adopting $m_{\mathrm{up}} = 50 - 100\,M_\odot$, the combination of the neon and $L_{\mathrm{IR}}/L_{\mathrm{Lyc}}$ ratios further suggests that most starbursts decay on short timescales of $\sim 10^6 - 10^7$ yr.

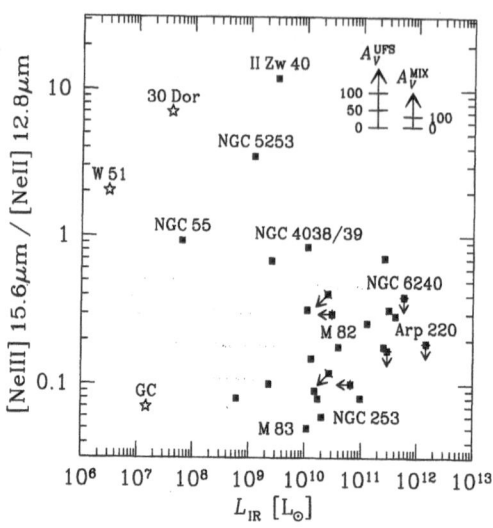

Fig. 1. Observed neon line ratio versus infrared luminosity for a sample of 27 starburst galaxies (filled squares) and template active star-forming regions (stars). Neon ratios for a larger sample of Galactic H II regions lie mainly between those of W 51 and the Galactic Center [2]. Extinction effects on the neon ratio are very small, as shown by the arrows for a uniform foreground screen (UFS) and a mixed (MIX) model.

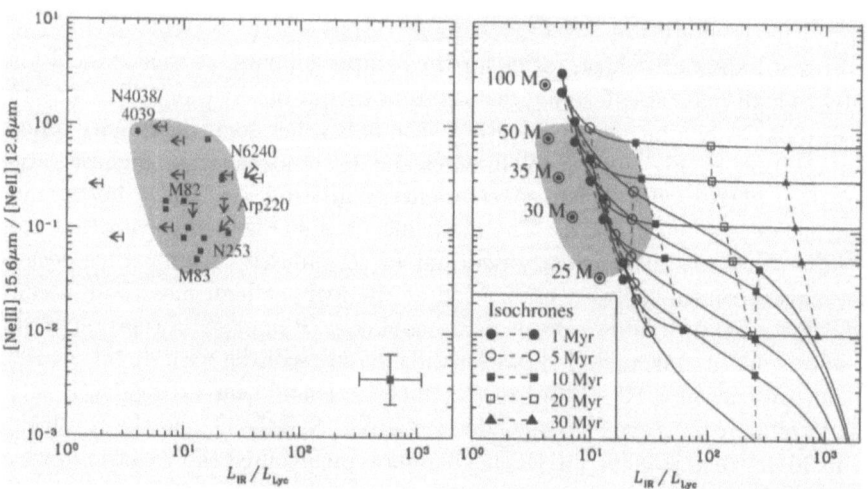

Fig. 2. Neon line versus L_{IR}/L_{Lyc} ratios for the solar-metallicity SWS starburst sample, compared to model predictions for various IMF upper mass cutoffs and burst ages (as labeled in the right-hand side panel), and burst timescales of 1 Myr and 5 Myr (steepest and flattest curves, respectively).

3   Case Study: M 82

M 82 is an ideal target for detailed investigations due to its proximity ($D = 3.3\,\mathrm{Mpc}$; [9]). Most of its infrared luminosity originates from a $500\,\mathrm{pc}$–diameter "starburst core" which has attracted considerable attention in the past (see e.g. [19], [15], and [17] for reviews and earlier starburst models). We have observed the central regions of M 82 with the near-infrared imaging spectrometer 3D at a spectral resolution of $R \sim 1000$ and angular resolution 1.5″. We complemented the 3D data with SWS spectroscopy and additional results at other wavelengths from the literature. These data provide an extensive set of diagnostics, notably for the hot massive stars (near-infrared H and He recombination lines and mid-infrared atomic fine-structure lines originating in H II regions) as well as for cool evolved stars (near-infrared continuum emission and absorption features produced in the atmosphere of red giants and supergiants).

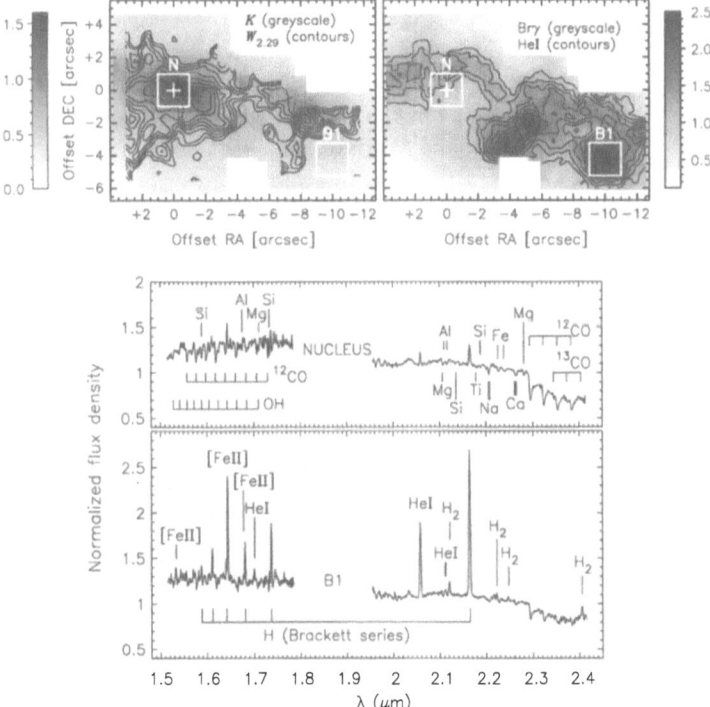

Fig. 3. 3D maps of M 82 and spectra of the $35\,\mathrm{pc}$ at the nucleus (cross, "N") and of region "B1." Greyscale levels are in $10^{-14}\,\mathrm{W\,m^{-2}\,\mu m^{-1}\,arcsec^{-2}}$ for the K-band map and $10^{-17}\,\mathrm{W\,m^{-2}\,arcsec^{-2}}$ for the Brγ map. Contours for the $W_{2.29}$ map range from $12\,\mathrm{\AA}$ to $16\,\mathrm{\AA}$ in steps of $0.5\,\mathrm{\AA}$, those for the HeI $2.058\,\mu m$ linemap range from 0.5 to 1.5 in steps of 0.1 and units of $10^{-17}\,\mathrm{W\,m^{-2}\,arcsec^{-2}}$. The spectra are normalized to unity in the range $2.2875 - 2.2910\,\mu m$ (normalizing factors in $\mathrm{W\,m^{-2}\mu m^{-1}\,arcsec^{-2}}$ are 6.52×10^{-14} for the nucleus and 1.78×10^{-14} for B1.

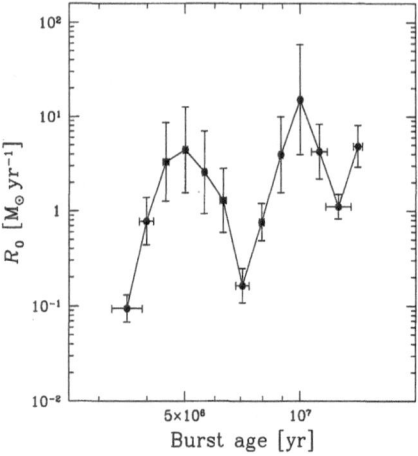

Fig. 4. Global star formation history in the central regions of M 82 reconstructed from the spatially-detailed modeling by integrating the burst intensity per age interval.

Figure 3 presents various 3D maps of M 82 together with the H- and K-band spectra of the central 35 pc at the nucleus and of the Brγ source $\approx 10''$ southwest of the nucleus (denoted "B1"). The 3D data show distinct burst sites as small as a few tens of parsecs and reveal important spatial variations in the strength of the absorption features and emission lines relative to the continuum. Application of the models to the starburst core, selected regions, and all individual resolution elements within the 3D field of view allowed us to constrain the detailed spatial and chronological evolution of starburst activity in the central regions of M 82. With an IMF extending up to $100\,M_\odot$, the results indicate burst timescales of a few million years on scales ranging from a few tens of parsecs to 500 pc. The various diagnostics can only be reconciled by invoking two successive bursts at each location, likely resulting from projection effects as the red supergiants are distributed in the disk and concentrated at the nucleus while the H II regions lie predominantly in a circumnuclear ring at radius ≈ 85 pc and along the ~ 1 kpc–long stellar bar (e.g. [20], [11], [1]). The derived integrated star formation history (Fig. 4) shows two distinct starburst episodes, and the spatial distributions of burst ages and intensities suggest the following scenario: (1) Following tidal interaction with M 81 $\sim 10^8$ yr ago (e.g. [23]), the interstellar medium (ISM) in M 82 experienced large-scale torques, loss of angular momentum, and infall towards the nuclear regions possibly channeled by the bar. In the absence of dynamical resonances, the ISM can reach the nucleus. (2) A starburst episode took place $8-15$ Myr ago throughout the central 500 pc of M 82, declined rapidly, and was particularly intense near the nucleus. (3) A subsequent episode occurred $4-6$ Myr ago predominantly in a circumnuclear ring and along the bar, was likely triggered by bar-induced dynamical resonances, and also decayed rapidly.

4   Conclusions

Our studies support that very massive stars can form in starburst environments, and suggest that starbursts can be episodic and relatively short-lived, lasting $\sim 10^6 - 10^7$ yr. Such short timescales can be easily understood on scales of a few tens of parsecs due to strong stellar winds from massive stars and subsequent supernova explosions rapidly disrupting the surrounding ISM. The modeling of the SWS galaxy sample and of the starburst core of M 82 suggests that starburst activity can have strong negative feedback effects on global scales as well.

References

1. J. M. Achtermann, J. H. Lacy: ApJ, **439**, 163 (1995)
2. P. Cox et al.: 'ISO Spectroscopy of Dense Regions.' In: *The Universe as seen by ISO*, ed. by P. Cox and M. F. Kessler, ESA SP-427 (ESA Publications Division, Noordwijk 1999), Vol. 2, pp. 631-637
3. T. de Graauw et al.: A&A, **315**, L49 (1996)
4. F. Eisenhauer, A. Quirrenbach, H. Zinnecker, R. Genzel: ApJ, **498**, 278 (1998)
5. G. J. Ferland: *Hazy, a Brief Introduction to Cloudy* (University of Kentucky Department of Physics and Astronomy Internal Report 1996)
6. D. F. Figer, F. Najarro, M. Morris, I. S. McLean, T. R. Geballe, A. M. Ghez, N. Langer: ApJ, **506**, 384 (1998)
7. N. M. Förster Schreiber, R. Genzel, D. Lutz, D. Kunze, A. Sternberg: ApJ, submitted (2000)
8. N. M. Förster Schreiber, R. Genzel, D. Lutz, A. Sternberg (in preparation)
9. W. L. Freedman, B. F. Madore: ApJ, **332**, L63 (1988)
10. M. F. Kessler et al.: A&A, **315**, L27 (1996)
11. J. E. Larkin, J. R. Graham, K. Matthews, B. T. Soifer, S. Beckwith, T. M. Herbst, A. C. Quillen: ApJ, **420**, 159 (1994)
12. P. Massey, D. A. Hunter: ApJ, **493**, 180 (1998)
13. F. Najarro, A. Krabbe, R. Genzel, D. Lutz, R.-P. Kudritzki, D. J. Hillier: A&A, **325**, 700 (1997)
14. A. W. A. Pauldrach, M. Lennon, T. L. Hoffmann, F. Sellmaier, R.-P. Kudritzki, J. Puls: 'Realistic Models for Expanding Atmospheres'. In: *Boulder-Munich II: Properties of Hot, Luminous Stars*, ed. by I. Howarth (ASP, San Francisco 1998), ASP Conf. Ser. vol. 131, p. 258
15. G. H. Rieke, K. Loken, M. J. Rieke, P. Tamblyn: ApJ, **412**, 99 (1993)
16. E. E. Salpeter: ApJ, **121**, 161 (1955)
17. S. Satyapal, D. M. Watson, J. L. Pipher, W. J. Forrest, M. A. Greenhouse, H. A. Smith, J. Fischer, C. E. Woodward: ApJ, **483**, 148 (1997)
18. A. Sternberg: ApJ, **506**, 721 (1998)
19. C. M. Telesco: ARA&A, **26**, 343 (1988)
20. C. M. Telesco, H. Campins, M. Joy, K. Dietz, R. Decher: ApJ, **369**, 135 (1991)
21. M. D. Thornley, N. M. Förster Schreiber, D. Lutz, R. Genzel, H. W. W. Spoon, D. Kunze, A. Sternberg: ApJ, **539**, 641 (2000)
22. L. Weitzel, A. Krabbe, H. Kroker, N. Thatte, L. E. Tacconi-Garman, M. Cameron, R. Genzel: A&AS, **119**, 531 (1996)
23. M. S. Yun, P. T. P. Ho, K. Y. Lo: Nature, **372**, 530 (1994)

Chemical Evolution and Starbursts

Uta Fritze-von Alvensleben

Universitätssternwarte Göttingen, Geismarlandstr. 11, 37083 Göttingen, Germany

Abstract. The first part of this paper deals with the impact of nonsolar and – for late-type, dwarf, and high redshift galaxies – generally subsolar abundances on the interpretation of observational data for starburst galaxies. It points out the differences in colors, luminosities, emission lines, etc. obtained from a model using low metallicity input physics for a starburst on top of the stellar population of a galaxy as compared to an otherwise identical model using solar metallicity input physics only.

The second part deals with the chemical evolution during a starburst and contrasts model predictions with observational clues.

1 Abundance Effects on Starbursts

The chemical abundances in the gas of a galaxy at the onset of a starburst set the initial abundances for the bulk of the burst stars. Only in long bursts, e.g. like those triggered by interactions between masive galaxies, may later generations of burst stars incorporate SN II products from earlier ones.

Starbursts in giant galaxies, e.g. triggered by an interaction or merger event, are the more spectacular the more gas is available. In the local Universe, a broad anticorrelation is observed between the size and the metallicity of a galaxy's gas reservoir. Hence, the strongest bursts can be excited in gas rich galaxies which genuinely are of low metallicity. Dwarf galaxies as well as young galaxies generally have lower abundances than today's giant galaxies.

We will show that for a burst of given strength, the spectrophotometric as well as the chemical evolution are significantly dependent on the metallicity. Interpretation of starburst galaxy observations therefore requires comparison with models of appropriate metallicity.

1.1 Abundances in Local and High Redshift Galaxies

While already in the Milky Way, the global average stellar and gas abundances both are subsolar ($\sim \frac{1}{2}$ solar), late type galaxies with their huge gas reservoirs show significantly lower abundances still [25,13,34,2,35].

So, strong starbursts in local *giant gas-rich galaxies*, e.g. triggered by interactions or mergers like in NGC 4038/39 or NGC 7252, are to be described by models accounting for the moderately subsolar metallicity of the gas in these objects. Burst durations in giant galaxies are typically of the order of the dynamical timescale $\tau_B \sim t_{dyn} \sim 10^8$ yr, i.e. long compared to the most massive stars' lifetimes. Hence self-enrichment in SN II products like O, Mg, and other α-elements during bursts may be important for giant galaxies. Depending on the cooling timescale, these SN II products may even be incorporated into burst stars that form after the first generation of massive burst stars has already died.

Dwarf galaxies in the local Universe, according to the luminosity – metallicity relations established both for the stellar metallicities of dwarf elliptical and spheroidal galaxies and for the gas metallicities of dwarf irregular galaxies, show significantly subsolar abundances that extend down to few percent solar [23,29]. E.g., Blue Compact Dwarf Galaxies (**BCDs**) typically feature $\langle Z \rangle \sim \frac{1}{10} Z_\odot$, extreme examples like IZw18 or SBS 0335-052 reach down to $\sim \frac{1}{40} Z_\odot$ [12]. Hence undoubtedly, the interpretation of starbursts occuring in dwarf galaxies requires models of appropriately low metallicity. In recent years, Tidal Dwarf Galaxies (**TDGs**) have been detected in rapidly increasing numbers, forming in the tidal tails of massive interacting spirals. Compared to dwarf galaxies of comparable luminosity they show enhanced metallicities, typically in the range $(\frac{1}{4} - \frac{1}{2}) Z_\odot$, as a result of being formed from stars and pre-enriched gas pulled out from their parent galaxies [1]. As for giant galaxies, typical burst durations in dwarf galaxies are of the order of the dynamical timescale, and, hence $\tau_B \sim 10^6$ yr for dwarf galaxies. These short burst durations – not longer than the lifetimes of massive stars – imply that self-enrichment during a burst will not be important. Mass loss due to SN-driven galactic winds, on the other hand, may be important for dwarf galaxies due to their shallower potential wells as compared to giant galaxies and may significantly affect their chemical evolution. The occurence and strength of galactic winds in dwarf galaxies, however, depend on the poorly constrained mass of their dark matter halos.

Galaxies in the early Universe, of course, have lower metallicity than their local counterparts. Lyman Break Galaxies at $3 \lesssim z \lesssim 4$, some or many of which are observed in phases of enhanced star formation (**SF**), show metallicities in the range $(0.1 - 1) Z_\odot$ as derived from the restframe UV stellar wind lines of their stellar populations [18,32] and from their [O III] emission lines [30]. The ample supply of neutral gas in Damped Lyα Absorbers at $0 \leq z \leq 4.4$ shows abundances from $10^{-3} Z_\odot$ to $\lesssim Z_\odot$ [21,22].

1.2   Chemical and Spectrophotometric Evolution Models

In the following we use evolutionary synthesis models to describe the spectrophotometric and chemical evolution of starbursts in various types of galaxies. The models for undisturbed galaxies of various types are parametrised by the respective appropriate star formation histories $\Psi(t)$ and constrained by the requirement to provide agreement after ~ 12 Gyr of evolution with average colors, luminosities (U ... K), emission and absorption line strengths, characteristic H II region abundances (= measured at R_{eff}), and gas content of local samples of the respective types and with template spectra. Then a burst of given strength $b := \frac{\Delta S}{S}$ (with S : stellar mass at the onset of the burst, ΔS : stellar mass added in the burst) and duration τ_B is assumed to start at some time t_B. Using sets of input physics (stellar evolutionary tracks, lifetimes, stellar yields, model atmosphere spectra, and absorption index calibrations) for a range of metallicities from 10^{-4} to 0.05, models follow the spectral evolution from UV – NIR, the chemical evolution of individual gas phase element abundances, as well as the metallicity distribution in the stellar population before, during, and after the burst. Models account for the finite lifetimes of the stars before they give back enriched material, include the contributions of type Ia SNe as described by [19] but they are otherwise kept as simple as possible in order to minimize the number of parameters. In particular, they are closed box 1-zone models and assume instantaneous and perfect mixing of the recycled gas.

1.3   Abundance Effects on Starbursts

At subsolar metallicities, bursts of a given strength in the same type of galaxy lead to higher peak luminosities, bluer optical colors, slower and weaker fading and reddening after the burst (Fig. 1), and higher mass loss from the burst population (due to shorter stellar lifetimes) as compared to models using solar metallicity input physics [14,16,26]. Furthermore, at lower metallicities, spectra look different, emission line ratios change [11], the emission contribution of the gas is higher to UBVR fluxes and lower in JHK bands (Fig. 2), UV stellar wind lines are weaker [17], and dust extinction less important.

Hence, in a low metallicity galaxy, the same blue optical colors imply a weaker or older burst than they would in a solar metallicity galaxy.

2   Chemical Evolution in a Starburst

2.1   Model Predictions

The chemical evolution – both in terms of ISM enrichment and metallicities of stars formed in the burst – depends on the burst strength, the size of the gas

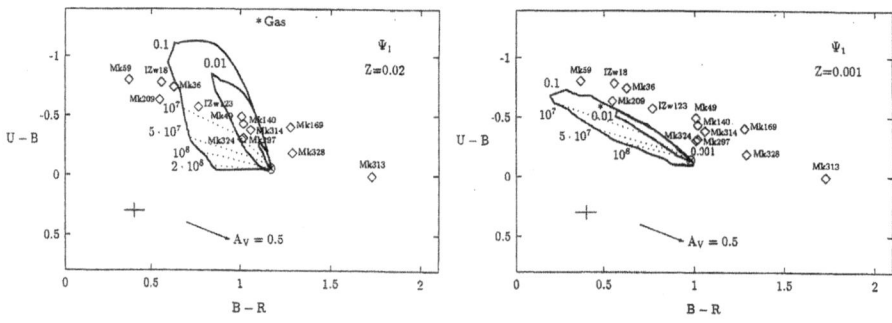

Fig. 1. Color-color diagram for starbust models at two different metallicities and 3 burst strengths each (b=0.1, 0.01, 0.001) on top of a galaxy with const. SF rate. ⊗ marks the galaxy before the burst, ∗ the color of the pure gas spectrum. BCD data are from [31].

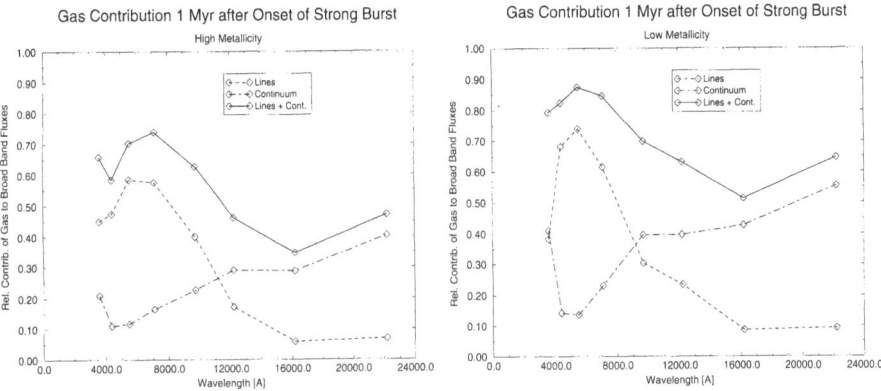

Fig. 2. Relative contribution of the gaseous emission to broad band fluxes UBVRIJHK 1 Myr after the onset of a strong burst (b=0.4). Left for solar metallicity, right for Z=0.001.

reservoir, and on the (time of) occurence or non-occurence of SN-driven galactic winds or superwinds. The latter, unfortunately, depends on several poorly known conditions, e.g. on the masses of DM halos, the geometry of the starburst region, etc. Note that the same number of burst stars can cause vastly different increases of the ISM metallicity depending on whether they shed their enrichment products into a small or a large gas reservoir. During a burst, individual element abundances in the ISM increase on individual timescales, depending on the nucleosynthetic origin of the respective element. Oxygen and other typical SN II products (α-elements) are restored on very short timescales given by the lifetimes of massive stars. Thus the abundances of oxygen and other α-elements start increasing shortly after the onset of the burst and do not continue for long after its end. Elements that are predominantly synthesized in intermediate mass stars, like C or N, as well as elements that have important contributions from type Ia SNe, such as Fe, only start enriching with certain time delays after the beginning of the burst and continue to do so for up to some Gyr after the end of the burst. This effect leads to strong changes in the element ratios between elements of different nucleosynthetic origin. As e.g. [20] showed, the large range of N/O-ratios at essentially all oxygen abundances among dwarf galaxies, Galactic and extragalactic HII regions may be explained by successive bursts of SF. During a short burst, the oxygen abundance increases and, hence N/O decreases while nitrogen is not changed yet. The nitrogen abundance only starts increasing after the end of the burst, leading to an increase of N/O at essentially constant oxygen abundance during up to a few Gyr.

A detailed understanding of the chemical evolution in the course of a starburst not only requires a consistent chemodynamical model but also the consideration of the multi-phase nature of the ISM, including a full description of all kinds of transition processes between the hot X-ray gas, the warm and neutral (HI) components and the cold and clumpy molecular gas. This latter component, often traced by CO, is increasingly difficult to observe in low metallicity dwarf galaxies. A consistent and complete model including all these phases and processes is still to be developed [24].

2.2 Observational Clues

ISM Abundances HII region abundances in BCDs are used to obtain the metallicity of the stars formed in the burst. Burst strengths in BCDs are weak as compared to those in massive gas rich interacting galaxies, increasing the stellar mass by typically few percent or less [15]. HI reservoirs, on the other hand, are large in BCDs. So, the metallicity increase in the burst and, hence, the metallicity difference between preburst and burst stars both are relatively small. This explains why models using input physics for the HII region metallicity observed in a BCD generally also allow for good agreement with its spectral energy distribution which, at short wavelengths, is dominated by burst stars while at long wavelengths (JHK) it is mostly due to the preburst component [15].

TDGs, on the other hand, with their observed HII region abundances enhanced over those of dwarf galaxies of comparable luminosity, in most cases turn out to be well describable with models including an old stellar population from a late type galaxy plus a burst star component of $\lesssim \frac{1}{2}$ solar metallicity [33]. Their HII region abundances indeed are $[O/H]_{TDG} \gtrsim [O/H]_{spiral}^{ISM}$.

Stellar and Star Cluster Abundances Stellar abundances for starburst populations are difficult to disentangle observationally from those of the preburst stars. Star clusters formed in starbursts in some dwarf galaxies and – in huge numbers – in the strong bursts accompanying gas rich galaxy mergers, however, provide a powerful tool to study the chemical enrichment process in these bursts. As opposed to field stars, they well separate from the preburst population and can be analysed individually.

Models predict their abundances on the basis of the progenitor galaxy ISM abundances and gas reservoirs, and of the strength and duration of the burst. E.g. for the young and very young star cluster populations in NGC 7252 and NGC 4038/39 average metallicties of $(\frac{1}{2} - 1)Z_\odot$ were predicted as well as some α-enhancement for the youngest clusters in the almost 1 Gyr old burst in NGC 7252 [7,6]. [16] and [5] show how significant the metallicity of a star cluster is in interpreting its photometric data, e.g. for age-dating and mass estimates. Young star clusters are bright – at ages of $\sim 10^8$ yr typically 4 mag brighter than old globular clusters (**GCs**) of the same mass. 10 m telescope spectroscopy thus directly gives access to metallicities, ages, velocity dispersions for kinematic mass estimates, ... [10,27,28].

For clusters older than few 10^8 yr which no longer feature emission lines, stellar absorption features in comparison with theoretical model calibrations at the appropriate young ages [16,26] give information about individual element abundances and abundance ratios. With spectroscopy of reasonable samples the metallicity and age distributions among young star clusters and their spatial variations will give information about the dynamics of the burst and its detailed enrichment process. The limiting factor for this kind of observations is the strong and spatially variable galaxy background.

MOS of old GCs is within reach of 10m telescopes out to Virgo cluster distances. The metallicity distributions this will reveal for the old GC populations of elliptical and S0 galaxies will teach us a lot about their formation processes [36,9].

3   Conclusions

Accounting for the appropriate metallicities in evolutionary synthesis models is essential for the interpretation of starburst galaxies as is the consideration of dust (see S. Charlot, *this vol.*).

In a burst, we see to first order star formation and ISM abundances at the metallicity of the gas in the preburst galaxy. Most clearly this is seen on star clusters rather than on field stars. Only to second order metallicity and, in particular, α-enhancements are expected in the long lasting bursts in massive interacting gas rich galaxies. Again these are best studied on individual star clusters.

Acknowledgement: I gratefully acknowledge partial financial support from the organisers.

References

1. P.A. Duc, I.F. Mirabel: IAUS **186**, 61 (1999)
2. A.M.N. Ferguson, J.S. Gallagher, R.F.G. Wyse: AJ **116**, 673 (1998)
3. U. Fritze-v. Alvensleben: A&A **336**, 83 (1998)
4. U. Fritze-v. Alvensleben: A&A **342**, L25 (1999)
5. U. Fritze-v. Alvensleben: in *Massive Stellar Clusters*, eds. A. Lançon, C. Boily, ASP Conf. Ser. 211 (ASP, San Francisco, 2000) p.3
6. U. Fritze-v. Alvensleben, A. Burkert: A&A **300**, 58 (1995)
7. U. Fritze-v. Alvensleben, O.E. Gerhard: A&A **285**, 751 (1994)
8. U. Fritze-v. Alvensleben, O.E. Gerhard: A&A **285**, 775 (1994)
9. K. Gebhardt, M. Kissler – Patig: AJ **118**, 1526 (1999)
10. L.C. Ho, A.V. Filippenko: ApJ **472**, 600 (1996)
11. Y.I. Izotov, T.X. Thuan: ApJ **511**, 639 (1999)
12. Y.I. Izotov et al.: ApJ **527**, 757 (1999)
13. J. Kilian - Montenbruck, T. Gehren, P.E. Nissen: A&A **291**, 757 (1994)
14. H. Krüger, U. Fritze-v. Alvensleben: A&A **284**, 793 (1994)
15. H. Krüger, U. Fritze-v. Alvensleben, H.-H. Loose: A&A **303**, 41 (1995)
16. O. Kurth, U. Fritze-v. Alvensleben, K.J. Fricke: A&AS **138**, 19 (1999)
17. C. Leitherer et al.: ApJS **123**, 3 (1999)
18. J.D. Lowenthal, D.C. Koo, R. Guzman, et al.: ApJ **481**, 673 (1997)
19. F. Matteucci, A. Tornambé: A&A **142**, 13 (1985)
20. F. Matteucci, M. Tosi: MNRAS **217**, 391 (1985)

21. M. Pettini, L.J. Smith, R.W. Hunstead, D.L. King: ApJ **426**, 79 (1994)
22. M. Pettini, S.L. Ellison, C.C. Steidel, D.V. Bowen: ApJ **510**, 576 (1999)
23. M.G. Richer, M. McCall: ApJ **445**, 642 (1995)
24. A. Rieschick, G. Hensler: AGM **17**, 72 (2000)
25. H.J. Rocha - Pinto, W.J. Maciel: A&A **339**, 791 (1998)
26. Schulz et al., in preparation
27. F. Schweizer, P. Seitzer: ApJ **417**, L29 (1993)
28. F. Schweizer, P. Seitzer: AJ **116**, 2206 (1998)
29. E.D. Skillman, R.C. Kennicutt, P.W. Hodge: ApJ **347**, 875 (1989)
30. H.I. Teplitz et al.: ApJ **542**, 18 (2000)
31. T.X. Thuan: ApJ **268**, 667 (1983)
32. S.C. Trager, S.M. Faber, A. Dressler, A. Oemler: ApJ **485**, 92 (1997)
33. P.M. Weilbacher, P.A. Duc, U. Fritze-v. Alvensleben, P. Martin, K.J. Fricke: A&A **358**, 819 (2000)
34. D. Zaritsky, R.C. Kennicutt, J.P. Huchra: ApJ **420**, 87 (1994)
35. L. van Zee, J.J. Salzer, M.P. Haynes et al.: AJ **116**, 2805 (1998)
36. S.E. Zepf, K.M. Ashman: MNRAS **264**, 611 (1993)

Chemical Abundances and Evolution in Nearby Starbursts

Michael A. Dopita, Lisa J. Kewley, Ralph S. Sutherland, and Charlene A. Heisler

Research School of Astronomy and Astrophysics, The Australian National University, Cotter Road, Weston Creek ACT 2611, Australia

Abstract. We have constructed self-consistent models of starburst emission line spectra using both the PEGASE v2.0 and STARBURST99 codes to generate the spectral energy distribution (SED) of the young star clusters as input to the MAPPINGS III code (with dust included) . We find that giant extragalactic H II regions can be well-modelled, provided that they are mostly youngy, but that the models of older starbursts where there is a large contribution of UV photons from Wolf-Rayet stars fit the observations less well. This is the result of fundamental uncertainties in atmospheric, and, to a lesser extent, evolutionary models of these stars.

1 Introduction

Observations of starburst galaxies can provide vital insights into the processes and spectral characteristics of massive star formation regions. In such regions the physical conditions are similar to those that existed at the time of collapse and formation of galaxies in the early universe, and they can provide an understanding of early galaxy evolution The emission line spectrum of the starburst, in particular, provides constraints upon the gas density, temperature and pressure, and the total rates of star formation. The theoretical tools required to interpret the spectra of such galaxies are now available. For example, detailed stellar population synthesis models have been developed for both instantaneous and continuous starbursts and using these models, one is able to derive parameters such as the starburst age and metallicity from the continuous spectrum. In such models, the stellar initial mass function (IMF), star formation rate (SFR) and stellar atmosphere formulations are all adjustable initial parameters.

Since the H II region emission line spectrum is very sensitive to the hardness of the ionizing EUV radiation, optical line ratio diagnostic diagrams provide an important constraint on the shape of the EUV spectrum and these may also be used to estimate the mean ionization parameter and metallicity of the galaxies. Such optical diagnostic diagrams were proposed by Baldwin et al. [2] and developed by Osterbrock & de Robertis [15] and Veilleux & Osterbrock [24] (VO87). In this paper, we review the results of our recent theoretical modelling of starbursts and H II regions in external galaxies, and show how further progress is critically dependent on improvements in the theory of Wolf-Rayet atmospheres.

2   Stellar Population Synthesis Models

A major uncertainty in the modelling of the H II region emission spectra concerns the reliability of models of the EUV spectra produced by evolving OB stars. Already, near the main sequence, there are important discrepancies between the EUV fluxes predicted by the COSTAR models of Stasinska & Schaerer [21] and those given by the plane parellel atmospheres of Kurucz [10] (*see also* [17]). These differences become even more marked for more evolved hot stars.

We have used both the PEGASE 2 [7] and the STARBURST99 [12] codes. The PEGASE 2 code uses the Lejeune et al. [13] grid of atmospheres covering the entire Hertzsprung-Russell diagram (HRD) plus the Clegg & Middlemass [4] planetary nebula nuclei (PNN) atmospheres for stars with high effective temperatives (T > 50000K). The STARBURST 99 code also uses the Lejeune grid. For stars with strong stellar winds it offers the choice of the Lejeune plane-parallel atmospheric grid or the Schmutz [20] extended model atmospheres (hereafter known as the Schmutz atmospheres). The prescription for the switch between extended and plane-parallel atmospheres is the same as in [11]. Since the Padova tracks [3] are used in PEGASE 2 while the Geneva tracks [18] are used in STARBURST99, there is sufficient flexibility to separate the effects of either the stellar atmospheres or the stellar evolutionary tracks on the theoretical H II region model line intensity ratios.

In these models, we distinguish between the *instantaneous* (zero-age) star formation case, and the *continuous* starburst models in which a balance between star birth and star death is set up for all stellar masses which contribute significantly to the EUV spectrum. The hydrogen-burning lifetime of massive stars is approximately $\tau = 4.5(M/40M_\odot)^{-0.43}$ Myr, so this condition is satisfied for any starburst which lasts longer than about 6 Myr. This allows a dynamical balance between star births and star deaths for all masses greater than about 20 $M_\odot$, and is also time enough for the Wolf Rayet stars to produce their full contribution to the EUV spectrum.

The EUV spectrum is very sensitive to age in both the PEGASE 2 and the STARBURST99 codes. However, after a few Myr of evolution, marked differences develop between the results two codes. The Schmutz extended atmospheres produce far more ionizing radiation at frequencies above the He II ionization limit than do the Lejeune model atmospheres, but diffences are much less important at lower energies. These differences are metallicity-dependent. At high metallicity, the EUV spectrum becomes harder more quickly as the starburst evolves and the high mass stars make a larger contribution to the EUV radiation field at younger burst ages. The most likely cause of these differences is to be found in the different atmospheric models - most important for the Wolf-Rayet stars.

The EUV spectrum emergent from a W-R model atmosphere depends critically on the fraction of ionizing photons which have been used up to maintain the ionization of the W-R wind region. This is determined by the size of the emission measure of the atmosphere; $\int n_e^2 dr$. This parameter is proportional to the product $(\dot{M}/v_\infty)^2 R_*^{-3}$, where $\dot{M}$ is the mass-loss rate, v_∞ the terminal velocity of the wind, and R_* is the photospheric radius of the star. This product is the Schmutz et al. [19] density parameter. Models with the same density parameter display very similar emission line equivalent widths, but the total scaling in luminosity depends on R_*^2. The density parameter can also be expressed in terms of a "transformed radius", R_t:

$$R_t = R_* \left[\frac{v_\infty \dot{M}_{ref}}{v_{ref} \dot{M}} \right]^{2/3}$$

where v_{ref} is a (normalising) reference velocity, and $\dot{M}_{ref}$ is a (normalising) reference mass-loss rate. Models with similar transformed radius give similar spectra. Stars which use a greater fraction of their EUV photons in maintaining the photoionization of their extended atmospheres would show a lower-intensity, harder, EUV spectrum below the He II ionization edge, and would be expected to exhibit more atmospheric blanketing by heavy elements. In principle, we would expect the Schmutz extended atmospheric modelling to be more appropriate to the modelling of our starburst galaxies. However, these assume a H + He atmosphere. The inclusion of the bound-free continuum opacity due to heavier elements would represent a major step forward in reducing the theoretical uncertainties in the shape of the EUV radiation field.

3   Photoionisation Modelling

The photoionization modelling used the computed EUV spectrum as input to the MAPPINGS III code ([23], *see also* [5]) to investigate the effects of stellar metallicity, age, and ionisation parameter q (cm s^{-1}) defined on the inner boundary of the nebula. This can be readily transformed to the more commonly used dimensionless ionisation parameter through the identity $\mathcal{U} \equiv q/c$. Dust physics is treated explicitly through the absorption of the radiation field on grains, grain charging and photoelectric heating by the grains. Standard depletion factors are adopted for each element. For non-solar metallicities both the dust model and the depletion factors are unchanged, since we have no way of estimating what they may be otherwise.

All elements except nitrogen and helium are taken to be primary nucleosynthesis elements. It is known that this assumption may be incorrect in systems where the time history of star formation in the galaxy is different, or where galactic winds are important. Nitrogen is assumed to be a secondary element above metallicities of 0.23 solar, while for Helium, a primary nucleosynthesis component is added to the primordial value derived from which is matched empirically to the observed abundances at SMC, LMC and solar abundances ([1], [16]) . The ionization parameter q was varied in the range $1 \times 10^7 - 3 \times 10^8$ cm s^{-1}, and the metallicities varied from $0.01 - 3$ solar for PEGASE and $0.05 - 2$ solar for STARBURST99. The metallicity values used are restricted by the stellar tracks used by the population synthesis models.

3.1 HII Region Models

Dopita et al. [5] find that the extragalactic H II region sequence is reproduced remarkably well, provided that the clusters which excite them are all rather young (< 2 Myr). The line ratio usually used for measuring the ionization parameter; [O III] 5007Å/ [O II] 3726,9Å, is indeed a good diagnostic, if somewhat sensitive to metallicity. Amongst other easily observed line ratios we found that the [N II] 6584Å/ [O II] 3726,9Å ratio gives the best diagnostic of abundance, as it is monotonic from 0.2 to over 3.0 times solar metallicity. The physical reasons for this is that nitrogen is a secondary element and its relative abundance increases at high metallicity, and that high metallicity H II regions are cool, so that the [O II] 3726,9Å lines are quenched. At low abundance, this ratio loses its sensitivity, because the N/O abundance ratio can vary wildly as the nucleogenic nature of nitrogen changes from primary to secondary. For these low abundances, it is probably better to use the more traditional R_{23} ([6], [14], [25]). For extragalactic H II regions in typical spiral and Magellanic Irregular galaxies, we find abundances 0.1 – 2.5 solar, ionization parameters in the range $10^7 < q < 10^8$ (or $\mathcal{U}$ in the range $-3.5 < \log \mathcal{U} < -2.5$). Observations of old evolved H II regions close to bright H II regions (for which the abundances can be accurately determined from these models) are likely to provide best discrimination between the rather different stellar atmospheric models and evolutionary tracks used by the stellar spectrum synthesis codes.

3.2 Starburst Models

When star formation continues over several Myr, the ionizing spectrum evolves until a dynamic balance between stellar births and stellar deaths is set up for all initial stellar masses which are important in producing EUV photons. This occurs after 6 - 8 Myr. In [9], we compared the theoretical grids for continuous star formation with the observations of southern warm IRAS galaxies, many of which are mergers or post-mergers, and for roughly 30% of which the line emission caused by an AGN dominated the starburst emission over the central kpc or so. The observations were of a large sample of 285 warm IRAS galaxies covering a wide range of infrared luminosities selected from the catalogue by Strauss [22] and consists of all objects south of Dec. = +10° satisfying the following criteria;

    1. Flux at 60 μm > 2.5 Jy with good detections at 25, 60 and 100 μm.
    2. cz < 8000 km s^{-1}; $\log(L_{FIR}) < 11$ and < 30000 km s^{-1}; $\log(L_{FIR}) > 11$.
    3. Galactic latitude $|b| > 15^0$, and declination $\delta < 0^0$
    4. Warm FIR colours; $8 > F_{60}/F_{25} > 0.5$ and $2 > F_{60}/F_{100} > 0.5$

where F_{25}, F_{60} and F_{100} are the IRAS fluxes at 25, 60 and 100 μm respectively. These selection criteria ensure that that the galaxies in the sample are well-resolved, and that the sample has a large dynamical range in luminosity, so that luminosity dependent effects can, in principle, be investigated.

    The distribution of points on the VO87 diagnostic diagram; [N II] 6584Å/Hα *vs.* [O III] 5007Å/Hβ , can be understood as a result of the "folding" of the abundance : ionization parameter surface of HII regions on these plots. Therefore, HII regions characterized by a rather wide range of abundance or ionization parameters project onto a narrow band located in the vicinity of this fold on these diagnostic plots. It is gratifying to note that the observed points indeed cluster in this area. However, we find that none of the continuous models can exactly reproduce the detailed shape of the fold. We can conclude that the Wolf-Rayet atmospheric models are at fault. The high-metallicity atmospheres need to be characterized by a relatively hard spectral index for energies below about 4 Ryd, but have to be cut off rather sharply above this energy. This problem could be resolved by the inclusion of metal bound-free continuum absorption, as discussed above.

We note that some galaxies optically classified as starbursts may also contain obscured AGN. However, very few galaxies with warm colours are found to have starburst-dominated line ratios in the optical [8]. Since warm colours usually mean that the galaxy is energetically dominated by an AGN we can conclude that the fraction of luminous obscured AGN must be low. Furthermore, we have also demonstrated that an AGN which contributes only 20% to the optical emission changes the line ratios sufficiently to move the starburst into the region where it would be classified as an AGN.

The PEGASE 2.0 grids use the hardest EUV spectrum, and therefore give a theoretical grid which is placed both highest and furthest to the right on the VO87 line ratio diagnostic diagrams. The lines dividing the theoretical starburst region from objects of other types of excitation can be parametrized by the following simple fitting formulae, which have the shape of rectangular hyperbolae;

$$\log\left(\frac{[\text{OIII}]\,\lambda5007}{\text{H}\beta}\right) = \frac{0.61}{\log([\text{NII}]/\text{H}\alpha) - 0.47} + 1.19 \qquad (1)$$

$$\log\left(\frac{[\text{OIII}]\,\lambda5007}{\text{H}\beta}\right) = \frac{0.72}{\log([\text{SII}]\,\lambda\lambda6717,31/\text{H}\alpha) - 0.32} + 1.30 \qquad (2)$$

$$\log\left(\frac{[\text{OIII}]\,\lambda5007}{\text{H}\beta}\right) = \frac{0.73}{\log([\text{OI}]\,\lambda6300/\text{H}\alpha) + 0.59} + 1.33 \qquad (3)$$

The shape and position of this maximum starburst line has not been previously established from theoretical models, and these theoretical upper limits for starburst galaxies have been used in [8] to classify the galaxies in our sample along with an 'extreme' mixing line produced using our shock modelling to classify galaxies into starburst, LINER and AGN types. The strength of our theoretical starburst classification line can be seen by observing the number of galaxies which have 'ambiguous' classifications. These galaxies are those which fall within the starburst region of one or two of the diagnostic diagram and the AGN region of the remaining diagram(s). We found that using our theoretical extreme starburst line, only 6% of our sample has ambiguous classifications compared with 16% ambiguous classifications using the standard VO87 classification lines. These results indicate that our theoretical starburst line is a reliable tool for optically classifying galaxies into starburst and AGN types, and is more consistent from diagram to diagram than the conventional VO87 method.

References

1. E. Anders, N. Grevesse: Geochim. & Cosmochim. Acta **53**, 197 (1989)
2. J.A. Baldwin, M.M. Phillips, R. Terlevich: PASP **93**, 5 (1981)
3. A. Bressan, F. Fagotto, G. Bertelli, C. Chiosi: A&AS **100**, 647 (1993)
4. R.E.S. Clegg, D. Middlemass: MNRAS **228**, 759 (1987)
5. M.A. Dopita, L.J. Kewley, C.A. Heisler, R.S. Sutherland: ApJ **542**, 224 (2000)
6. M.G. Edmunds, B.E.J. Pagel: MNRAS **246**, 678 (1984)
7. M. Fioc, B. Rocca-Volmerange: A&A **326**, 950 (1997)
8. L.J. Kewley, C.A. Heisler, M.A. Dopita, S. Lumsden: ApJS, 2000, in press
9. L.J. Kewley, M.A. Dopita, C.A. Heisler, R.S. Sutherland, J. Tevena: ApJ, 2000, in press
10. R.L. Kurucz: in IAU Symp. 149, *The Stellar Populations of Galaxies*, ed. B. Barbuy & A. Renzini (Dordrecht, Kluwer 1992), p.225
11. C. Leitherer, T.M. Heckman: ApJS **96**, 9 (1995)
12. C. Leitherer, D. Schaerer, J.D. Goldader, R.M. González Delgado, R.C. Kune, D.F de Mello, D. Devost, T.M. Heckman: ApJS **123**, 3 (1999)
13. Th. Lejeune, F. Cuisinier, R. Buser: A&AS **125**, 229 (1997)
14. S.S. McGaugh: ApJ **380**, 140 (1991)
15. D.E. Osterbrock, M.M. de Robertis: PASP **97**, 1129 (1985)
16. S.C. Russell, M.A. Dopita: ApJ **384**, 508 (1992)
17. D. Schaerer: in *Properties of Hot, Luminous Stars*, ASP Conf Ser. v131, ed. D. Howarth, p.310 (1998)
18. G. Schaller, D. Schaerer, G. Meynet, A. Maeder: A&AS **96**, 269 (1992)
19. W. Schmutz, W.-R. Hamann, U. Wessolowski: A&A **210**, 236 (1989)
20. W. Schmutz, C. Leitherer, R.B. Gruenwald: PASP **104**, 1164 (1992)
21. G. Stasińska, D. Schaerer: A&A **322**, 615 (1997)
22. M.A. Strauss, J.P. Huchra, M. Davis, A. Yahil, K.B. Fisher, J. Tonry: ApJS **83**, 29 (1992)
23. R.S. Sutherland, M.A. Dopita: ApJS **88**, 253 (1993)
24. S. Veilleux, D.E. Osterbrock: ApJS **63**, 295 (1987) (VO87)
25. D. Zaritsky, R.G. Kennicutt, J.P. Huchra: ApJ **420**, 87 (1994)

Can We Date Starbursts?

Ariane Lançon

Observatoire de Strasbourg, 11, rue de l'Université, 67000 Strasbourg, France

Abstract. Age dating starbursts is an exercise with many caveats. We attempt to summarise a discussion session that was lead along a rather optimistic guideline: the aim was to highlight that current age estimates, despite undeniable uncertainties, do provide constraints on the physics of starbursts. In many cases, better starburst theories will be needed before the improvement of empirical timelines becomes crucial.

1 Introduction

Many questions can be asked about our ability to trace the history of star formation (SF) in starbursts. The phrasing chosen by the Organizing Committee of this workshop was "Can we date starbursts?". This formulation calls for one of only two answers: *yes*, or *no*... When hearing the question, one automatically recalls one's most recent conversation about the complexity of starburst galaxies or about uncertainties in stellar population synthesis models. Is there any chance for a positive answer? In preparing guidelines for the discussion, we took the optimistic approach of attempting to defend a *yes*. Of course the final answer ended up not being as clear-cut, but some negative intuitions were countered.

Clearly, our degree of satisfaction with starburst age or duration measurements depends on the intended application. The initial question really holds two: how accurately and reliably can we date starbursts? and is that sufficient to make astrophysics progress?

Starburst galaxies are composite objects. The SF may occur both in a diffuse mode and in clusters [22]. The global duration of active episodes can approach 10^9 yrs, while individual starburst clusters are often thought to form instantaneously ($< 10^6$ yrs). To avoid confusion in the meaning of the word "starburst", the following pages deal successively with (i) individual young starburst clusters, (ii) individual intermediate age "post-starburst" clusters that trace starburst activity of the recent past, and (iii) starburst galaxies as a whole. More extensive reviews and references regarding the age dating of stellar populations can be found in [18,11,13] and in this volume.

2 Individual Young Clusters

This section focuses on starburst clusters with ages below 10^7 yrs, as observed in large numbers in the main body of starburst galaxies [6,28,1] or in tidal tails of interacting objects [28,7].

The conditions for cluster age determination are most favourable when the spectroscopic study of individual stars is possible. Until now such studies have been limited to the local neighbourhood of the Milky Way, where many young OB associations exist but massive compact young clusters (as seen in starburst galaxies) are rare/non-existent; 30 Doradus in the LMC and NGC 3603 in the Milky Way are the most relevant accessible targets. Nevertheless, the nearby objects highlight some of the difficulties:

- Samples of cluster stars with spectroscopically confirmed positions in the Hertzsprung-Russell (HR) diagram are small and strongly affected by stochastic fluctuations or spatial variation in the extinction; they are potentially contaminated by field stars.
- Massive star main sequence lifetimes vary between authors by up to $\sim 25\%$.
- Rotation is poorly understood, but rotational velocities above $100\,\mathrm{km/s}$ are the rule in early type stars. Meynet (in [11]) shows that the main sequence lifetime of a massive star may be extended by 20–25 % in case of rotation.
- The proportion of double stars and the effect of binarity on evolution are unknown. Binaries are usually neglected in predictions of frequently used properties such as the number fractions of various types of Wolf-Rayet stars.

In more distant starburst clusters, one integrates the cluster light. The photospheric and wind features in the UV spectrum are considered the most sensitive age indicators and in principle give instantaneous burst ages to within a few Myr [19]. The study of line equivalent widths allows similar formal age accuracies if the light of the whole H II region surrounding the cluster can be summed, the fraction of escaping Lyman continuum photons considered negligible and the continuum contamination by background stars subtracted. The above-mentioned problems associated with rotation, binarity and stellar tracks remain. Charlot (in [18]) for instance points out a delay of about 0.1 dex (25 %) between the appearance of the first red star contributions in two sets of commonly used evolutionary tracks. The risk of stochastic fluctuations between the properties of clusters with identical ages also persists because of fluctuations in the small numbers of very luminous stars. Monte Carlo simulations [4] indicate that these fluctuations contribute less than ~ 1 Myr additional uncertainty to the age estimate as long as clusters more massive than $10^4\,\mathrm{M_\odot}$ are considered.

How the described sources of uncertainties add up or compensate each other is not known. Today, if telescope time is not a limiting factor, a detailed multiwavelength study of a young cluster can be thought to provide an age estimate to better than $\sim 50\%$. Opinions in the workshop audience varied from 30 % (which I would support at least in favourable cases), to a provocative 0.3 dex (which are probably realistic at extreme metallicities or in environments of particularly complex structure).

Can astrophysical questions be addressed with a 50 % accuracy in young starburst cluster ages? Problems of physical interest include SF processes themselves (delay between an external trigger and the onset of SF, formation timescale for massive clusters, propagation of SF within a galaxy) and their effects on the environment (survival times of molecular clouds around starburst clusters, bubble expansion timescales).

Many examples illustrate that spectrophotometric ages, despite the uncertainties, provide interesting constraints. Age spreads of several Myr have been found in OB associations [29,2,25], showing that a unique number does not suffice to describe their age. WC/WN star number ratios indicate that spreads of a few Myr may also be relevant to clusters in starburst environments [26]. Rather complex age structures are seen in NGC 3606 and 30 Dor. In both cases the massive stars of the youngest, 2-3 Myr old component, are concentrated in the central few parsecs and surrounded with significantly older components [9,27]. This situation remains to be convincingly explained by cluster formation models (what are the relative roles of a progressive onset of SF [25], mass segregation [23,12] propagation, merging?). Oey & Massey [24] studied the LH 47/48 and the surrounding superbubble in the LMC, and found a significant disagreement between the stellar ages and the bubble properties predicted from a simple dynamical model, calling for more detailed modelling of the reactions of the ISM. Uncertainties in the identification of external triggers and in *their* onset time dominate in many studies of the initiation of SF in young clusters. Clearly, spectrophotometric dating has been successful in providing other fields of starburst cluster research with new problems.

3   Individual Post-Starburst Clusters

Star clusters with ages between 10^7 and 10^9 yrs are useful to relate current SF activity to potential starbursts of the recent past. They are often found together with the young clusters discussed previously. As they do not ionize their surroundings and have already faded at optical wavelengths, they have not yet been searched for and studied as systematically as their younger counterparts.

Post-starburst clusters are dominated by B then A type stars in the optical/near-UV, by red supergiants (RSG) and then giants of the upper asymptotic giant branch (TP-AGB) in the near-IR. The effects of mixing processes, due e.g. to rotation, appear essential to explain the location of B stars in the HR diagram [20]; Figueras & Blasi [5] use simulations of the Strömgren photometry of stellar populations with reasonable rotation velocity distribution to conclude that photometric ages are affected at the 30–50 % level. More consistent approaches combining the effects of rotation on internal structure and on observable properties have not yet been systematically applied to age studies. Supergiant counts should be used with caution at non-solar metallicities (Z) as the Z-dependance of the blue/red number ratio is not predicted correctly by models [17]. It seems that at $Z_{M31} \simeq Z_{\odot}$ the RSGs have later spectral types but are only produced for $m < 15 \, M_{\odot}$ (age $\sim$12 Myr) as opposed to $m < 25 - 30 \, M_{\odot}$ (age $\sim$7 Myr) at $Z_{NGC6822} \simeq Z_{\odot}/3$ [21]. Modelling the thermal pulses and the Mira-type pulsation along the TP-AGB, in addition to the early AGB, is essential when studying stars in $10^8 - 10^9$ yr old clusters. Number counts that separate C-rich stars from O-rich stars of various subtypes then are potential age-indicators (Lançon & Mouhcine, this volume, and references therein).

For unresolved solar metallicity clusters younger than $\sim 50\,$Myr, well-isolated from the host galaxy background, the UV features give ages to within $\sim 20\,\%$ [3]. Effects of metallicity are uncertain, but empirical calibrations are being attempted (Tremonti, this workshop). The absorption line spectrum (HI and metals) *together* with the energy distribution in the Balmer region gives ages to within $\pm 30\,\%$ [6,10]. Reddening-independent colour-indices in [28] are efficient and could be generalised to include near-IR fluxes. Gilbert (this workshop) showed that, at a given metallicity, near-IR spectra of synthetic clusters with ages of $10 - 25\,$Myr (RSG-dominated) and age differences of $2 - 3\,$Myr can be distinguished and sorted. TP-AGB stars leave potentially useful spectral signatures in integrated spectra of slightly older objects [15]. Stochastic fluctuations in the integrated spectrophotometry, that are dominated by the most luminous red stars, add negligible amounts to the other dating errors as long as the clusters contain more than $10^4\,M_\odot$ of stars [14].

Again, when enough telescope time can be obtained to combine several of the above approaches, ages can be expected to within a conservative $\pm 50\,\%$ (25 % in favourable cases, 0.3 dex for sceptical attendees).

The ages discussed here are comparable to galaxy interaction timescales and more generally to the duration of starburst activity on galaxy scales. Mihos (this workshop) reminded us that the treatment of the transition from a dynamical perturbation to star formation in dynamical models is simplistic; delays of 100 to 500 Myr are found to be typical before onset of starburst activity. Obvious morphological signatures of an interaction fade away over similar timescales; in the case of NGC 4038/39 the spectrophotometric age distribution of the clusters is probably a safer indication of a second encounter than model adjustments to the projected system structure. In NGC 1614 and IC 342 (Rieke, Genzel, this session) starburst knots form a $\sim 0.5\,$kpc nuclear ring, with younger knots (Hα sources, $\leq 6\,$Myr old) located at larger galactic radii than older ones (RSG hosts, $\leq 7\,$Myr old). No dynamical models are as yet available to explain this situation well enough to require improved spectrophotometric ages.

Is the formation of generations of starburst clusters a recurrent phenomenon? When cluster ages become comparable to the dynamical timescales of a galaxy, age differences much shorter than this time cannot be interpreted as separate SF episodes, but rather as one extended one. Therefore a 50 % precision on the age is sufficient to detect potential separate episodes. Then, attempts to the compare properties of the starburst clusters of the current and the previous active phases must deal with a large variety of dynamical effects that rule the survival/destruction of starburst clusters over timescales of $10^8 - 10^9\,$yrs [8]. Uncertainties in those are likely to wash out 50 % age errors.

In this section again, our (biased) approach demonstrates that current age estimates pose challenging astrophysical problems that are far from being resolved to the point of necessitating better timelines.

4   Starburst Galaxies at Low Spatial Resolution

Let us finally question the dating of starburst galaxies observed at a spatial resolution no better than a few 100 pc, or completely unresolved. Partial spatial resolution has obvious advantages but also has some dangers: aperture mismatch between wavelengths, the likelihood that wavelength-dependent photon-exchanges with regions outside the line of sight (through scattering) falsify energy balances, the possibility that average obscuration curves don't apply, etc. The youngest and/or least reddened stellar component is usually dominant at UV wavelengths; but underlying "evolved" populations have been found in all starbursts. Age studies must also aim at determining whether these are part of an extended starburst episode that is still going on, or whether they are remnants of previous, dynamically unrelated star formation.

The nuclear starburst in the interacting spiral galaxy NGC 7714 will be used here for illustration. Integrated photometry is available over the whole electromagnetic spectrum. Extinction is very inhomogeneous and typically $A_v \sim 0.8$. A recent study [16] addresses the photometry and the UV+V+near-IR spectra of the central 300 pc. There, the UV is dominated by a young (~ 5 Myr old) burst, obviously seen through a hole in the dust distribution; the short wavelengths thus contain no information on putative other young populations, including those required to explain the far-IR emission. The broad band photometry can be adjusted satisfactorily with many models: continuous SF over as little as a few 10^7 yrs or as long as $\sim 10^9$ yrs, or a succession of brief bursts. Dust distributions provide more than enough degrees of freedom. More stringent constraints come from spectroscopy : the Hubble Space Telescope UV spectrum favours the presence of at least one instantaneous 5 Myr burst ; the Balmer line region rejects the optical predominance of populations younger than ~ 300 Myr or older than ~ 900 Myr (note that the continuum shape had to be used in addition to the line profiles of the rectified spectrum in order to reach this conclusion); the K band spectrum suggests mixed contributions, as opposed to a population purely dominated by RSG or by TP-AGB stars. The far-IR flux sets a loose upper limit on the amount of heavily obscured young stars, and the reddened Balmer ratio a lower one. The study concludes that starburst activity has been going on with ups and downs over an extended time, and that durations between ~ 300 and ~ 900 Myr are consistent with the data. This is an age to ± 50 %.

The observational constraints on starburst studies can and must still be improved, using available instruments; but on the other hand, more dust configurations and the effects of chemical evolution must be explored systematically, adding even more free parameters. We will thus probably have to bear with ± 50 % estimates for a while.

Is that enough? In the case of NGC 7714, it is at least sufficient to point out an astrophysical problem: the comparison of the system morphology with dynamical simulations indicates that the closest encounter with NGC 7715 occured about 100 Myr ago. The model parameters would allow an increase in the time since interaction by about a factor of 2, but it seems difficult to reconcile this dynamical timescale with the starburst timescales derived from spectrophotometry.

5 Conclusions

Although some workshop participants never accepted age uncertainty estimates below 0.3 dex, we believe that detailed multiwavelength studies, as possible with current instruments (when access to them is not a limiting factor), allow one to reach ±50%, or even better in particularly favourable configurations. The session has allowed many examples to be discussed, and we hope it has conveyed the positive impression that current age determinations, despite their uncertainties, are indeed providing essential constraints on theoretical issues related to starbursts.

References

1. T. Böker: In [13], p. 227 (2000)
2. V. Caloi, A. Cassatella: A&A **330**, 492 (1998)
3. D. de Mello, C. Leitherer, T.M. Heckman: ApJ **530**, 251 (2000)
4. M. Cerviño, V. Luridiana, F.J. Castander: A&A **360**, L5 (2000)
5. F. Figueras, F. Blasi: A&A **329**, 957 (1998)
6. J.S. Gallagher, L.J. Smith: MNRAS **304**, 540 (1999)
7. S.C. Gallagher, S.D. Hunsberger, J.C. Charlton, D. Zaritsky: In [13], p. 247 (2000)
8. O. Gerhard: In [13], p. 12 (2000)
9. E.K. Grebel, W. Brandner, Y.-H. Chu: Bull. AAS **194**, 6801 (1999)
10. R.M. González Delgado, C. Leitherer, T.M. Heckman: ApJS **125**, 489 (1999)
11. I. Hubeny, S.R. Heap, R.H. Cornett (eds.): *Spectrophotometric Dating of Stars and Galaxies*, ASP Conf. Ser. **192** (ASP, San Francisco 1999)
12. P. Kroupa: In [13], p. 233 (2000)
13. A. Lançon, C. Boily (eds.): *Massive Stellar Clusters*, ASP Conf. Ser. **211** (ASP, San Francisco 2000)
14. A. Lançon, M. Mouhcine. In [13], p. 34 (2000)
15. A. Lançon, M. Mouhcine, M. Fioc, D. Silva: A&A **344**, L21 (1999)
16. A. Lançon, J.D. Goldader, C. Leitherer, R. Gonzalez Delgado: ApJ, submitted
17. N. Langer, A. Maeder: A&A **295**, 685 (1995)
18. C. Leitherer, U. Fritze-von Alvensleben, J. Huchra (eds.): *From Stars to Galaxies: the Impact of Stellar Physics on Galaxy Evolution*, ASP Conf. Ser. **98** (ASP, San Francisco 1996)
19. C. Leitherer, D. Schaerer, J.D. Goldader et al.: ApJS **123**, 3 (1999)
20. D.J. Lennon: In [11], p.24 (1999)
21. P. Massey: AJ **501**, 153 (1998)
22. G.R. Meurer, T.M. Heckman, C. Leitherer et al.: AJ **110**, 2665 (1995)
23. G. Meylan: In [13], p. 215 (2000)
24. M.S. Oey, P. Massey: ApJ **452**, 210 (1995)
25. F. Palla, S.W. Stahler: ApJ **540**, 255 (2000)
26. D. Schaerer, T. Contini, D. Kunth: A&A **341**, 399 (1999)
27. F. Selman, J. Melnick, G. Bosch, R. Terlevich: A&A **347**, 532 (1999)
28. B.C. Whitmore, Q. Zhang, C. Leitherer et al.: AJ **118**, 1551 (1999)
29. J.-M. Will, D.J. Bomans, A. Vallenari, J.H.K. Schmidt, K.S. de Boer: A&A **315**, 125 (1996)

Investigating ULIRGs in the Near-Infrared: Imaging and Spectroscopy

Ric Davies[1], Amanda Burston[2], and Martin Ward[2]

[1] Max-Planck-Institut für extraterrestrische Physik, 85741 Garching, Germany
[2] X-ray Astronomy Group, University of Leicester, Leicester, LE1 7RH, UK

Abstract. We present imaging and spectroscopic observations of two nearby ($z <$ 0.1) ULIRGs from a larger sample, and address the question of whether the JHK continuum colours and slope might be effective probes of the nuclear region in searches for AGN. Certainly there is evidence for significant quantities of hot dust emission at temperatures $\gtrsim 1000$ K; but it may be that rather than pointing to an AGN, this instead tells us more about the environment and evolution of the star formation.

1 Introduction

Studies of ULIRGs at different wavelengths face a variety of challenges. These include attenuation at shorter wavelengths (soft X-ray to optical) due to high extinction, and poor spatial resolution at longer wavelengths (mid- to far-infrared). A good compromise can be found in the near-infrared.

Although there have been a number of JHK imaging studies of ULIRGs [9,3], colours have tended to be examined in large apertures for which dilution from extended emission can be significant. We present data on 2 objects from a larger sample, with a resolution $<0.5''$ allowing us to probe much closer into the nucleus. Furthermore, combining imaging data with spectroscopy provides much tighter constraints on the origin of the continuum, which cannot be achieved with K-band spectra alone.

2 IRAS 23365 +3604

Figure 1 shows the K-band image of IRAS 23365, which displays signs of interaction although with no obvious nearby candidates, at a resolution of 0.40''. The colours in incremental annuli out to 5'' are drawn in the JH-HK colour diagram. The nucleus (classified as a LINER [15] or composite [1]) is reddest; colours at larger radii become bluer until beyond a radius of ~1.5'' they stabilise. For comparison the diagram shows at the redshift of the galaxy, the colours of stellar populations from [6] for instantaneous and continuous star formation, as well as the effect of extinction and hot dust.

We have fitted models to the nuclear colours (in a 1'' aperture), consisting of a reddened 100 Myr old stellar population and a hot dust component. Except in the extreme case of a very young starburst <2.5 Myr old – which is ruled out by the Brγ equivalent width – the nuclear colour cannot be reproduced by reddened stellar light. Therefore including hot dust (which contributes

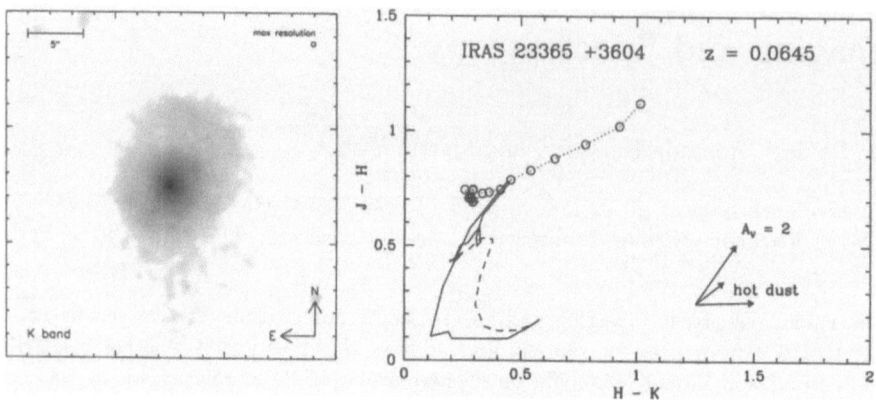

Fig. 1. Left: K-band image of IRAS 23365 with a resolution of 0.5″. Right: JHK colour diagram showing the colours in annuli of increasing radius; the nucleus is reddest. Also shown are colours of instantaneous (solid line) and continuous (dashed line) star formation from Starburst 99 [6], and the effects of extinction and dust emission at 500 K and 1500 K.

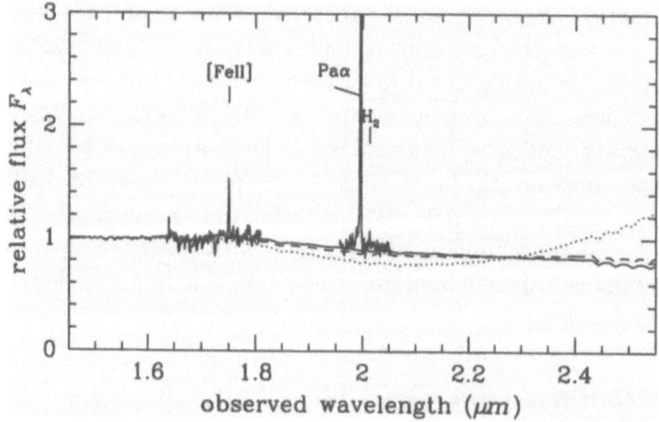

Fig. 2. Spectrum of IRAS 23365 extracted in a 0.58″×0.58″ aperture, kindly provided by R. Genzel [5]. Overplotted are the 3 models (at low spectral resolution) derived from the JHK colours above for dust emission at 1500 K (solid line), 1000 K (dashed line), and 500 K (dotted line). These show that dust emission below ~1000 K has a characteristic shape which is inconsistent with the data.

a fraction f_{Kd} of the K-band emission) is mandatory. But since its temperature cannot be constrained from JHK colour data alone, we consider three particular cases: $T_D=1500$ K, close to the sublimation temperature of grains is an upper limit, 1000 K is intermediate, while 500 K represents the coolest dust that can still significantly affect the K-band. With these we find that for

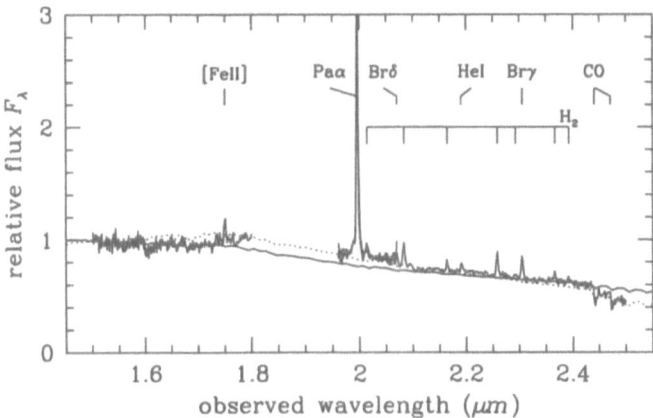

Fig. 3. Spectrum of IRAS 23365 extracted in a $0.6'' \times 3.0''$ aperture, from [2]. Overplotted are models derived from the JHK colours in a $3''$ aperture for: solid line – a reddened 100 Myr starburst and hot dust emission at 1500 K. Dotted line – a highly reddened starburst with no extra dust emission.

$T_D = 1500$ K, $f_{Kd} = 0.59$ and $A_V = 2.85$; for $T_D = 1000$ K, $f_{Kd} = 0.49$ and $A_V = 3.77$; and for $T_D = 500$ K, $f_{Kd} = 0.50$ and $A_V = 4.36$.

The spectrum associated with each of these models is overplotted on a spectrum of the galaxy nucleus [5] in Fig. 2. They show that the model matches the spectrum extremely well, and constrains the temperature of the dust emssion to be $\gtrsim 1000$ K. Another spectrum of IRAS 23365 in Fig. 3, discussed in more detail in [2], was extracted in a larger aperture and covers a longer wavelength range. The line emission shows none of the typical signs of an AGN: the Paα and Brγ lines are narrow (300–500 km s^{-1}), and there is no detectable coronal [Si VI]. Here we have overplotted models similar to those above, but derived from colours extracted in $3''$ apertures; the difference from the previous figure emphasises the importance of using consistent apertures when comparing different data. The model shown (solid line) uses $T_D = 1500$ K and matches the shape of the continuum quite well. To evaluate whether a model with no dust emission is feasible, we also show a pure stellar (and nebula) continuum with $A_V = 6$ (dotted line). Although this provides a reasonable match to the spectra, it can be ruled out because its JH colour is too red by 0.32 mags. Thus in order to replicate both the JHK colours and the HK spectral shape, we find that a contribution to the K-band emission from dust at $T_D \gtrsim 1000$ K is required.

3   IRAS 20210 +1121

In the K-band image in Fig. 4 (resolution $0.48''$), IRAS 20210 has two nuclei with a ridge of emission between them. The JH-HK diagram shows that the colours of annuli centered on the north nucleus (squares) are consistent with a stellar population reddened by $A_V \sim 1$. The circles denote the colours centered on the south

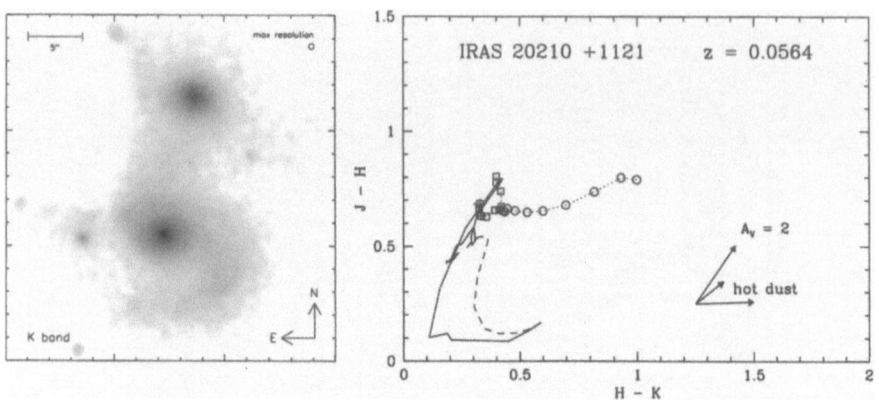

Fig. 4. Left: K-band image of IRAS 20210 at a resolution of 0.48″. Right: JHK colour diagram of the two nuclei (north, squares; south, circles) in incremental annuli. Shown are the locus of star formation models, and the effect of extinction and dust emission at 500 K and 1500 K.

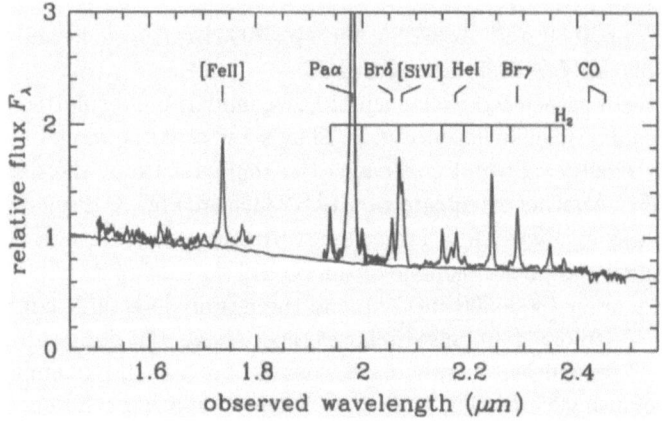

Fig. 5. Spectrum of IRAS 20210 extracted in a 0.6″×3.0″ aperture, from [2]. Overplotted is a model derived from the nuclear JHK colours for a reddened starburst and hot dust emission at 1500 K. It provides a remarkably good fit to the continuum shape.

nucleus (classified as a Seyfert 2 [8,14]). As before, at radii larger than ∼1.5″ these are typical of a moderately reddened stellar population, but for the nucleus itself hot dust is required. Fitting models to the colours of the south nucleus in a 1″ aperture yields: for T_D=1500 K, f_{Kd}=0.70 and A_V=0.04; for T_D=1000 K, f_{Kd}=0.60 and A_V=1.38; and for T_D=500 K, f_{Kd}=0.60 and A_V=2.20.

Although only 2 examples are presented in this contribution, we have looked at 8 nuclei in 6 ULIRGs, the analysis of which is given in [4]. For the sample as a whole, varying the age or metallicity of the stellar population does not affect the results much because, except for ages less than 10–20 Myr, the JHK colours hardly change. It is although – and because – the stellar population cannot be constrained, that the result we find here about significant hot dust emission being mandatory is robust. Our results are similar to those found by [13].

Figure 5 shows spectra for the southern nucleus of IRAS 20210 (see [2]), which do exhibit typical signs of an AGN: the Paα is broad as expected for a narrow line region ($\sim$700 km s^{-1}), and there is easily detectable [Si VI] – which is stronger than H$_2$ 1-0 S(1) and even twice as strong as Brγ. A model derived from the broadband colours as before (but in a 2'' aperture) is over-drawn. It matches the continuum extremely well through both the H- and K-bands. As for IRAS 23365 we find that not only is a dust component required but that we can now constrain the temperature to be close to the top end of the permissible range, $T_D \gtrsim 1000$ K.

4 Starburst or AGN ?

An important question is whether the presence of dust at $\gtrsim 1000$ K implies an AGN exists. Typically, heating dust to its sublimation temperature requires a very intense UV radiation field that only occurs within a few parsecs of an AGN, at the inner edge of a putative torus. Indeed, ISO spectroscopy of classic 'template' starburst and AGN galaxies [12,7] suggests that it is only AGN which have spectral shapes indicative of dust emission at these temperatures. However, the ISO-SWS 2.4–12 μm aperture is large enough to include a significant fraction of the more extended emission from a galaxy bulge or disk. So late-type stars could easily dominate the emission observed by ISO and hide any relatively weaker hot dust emission that might originate only in younger star clusters.

Evidence that stellar processes can produce dust hotter than 1000 K arises from 3 reflection nebulae that were observed to have 2–5 μm continua charac-terised by a colour temperature of $\sim$1000 K [10]. This is explained in terms of stochastic heating of very small grains of radius 5–10Å [11], with the result that the fraction of the dust mass in such grains is 0.002, and the fraction of stellar luminosity absorbed and re-radiated by them is 0.004 of that absorbed and re-radiated by all grains. A simple calculation comparing the luminosities for dust emission at 2.2 μm and 60 μm yields ratios of $\sim$0.01, consistent with this model.

A possible problem with applying this model to ULIRGs is that the reflec-tion nebulae contain non-ionising stars; whereas ULIRGs must host significant numbers of ionising stars, from which the UV radiation could destroy very small grains and the thermal nebula emission could mask small grain emission. How-ever, one plausible hypothesis for the star formation history in ULIRGs is that individual star clusters (or groups of clusters) have formed in multiple episodes spread over a timescale of several 100 Myr – effectively continuous star formation. In such a scheme only the youngest clusters, in which star formation occured within the last 10 Myr, host ionising stars. Most of the far-infrared luminosity still originates in these clusters through thermal heating of dust grains; but the bulk of the starburst population resides in older clusters without HII regions, and could have an observable small grain population. The 1000Å luminosity of these 10–100 Myr clusters is an order of magnitude greater than that at $2\,\mu$m, so there are enough energetic photons to heat the small grains and hence account for the equal contributions from dust and stars to the $2\,\mu$m emission. One could also speculate that in the turbulent environment expected in the nuclear region of a ULIRG, grain destruction through shocks – via the merging process as well as the high supernova rate – cause an over-density of very small grains. Thus it is certainly possible that in a ULIRG the necessary prerequisits are met so that stochastic excitation of small grains becomes an observable phenomenon.

5   Conclusions

As part of a larger study, we present an analysis which combines JHK imaging and HK spectroscopy of 2 ULIRGs, one that harbours a Seyfert 2 nucleus and one which shows none of the typical signs of an AGN. The crucial result is that models of reddened star formation cannot explain the JHK colours of the nuclei; additional hot dust emission is needed. Although the temperature is unconstrained by colours alone, comparison to the spectra show it is $\gtrsim 1000$ K. Hot dust emission in the K-band is usually associated with the inner edge of a torus around an AGN, but we have presented evidence that in ULIRGs it could also arise through stochastic heating of very small grains by non-ionising stars.

Acknowledgements: RID thanks the many people at the meeting, particularly J. Graham and M. Lehnert, who provided helpful input to this work. Some of the data here were obtained as part of the UKIRT Service Programme, and we thank all those who were involved. UKIRT is operated by the JAC on behalf of the U.K. PPARC.

References

1. W. Baan, J. Salzer, R. Lewinter: ApJ **509**, 633 (1998)
2. A. Burston, M. Ward, R. Davies: MNRAS, 2001, submitted
3. D. Carico, et al.: ApJL **349**, 39 (1990)
4. R. Davies, A. Burston, M. Ward: MNRAS, 2001, submitted
5. R. Genzel, L. Tacconi, D. Rigopoulou: 2001, in preparation
6. C. Leitherer, et al.: ApJS **123**, 3 (1999)
7. D. Lutz: New Astronomy Reviews **44**, 241 (2000)
8. E. Perez, A. Manchado, P. Garcia-Lario, S. Pottasch: A&A **227**, 407 (1990)
9. D. Sanders, et al.: ApJ **325**, 74 (1988)
10. K. Sellgren, M. Werner, H. Dinerstein: ApJ **271**, L13 (1983)
11. K. Sellgren: ApJ **277**, 623 (1984)
12. E. Sturm, et al.: A&A **358**, 481 (2000)
13. J. Surace, D. Sanders, A. Evans: ApJ **529**, 170 (2000)
14. J. Vader, J. Frogel, D. Terndrup, C. Heisler: AJ **106**, 1743 (1993)
15. S. Veilleux, D.-C. Kim, D. Sanders, J. Mazzarella, B. Soifer: ApJS **98**, 171 (1995)

Post-Starburst Populations Near and Far –
The Potential of Near-IR Spectroscopy

Ariane Lançon and Mustapha Mouhcine

Observatoire de Strasbourg (UMR 7550), 11, rue de l'Université, 67000 Strasbourg, France

Abstract. The efficient use of near-IR data in studies of external stellar populations depends on our ability to recognize the nature of the predominant sources of light, and to interpret these findings in terms of age and metallicity. Here we focus on elderly post-starburst populations, with ages of $10^8 - 10^9$ yrs. New models confirm that they are indeed expected to display specific spectral signatures in the near-IR, due to variable M stars of the asymptotic giant branch and to carbon stars. The signatures depend on age and metallicity. We summarize the status of current quantitative predictions and emphasize the importance of an empirical calibration of the spectral synthesis models.

1 Overview

Near-IR data has been used very often to confirm the presence of an "evolved stellar component" in starburst galaxies. The need to study stellar populations deeply embedded in dust is also frequently invoked to justify near-IR observations. But what exactly are these cool stellar populations seen between 1 and $3\,\mu$m? Population synthesis models tell us that red supergiants are the main sources $10^7 - 10^8$ yrs after a star formation episode; during the following 10^9 yrs, stars of the upper, thermally pulsing asymptotic giant branch (TP-AGB) become predominant; and finally giants of the early AGB (E-AGB) and of the first red giant branch (RGB) outweigh other populations. The efficient use of near-IR data depends on our ability to distinguish these categories from each other and to interpret them in terms of age and metallicity.

Very strong CO absorption ($2.3\,\mu$m) is a clear signature of a red supergiant population [1] initiated the systematic exploration of CO in cool stars; [18] analysed early observations in starbursts; [8] discussed the diagnostic power of the feature and its limitations). But the dependence of the red supergiant population on metallicity is strong [13] and not reproduced by stellar evolution models unless ad hoc corrections are invoked [11,16]. Therefore the threshold above which the CO absorption becomes an unambiguous signature of a supergiant population is uncertain. On the other hand, it has recently been claimed that high quality, high spectral resolution near-IR observations allow for *relative* age-dating of young post-starburst populations to within $2 - 3\,$Myr, at least at solar metallicity (Gilbert, this workshop).

The focus of our work is on somewhat older post-starburst populations, such as we are likely to find in evolved mergers or "E+A" galaxies, i.e. after the most obvious signs of a recent starburst have faded away. Between 10^8 and 10^9 yrs

of age, of the order of 50 % of the K band light of a coeval population originates from TP-AGB stars. Large stellar samples show that essentially all the
TP-AGB stars are long period variables [20]. Spectral libraries as well as model
atmospheres demonstrate that this pulsation results in deeper H_2O absorption
bands than seen in E-AGB or RGB stars. This provides us with a means of identifying strong contributions of oxygen-rich (e.g. M type) TP-AGB stars. In addition, the convective dredge-up of carbon-rich material from the stellar core into
the envelope progressively turns some TP-AGB stars into carbon stars. Those
again have unmistakable spectral signatures in the near-IR. Finally, TP-AGB
stars may become dust enshrouded mid-IR sources, and it has been suggested
that the presence of these modifies the location of stellar populations in near-
IR versus mid-IR two-colour diagrams [3]. All the mentioned processes depend
sensitively on the initial stellar mass (and thus on post-starburst age) as well as
on metallicity. This was the motivation for our work. We summarize its status
below.

2   Model Construction and Predictions

Our population synthesis models are based on the code structure of PÉGASE [5],
the stellar tracks of the "Padova Group" [2,4] up to the end of the E-AGB, and
the spectral library of Lejeune et al. [12]. We have completed this set of inputs
with synthetic evolutionary tracks for the TP-AGB that incorporate the effects
of mass loss, dredge-up and envelope burning of dredged-up materials [14,15];
we also use a new library of cool stellar spectra [10].

Various global properties can be determined without invoking spectral libraries, on the basis of the TP-AGB tracks alone [15]. The values given here
are based on our currently prefered internal model parameters (e.g. mass loss
prescription, convective mixing length, dredge-up efficiency), which have been
tested against statistical properties in nearby stellar populations (see below).
The stars with the longest TP-AGB durations (and thus the largest luminous
energy output, integrated along the TP-AGB) are found to have initial masses
(M_i) between 2 and $2.5 \, M_\odot$ (i.e. ages of 0.7 to 1.1 Gyr); stars with M_i in the
$1.7 - 3 \, M_\odot$ range last $\sim 10^6$ yrs or more in that phase. Mass loss has the strongest
direct influence on these durations, and itself depends on various other model
parameters through the instantaneous masses, luminosities and radii. As the
TP-AGB stars are the coolest luminous stars present, the duration of that phase
essentially controls the (V−K) colour of the stellar population [6]. The fraction
of the TP-AGB duration spent as a carbon star is found to be maximal for objects with initial masses between 2.5 and $3 \, M_\odot$ (i.e. ages of 0.4 to 0.8 Gyr), where
it typically amounts 45 % at solar metallicity and about 80 % at the metallicity
of the Large Magellanic Cloud (Fig. 1). Due to strong mass loss, stars above
$M_i = 3.5 \, M_\odot$ have relatively short TP-AGB durations ($< 5 \times 10^5$ yrs) and are
embedded in dust for a significant fraction of this time. This reduces the predicted contribution of TP-AGB stars to the light at ages of $0.1 - 0.4$ Gyr, when
compared to our earlier models [9].

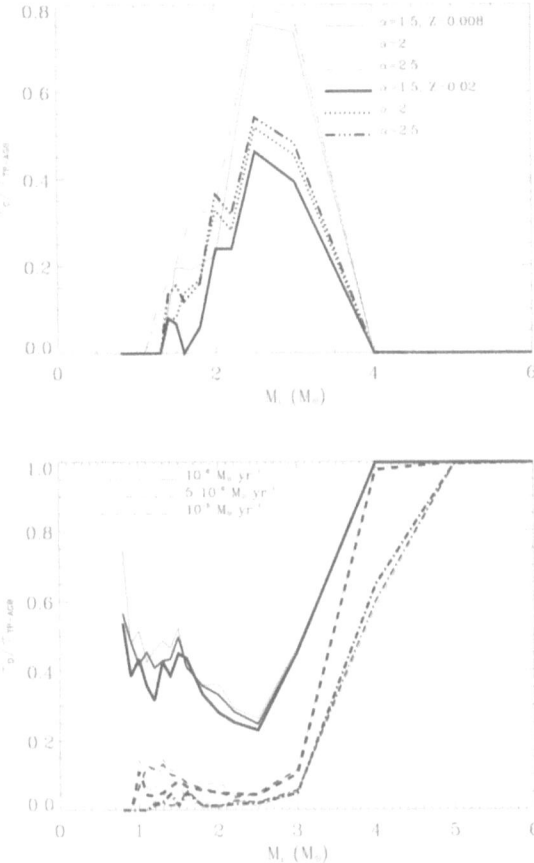

Fig. 1. *Top*: Fraction of the TP-AGB duration spent as a C-star, as a function of initial stellar mass. Two metallicities and various values of the mixing length parameter are shown. *Bottom*: Fraction of the TP-AGB duration spent as a star with mass loss above 10^{-6}, 5×10^{-6} and $10^{-5}\,M_\odot/\mathrm{yr}$, at solar metallicity.

As mentioned, our trust in the model prediction is based on a variety of tests. For instance, the new evolutionary models for the TP-AGB successfully reproduce the empirical initial to final mass relation for stars in the solar neighbourhood. The predicted contributions of C-stars to the near-IR light of intermediate age populations are consistent with those observed in the Magellanic Clouds (e.g. [17]). One of us (M.M.) has coupled the TP-AGB models with chemical evolution models for galaxies (including the yields of intermediate mass stars as given by the new tracks). The tight relations observed in the Local Group galaxies between metallicity and the mean luminosity of C-stars, or between metallicity and the C-star/M-star number ratio can be reproduced very naturally. These results will be discussed in detail elsewhere.

Stellar libraries are required for the prediction of near-IR colours or spectral features. Based on the extremely diverse AGB star spectra of [10] ($\lambda/\delta\lambda \simeq$ 1100 between 1 and 2.5 μm), we have constructed a regular sequence of averages that can be more conveniently coupled with population synthesis models [7]. As a result, we predict that the signatures of upper AGB stars are seen in the integrated light of a stellar populations between ~ 0.2 and ~ 1.2 Gyr after a starburst [15,9]. The main signatures are extremely broad, and best detected with low resolution data with very wide spectral coverage (the K band spectrum alone is not sufficient). As the features reach across telluric absorption bands, observations at good IR sites are essential. The features are due to H_2O if the TP-AGB stars are O-rich, to C_2 and CN if they are C-rich. Narrow band indices can be defined that detect TP-AGB stars independently of their chemical nature, and others that then separate the two classes [9]. Photometry with an accuracy better than 5 % must be aimed for. We are exploring the feasibility of surveys with the available filters on the NICMOS camera onboard the Hubble Space Telescope.

3   The Need for Empirical Calibration

Once TP-AGB evolutionary tracks have been constructed and tested, only a part of the work is done: associating a spectrum with a given point of the theoretical HR diagram is difficult, as illustrated in Fig. 2. Individual stars vary significantly in luminosity and colour temperature. While optical colours and temperature indicators such as (V−K) correlate well, the near-IR properties (colours, strength of the H_2O or CO bands) are very dispersed at a given colour temperature. In addition, no model atmospheres are available to reliably relate an instantaneous colour temperature to the effective temperature of the simple static stars

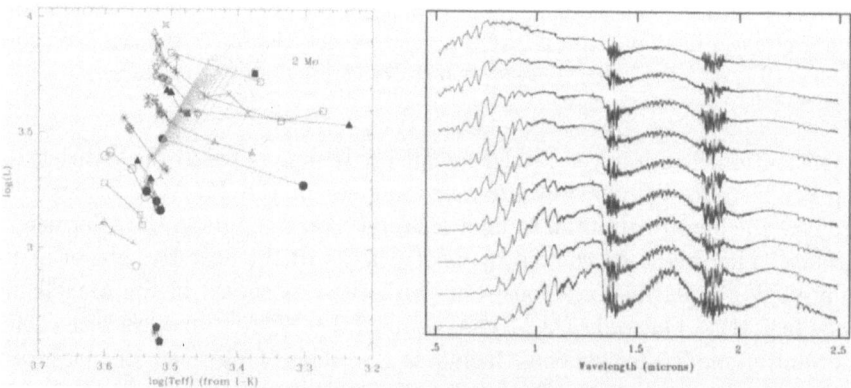

Fig. 2. *Left*: Estimated location of individual spectra of [10] in the HR diagram. Multiple observations of individual stars are connected with lines. An illustrative TP-AGB track for a 2 $M_\odot$ star is also shown. *Right*: Empirical sequence of average O-rich TP-AGB star spectra [7].

on which evolutionary tracks are based. The regularity of the spectral sequence obtained by [7] gives some confidence in the evolution of the spectra with temperature; but only empirical calibrations, based on stellar populations with ages and metallicities known from optical observations, can constrain absolute relationships.

Despite the more numerous near-IR instruments now available on large telescopes, appropriate targets for the model calibration remain hard to find. The LMC and M33 clusters, of which several have adequate ages, tend to be too small: they contain a handful of TP-AGB stars and their integrated properties are affected by the stochastic nature of this subpopulation. In [10], it was shown that stellar populations with total masses of $\sim 10^5 \, M_\odot$ are needed to obtain significant constraints on the TP-AGB models. Massive stellar clusters around merging galaxies are well suited, but only a few are bright enough for near-IR spectroscopy. We have observed the 500 Myr old cluster W3 in NGC 7252 with SOFI (NTT, ESO), between 1 and 2.4 μm. Illustrative adjustments with models are shown in Fig. 3. The figure illustrates the need for broad wavelength coverage and excellent relative calibrations of the various atmospheric windows (with SOFI the overlap between grisms ensures this). Our preliminary conclusions are the following. The data reject models that have few C-stars *and* use a cool effective temperature scale for the TP-AGB spectra: the water bands in those are too strong. This suggests that the metallicity is not more than solar and that the coolest of the Mira temperature scales in the literature is inappropriate for our purpose. Good fits to the spectral shape are obtained at solar and LMC metallicities, with extinction values compatible with optical estimates [19]. More calibration points will be required, but this first test already restricts the allowed range of parameters and gives us confidence in the new tool.

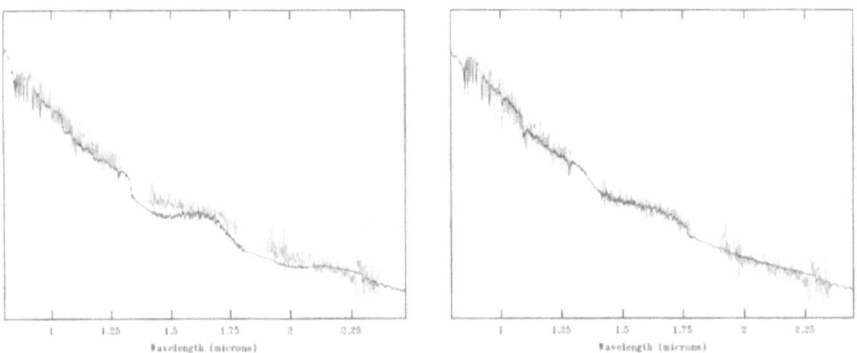

Fig. 3. The observed spectrum of W3 and two illustrative models. *Left*: solar metallicity, formation of C-stars switched off, cool TP-AGB temperature scale. This model is ruled out. *Right*: LMC metallicity, formation of C-stars allowed. This model matches the data well.

4   Conclusions

Based on new synthetic evolution models for the TP-AGB and on a purpose-designed library of stellar spectra, we have shown that near-IR spectra of post-starburst populations indeed carry information about post-starburst age and metallicity. Elderly post-starbursts with ages of 10^8 to 10^9 yrs can be distinguished from younger or older ones, using the specific spectral signatures of O-rich and C-rich TP-AGB stars. The chemical nature of the predominant TP-AGB stars is a metallicity indicator. The population of dust-obscured TP-AGB stars depends both on age and metallicity. We have pointed out that the relevant spectral features can be detected at low spectral resolution as long as a broad continuous spectral coverage is obtained. High quality narrow band photometry can also be used. The quantitative prediction of the strength of the near-IR features must be based on empirical calibrations. We have described the current status of these.

Applications of this work include studies of the duration of star formation episodes in starburst galaxies or mergers, of the survival timescale of young starburst clusters in the galaxy environment, and of the nature of "E+A" galaxies. First observations of "E+A" galaxies will be available to us soon.

References

1. J.R. Baldwin, J.A. Frogel, S.E. Persson: ApJ **184**, 427 (1973)
2. A. Bressan, F. Fagotto, G. Bertelli, C. Chiosi: A&AS **100**, 647 (1993)
3. A. Bressan, G.L. Granato, L. Silva: A&A **332**, 135 (1998)
4. F. Fagotto, A. Bressan, G. Bertelli, C. Chiosi: A&AS **105**, 29 (1994)
5. M. Fioc, B. Rocca-Volmerange: A&A **326**, 950 (1997)
6. L. Girardi, G. Bertelli: MNRAS **300**, 533 (1998)
7. A. Lançon, M. Mouhcine: In preparation (2001)
8. A. Lançon, B. Rocca-Volmerange: New Astron. **1**, 215 (1996)
9. A. Lançon, M. Mouhcine, M. Fioc, D. Silva: A&A **344**, L21 (1999)
10. A. Lançon, P.G. Wood: A&AS **146**, 217 (2000)
11. N. Langer, A. Maeder: A&A **295**, 685 (1995)
12. T. Lejeune, F. Cuisinier, R. Buser: A&AS **125**, 229 (1997)
13. P. Massey: AJ **501**, 153 (1998)
14. M. Mouhcine, A. Lançon: In *Massive Stellar Clusters*, ed. by A. Lançon, C. Boily (ASP, San Francisco 2000), ASP Conf. Ser. **211**, p. 144
15. M. Mouhcine, A. Lançon: In preparation (2001)
16. L. Origlia, J.D. Goldader, C. Leitherer, D. Schaerer, E. Oliva: ApJ **514**, 96 (1999)
17. S.E. Persson, M. Aaronson, J.G. Cohen, J.A. Frogel, K. Matthews: ApJ **266**, 105 (1983)
18. G.H. Rieke, M.J. Lebofsky, R.I. Thompson, F.J. Low, A.T. Tokunaga: ApJ **238**, 24 (1980)
19. F. Schweizer, P. Seitzer: AJ **116**, 2206 (1998)
20. P.G. Wood: In *AGB stars*, IAU Symp. 192.

Optical Spectra of Dusty Starbursts

Bianca M. Poggianti

Osservatorio Astronomico, 35122 Padova, Italy

Abstract. This contribution presents the optical spectral properties of FIR-luminous galaxies, whose distinctive feature is often the simultaneous presence in the spectra of a strong Hδ line in absorption and of emission lines (e(a) spectra). A discrepancy between the star formation rate estimated from the FIR luminosity and that derived from the Hα luminosity persists even after having corrected the Hα flux for dust according to the observed Balmer decrement. It is shown that the e(a) spectrum can be reproduced assuming a current starburst and a dust extinction affecting the youngest stellar populations much more than the older stars.

1 Introduction

is not the equivalent width of the emission lines[1], but the equivalent width of the Hδ line in absorption, which is generally stronger than in the spectra of optically selected galaxies. In fact, emission line spectra with EW(Hδ) > 4 Å (e(a) spectra) are frequent among FIR galaxies, while the great majority of nearby spirals in optical samples have EW(Hδ) < 4 Å (e(c) spectra) [5,6]. The e(a) spectral class differs from the so-called "k+a" (or "E+A") galaxies which only display a strong Hδ line in absorption, but no emission lines.

What is the origin of the difference in the Hδ strength between dusty starburst galaxies and quiescent spirals? What star formation and dust properties generate the peculiar spectral combination found in e(a) galaxies? In the following I will address these issues presenting the analysis of a spectroscopic survey of luminous IRAS galaxies [6] and the results of an effort to model the e(a) spectrum in detail [7].

2 FIR-Luminous Galaxies: Spectral Features and Star Formation Properties

In [6] we analyzed the spectral characteristics of a complete sample of IRAS galaxies from Wu et al. [9,10] (W98) comprising 73 Very Luminous Infrared galaxies (VLIRGs, $log(L_{IR}/L_{\odot}) > 11.5$) with a median $log(L_{IR}/L_{\odot}) = 11.72$ and 40 companion galaxies, selected from the 2 Jy redshift survey of Strauss et al. [8]. The great majority of these galaxies show evidence for a strong interaction or a merger [10]. The spectra typically cover the central 2 kpc of the galaxies at a $\sim$ 10 Å resolution.

---

[1] These are usually lower than those in HII galaxies and UV-bright starbursts.

Table 1. Fraction of galaxies and E(B-V) as a function of the spectral class.

Class	fraction	Median E(B-V)	Comments (see PW00 for details)
e(a)	0.56±0.10	1.11	Strong Balmer absorption plus emission
e(c)	0.25±0.07	0.68	Weak/moderate Balmer absorption plus emission
e(b)	0.10±0.04	0.62	Strong emission (EW([OII])> 40 Å)
sey1	0.10±0.04	–	Seyfert 1 from broad hydrogen lines in emission

Table 1 shows the fraction of VLIRGs as a function of the spectral class; a detailed description of each class can be found in [6]. More than half of these galaxies have an e(a) spectrum, about 1 out of 4 has an e(c) spectrum, and only 1 out of 10 has very strong [OII] emission (e(b) type). None of the e(a) spectra would be classified as a Seyfert 1 or 2 according to the standard diagnostic diagrams of line intensity ratios. Notably no k+a spectrum is found among the FIR luminous galaxies, neither among their companions: all the VLIRGs display emission lines in their spectra.

A general characteristic of the e(a) spectra in all environments and at all redshifts appears to be the low [OII]/Hα ratio (see Fig. 1): both the equivalent width and the flux ratios of these lines are a factor of 2 lower than the median ratios observed in nearby spirals. Though such low ratios may be caused by various reasons, the most likely explanation is reddening by dust: the color excess E(B-V) derived from the observed Balmer decrement (see Table 1) is consistent with – and fully accounts for – the observed [OII]/Hα values. It is noteworthy that the difference in the *equivalent width* ratio between this sample and optically selected nearby spirals is entirely due to the difference in the average *flux* ratio of these two lines. This can be interpreted as an effect of *selective dust extinction* which affects the youngest stellar generations (still deeply embedded in large amounts of dust and responsible for the ionizing photons producing the emission lines) much more than older, less extincted stellar populations producing the continuum underlying the lines: the net result is a low EW ratio. Additional evidence for a selective extinction will be discussed in Sec. 3.

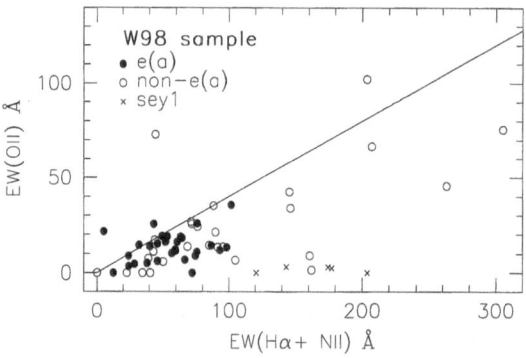

Fig. 1. The line represents the fit for normal field galaxies in the local Universe (EW([OII])=0.4 EW(Hα + NII)).

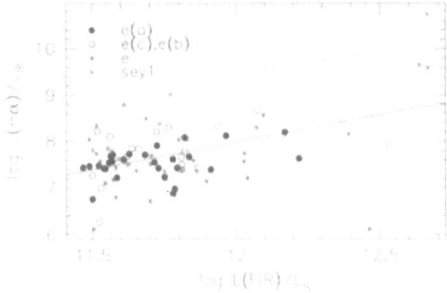

Fig. 2. FIR and Hα luminosities in solar units. No dust correction has been applied. The best fit to the datapoints is shown as a solid line. The relation found by Devereux & Young [2] for a sample of field spiral galaxies in the local Universe is extrapolated to the FIR luminosities of the present dataset and is shown as a dotted line. "e" spectra are those with at least one detected emission line, but Hδ unmeasurable.

As in all FIR luminous samples, there is a deficiency of Hα luminosity at a given FIR luminosity as compared to optically selected spirals (Fig. 2) and this scarcity of Hα flux translates directly into a difference in the SFR estimate: even applying to the Hα flux the dust correction derived from the Balmer decrement is not sufficient to reconcile the two estimates of current star formation obtained from the FIR and Hα luminosities (Fig. 3). After dust correction, the SFR$_{H\alpha}$ is still a factor of ~ 3 lower than the SFR$_{FIR}$; it is unlikely this discrepancy is entirely due to the limited aperture of the spectroscopic slit and I will come back to this point in Sec. 3.

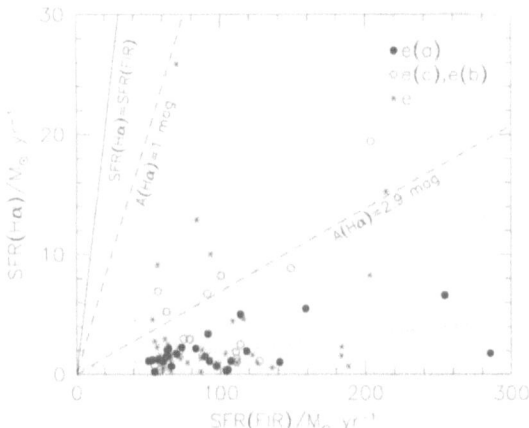

Fig. 3. The SFR derived from the observed Hα flux versus the FIR–based estimate. The dotted line is the fit to the e(a) population. The solid line shows the relation $SFR_{H\alpha} = SFR_{FIR}$, and the dashed lines are found for 1 mag extinction at Hα (average extinction in nearby spirals) and for the average extinction in e(a)s in the W98 sample as determined by the Balmer decrement (E(B-V)=1.1, A(Hα) = 2.9 mag).

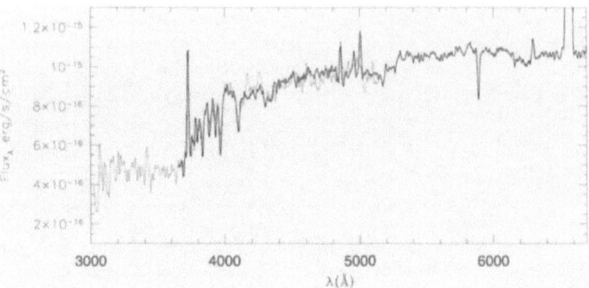

Fig. 4. A rest-frame comparison of the average e(a) spectrum of Very Luminous Infrared galaxies (thick line) and the spectrum of the ISO starburst galaxies at $< z > \sim 0.5$ (thin line). The latter is taken from Fig. 12 in Flores et al. [3] and has been normalized to the VLIRGs spectrum over the wavelength range in common. This is the average spectrum of 5 ISO-detected galaxies at $z < 0.7$ whose spectral energy distributions at visible, near-IR, MIR and radio wavelengths resemble those of highly reddened starbursts in the local universe. The spectral resolution is 10 Å (Wu) and 40 Å (Flores).

The analysis presented so far refers to a well defined sample of IRAS luminous galaxies, but the e(a) phenomenon appears to be widespread among IR luminous galaxies at any redshift (see [6] for a census of the e(a) occurrence). As an example, Fig. 4 shows that the average e(a) spectrum of the W98 sample is very similar to the average spectrum of distant starburst galaxies detected at 15 micron by ISO, both in the continuum shape and in the strength of the [OII], Hδ and Hβ lines.

3   Physical Origin: Modelling e(a) Spectra

A selective dust extinction has been proposed as the physical origin of the e(a) spectra [5,6]: dust obscuration affects the youngest stellar generations more than the older stars. This is expected to explain both the observed [OII]/Hα ratio, as discussed in the previous section, and the Hδ strength because the stellar populations responsible for this line (with ages a few times 10^7–$1.5 \cdot 10^9$ yr) have had time to drift away from or disperse the dusty molecular clouds where they were born and their emission can dominate the integrated spectrum at 4000 Å if younger stars are more heavily obscured. Furthermore, a selective extinction appears to be the most plausible explanation for the fact that, even within the same spectrum, different values of extinction are measured depending on the spectral region/feature used to estimate it: for example, the dust attenuation of the UV/optical stellar continuum is often measured to be lower than the obscuration of the emission lines (see [6] for a reference list).

On the basis of our models we find the following results:

In order to verify whether the hypothesis of selective extinction in a starburst galaxy can indeed account for the e(a) spectrum, a simplified spectrophotometric model, including only 10 stellar populations, has been developed [7]. This

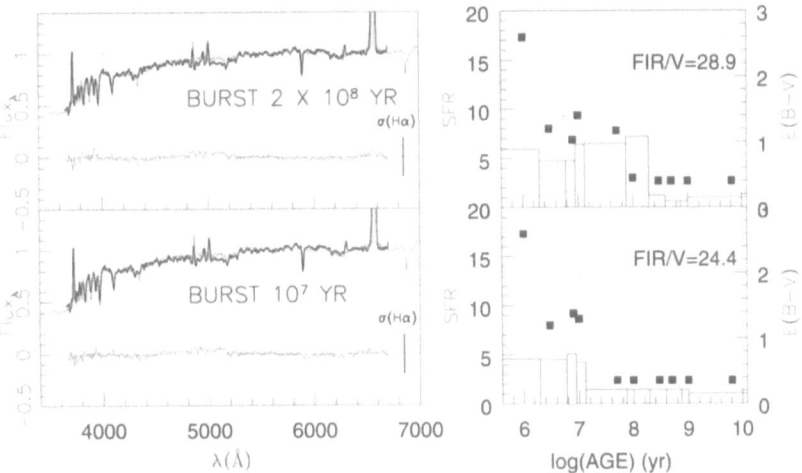

Fig. 5. *Left:* Comparison between the average observed spectrum of the e(a) galaxies (thick line) and the spectrum of two starburst models (thin lines), normalized at 5500 Å. The emission lines included in the models are those of the Balmer series and the [OII] line. The difference between the model and the observed spectrum is also shown (at Flux∼ 0). The vertical segment on the right side represents the 1σ error in $H\alpha$. *Right:* SFR (histogram, normalized to 1 in the old population) and E(B-V) (dots) of the models whose spectrum is shown in the left panel. The observed FIR/V ratio is 88.0.

represents the minimum set of stellar populations that are known to be essential because they affect the spectral features that we wish to reproduce: four young generations ($10^6, 3 \cdot 10^6, 8 \cdot 10^6, 10^7$ yr) responsible for the ionizing photons that produce the emission lines; five intermediate populations ($5 \cdot 10^7, 10^8, 3 \cdot 10^8, 5 \cdot 10^8, 10^9$ yr) with the strongest Balmer lines in absorption, and older stars modelled as a constant star formation rate between 1 and 12 Gyr before the moment of the observation which can give a significant contribution to the continuum. The spectrum of each stellar generation is found from a spectrophotometric model that includes both the stellar and the nebular contribution [1] and it is extincted with its own extinction value – that is allowed to vary from one stellar population to another – assuming a dust screen with a standard Galactic extinction law and a Salpeter IMF between 0.1 and 100 $M_\odot$. The results of this model are compared with the average e(a) spectrum of the VLIRGs discussed in Sec. 2; the quality of the fit is assessed considering the differences between the model and the observed spectrum in the equivalent widths of four lines ([OII]λ3727, $H\delta$, $H\beta$ and $H\alpha$) and the continuum flux in eight spectral windows in the range 3770–6460 Å.

1) the e(a) spectrum is consistent with the starburst/selective extinction scenario. The upper left panel of Fig. 5 presents the comparison between the average e(a) spectrum and a model of a burst that began 2×10^8 yr ago; the extinction of the starburst populations is significantly higher than that of the

older generations (see the right panel in Fig. 5). The fit is remarkable, both as far as the line equivalent widths and the continuum are concerned[2].

2) The model described above can only account for about 1/3 of the observed FIR luminosity. Different effects can contribute to this discrepancy:

a) Slit aperture effects i.e. a mismatch in the galactic area sampled by the optical spectrum (central 2 kpc) and by IRAS (integrated). Given that the IR emissivity is usually concentrated in the central regions of luminous infrared galaxies, it is hard to envisage how slit effects can account for the discrepancy between the modelled and the observed IR flux.

b) Dusty starburst galaxies often have star forming regions which are completely obscured at optical wavelengths, hence give no contribution to the spectrum but produce a significant fraction of the FIR luminosity (e.g. [4]). The observed FIR/V ratio can be reproduced by starburst models with regions that are highly obscured by dust with an E(B-V) even greater than in the models of Fig. 5.

3) No strong constraint can be placed on the burst duration. Models with a starburst as short as a few times 10^6 yr and as long as 10^9 yr are able to fit the observed e(a) spectrum as long as: a) the youngest populations are highly extincted and b) the old population provides a contribution but does not overwhelm the intermediate age contribution at 4000 Å. As an example, the lower left panel in Fig. 5 shows the excellent fit obtained with a current burst that began $\sim 10^7$ yr ago. In the case of short bursts ($< 5 \times 10^7$ yr), the strong Hδ line is not produced by the stars born during the starburst event, but by the previous stellar populations.

4) We also considered a family of "post-starburst models" with extinction, assuming a galaxy is seen *after* a strong starburst phase when a small amount of residual star formation activity is still ongoing. This type of model – besides accounting for not more than $\sim 1/10$ of the observed FIR luminosity – fails to reproduce simultaneously the Hα and the Hβ lines, either underestimating the former or overestimating the latter.

References

1. G. Barbaro & B.M. Poggianti: A&A **490**, 504 (1997)
2. N.A. Devereux, J.S. Young: ApJ **350**, L25 (1990)
3. H. Flores, et al.: ApJ **517**, 148 (1999)
4. I.F. Mirabel, et al.: A&A **333**, L1 (1998)
5. B.M. Poggianti, et al.: ApJ **518**, 576 (1999)
6. B.M. Poggianti, H. Wu: ApJ **529**, 157 (2000)
7. B.M. Poggianti, A. Bressan, A. Franceschini: ApJ, in press (2001) astro-ph/0011160
8. M.A. Strauss, et al.: ApJS **83**, 29 (1992)
9. H. Wu, Z.L. Zou, X.Y. Xia, Z.G. Deng: A&AS **127**, 521 (1998a)
10. H. Wu, Z.L. Zou, X.Y. Xia, Z.G. Deng: A&AS **132**, 181 (1998b)

---

[2] There is a small discrepancy around 5000 Å but this is only at the 4% level.

The Nature of ISOCAM Galaxies
in the Hubble Deep Field South

Dimitra Rigopoulou[1], Alberto Franceschini[2], and Reinhard Genzel[1]

[1] Max-Planck-Institut für extraterrestrische Physik, Postfach 1312, 85741 Garching, Germany

[2] Dipartimento di Astronomia, Universitá di Padova, Vicolo Osservatorio 5, 35122 Padova, Italy

Abstract. The various deep ISOCAM surveys revealed a new class of infrared luminous galaxies which are characterized by a high rate of evolution and are found at redshifts of z~1. Here we report results of our near-infrared VLT-ISAAC spectroscopic survey aimed at characterising the nature of these sources. We find that the rest-frame R-band spectra resemble those of powerful dust enshrouded starbursts. More recent detailed studies revealed that some of these systems are in fact extremely massive galaxies.

1 Introduction

The COBE detection of an extragalactic far-infrared/submm background [13], with an integrated intensity similar to or greater than that of optical light (e.g. [7]), strongly suggests that a significant fraction of the cosmic star formation in the Universe is obscured by dust and thus missed by the various optical surveys.

With the advent of the Infrared Space Observatory (ISO, [8]) deep mid-IR surveys for distant galaxies, have been successfully carried out for the first time. Operating in the 5–18 μm band, ISOCAM on board ISO was more than 1000 times more sensitive than IRAS and thus had the potential to study infrared bright galaxies at redshifts beyond 0.5. A number of cosmological surveys have been performed with ISOCAM especially in the LW3 filter (12–18 μm). The source counts from all ISOCAM surveys combined with those of IRAS are in good agreement with a no or moderate evolution model up to a flux level of 100 mJy. However at fainter flux levels the situation rapidly changes: the counts steepen considerably and at ~200–600 μJy they are about an order of magnitude greater than the predictions of no evolution models. This steepening in the log N–log S plot and a pronounced maximum in the differential number counts at ~400 μJy implies that ISOCAM surveys have probably revealed a new population of strongly evolving galaxies [3]. Among the deepest ISOCAM surveys are the observations of the Hubble Deep Field (HDF) North (N) and South (S) regions resulting in the detection of ~150 sources down to μJy levels. The results we discuss here focus on a sample of galaxies drawn from the ISOCAM survey of the HDF-S. We present our detailed near-infrared spectroscopic followup carried out using the Very Large Telescope (VLT) aiming at characterizing the nature of this new population. As an additional step we investigate the dynamical stage of the ISOCAM galaxies.

2   Observations of HDF-S

2.1   The ISOCAM HDF-S Sample

The HDF-S was observed by ISOCAM as part of the European Large Area ISO Survey (ELAIS, [11]). The observations were carried out at two wavelengths, LW2 (6.75 μm) and LW3 (15 μm). The data have been analysed independently by Oliver et al. [11] and Aussel et al. [1]. The latter analysis was carried out using the PRETI method and resulted in the detection of 63 sources in the LW3 band. The results presented here are based on the Aussel et al. analysis.

2.2   The VLT-ISAAC Sample and Observations

The sample discussed here was selected from the ISOCAM LW3 detections of the HDF-S (hereafter ISOHDFS sample). All of the selected galaxies had secure LW3 detections and I and/or K-band image counterparts (I-band data from Dennefeld and collaborators, K-band from ESO-EIS Deep). We did not apply any selection based on colours. Our sample (hereafter VLT ISOHDFS sample) is thus a fair representation of the strongly evolving ISOCAM population near the peak of the differential source counts [3]. Our VLT ISOHDFS sample contains about 25 galaxies. The LW3 flux ranges between 100–400 μJy.

In total we observed 12 galaxies and H$_\alpha$ was succesfully detected in all but one of them. [NII] emission is also seen in some spectra. Figure 1 shows selected representative spectra [15].

3   The Nature of the ISOCAM Faint Galaxies

Prior to our VLT-ISAAC observations no near-infrared (rest-frame R-band) spectroscopy had been carried out for the ISOCAM population, mostly due to the faintness of the galaxies. Optical spectroscopy (rest-frame B-band) has been done for Hubble Deep Field North [2] and the Canada France Redshift Survey (CFRS) field [4]. In both of these samples the median redshift is about 0.7–0.8. Our VLT ISOHDFS sample contains 7 galaxies with 0.4<z<0.7 and 5 galaxies with 0.7<z<1.4. Thus our sample has a z-distribution very similar to the HDF-N [1] and CFRS [4] samples.

Based on rest-frame B-band spectra and the presence of prominent deep Balmer absorption features and a moderate [OII] emission, Flores et al. [4] proposed that the ISOCAM galaxies in the CFRS field look like post-starburst systems. However, our recently acquired rest-frame R-band spectra (Fig. 1) displaying strong H$_\alpha$ emission with large equivalent widths imply that these galaxies have active on-going star formation. Surprisingly, the well-known starburst galaxy M82 displays exactly the same characteristics (Fig. 2). The answer to

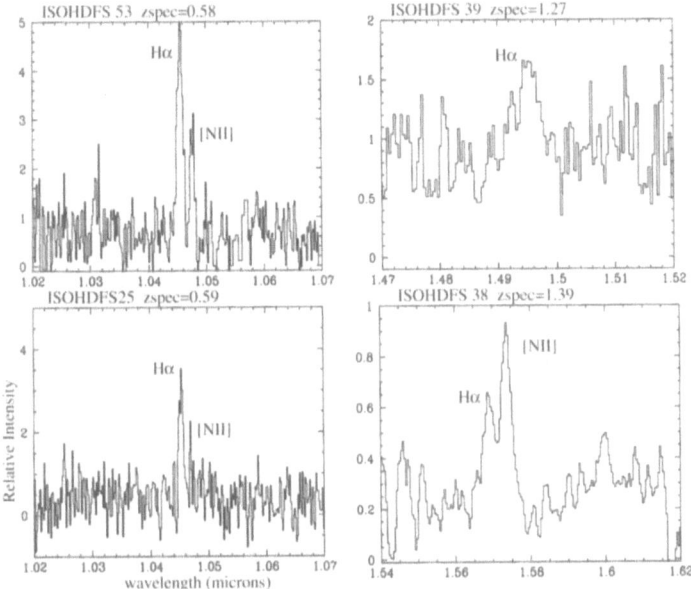

Fig. 1. ISAAC-VLT spectra of ISOCAM galaxies, showing two low-z and two high-z sources. The resolution of 600 corresponds to a resolution of about 12 Å at the source distance. The H$_\alpha$ and [N$_{II}$] lines are resolved in three of the spectra. In source ISOHDFS-38 the H$_\alpha$/[N$_{II}$]<1 implies that this source is an AGN (spectra from [15]).

this phenomenal mismatch is differential extinction: large amounts of dust exist within the H$_{II}$ regions where the H$_\alpha$ and [O$_{II}$] line emission originates. [O$_{II}$] emission is affected more than H$_\alpha$ simply because of its shorter wavelength. The continuum is due to A-stars. This A-star signature comes from earlier (0.1–1.0 Gyr) star formation activity that is not energetically dominant, in fact it plays a small role once the dusty starburst is dereddened. Such a scenario implies that these galaxies undergo multiple burst events: the less extincted population is due to an older burst while in the heavily dust enshrouded H$_{II}$ regions there is ongoing star formation. We thus suggest that ISOCAM galaxies are actively star forming, dust enshrouded galaxies, akin to local LIRGs (e.g. NGC 3256, [14]).

4   Star Formation Rates

Being the strongest of all Balmer lines, the H$_\alpha$ emission line has been traditionally used to derive quantitative star formation rates in galaxies. The H$_\alpha$ luminosity scales directly with the ionizing luminosity of the embedded stars and is thus proportional to the star formation rate (SFR). The conversion factor between ionizing luminosity and SFR is computed with the aid of an evolutionary synthesis model. The prime contributors to the integrated ionizing flux are massive stars (M>20 M$_\odot$) with relatively short lifetimes ($\leq 10^7$ yrs). Using the evolution-

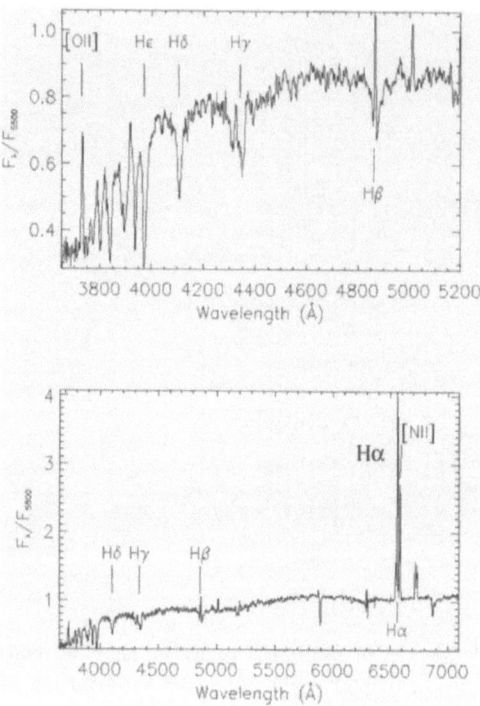

Fig. 2. Top panel: B-band spectrum of the well known starburst galaxy M82. Note the prominent Balmer H_δ and H_ϵ absorption lines as well as the moderate [OII] emission. Lower panel: R-band spectrum of M82. The spectrum is dominated by the very strong H_α emission (spectra from [9]).

ary synthesis code of Kovo and Sternberg [18] for solar abundances, a Salpeter IMF (1–100 $M_\odot$) and for a short duration burst (t $\sim \Delta$t $\sim$ a few 10^7 yrs) we obtain:

$$SFR \ (M_\odot/yr) = 5 \times 10^{-42} L(H_\alpha) \ (ergs^{-1}). \tag{1}$$

Using this formula we estimate that the SFR rates in our ISOHDFS galaxies range between 1–20 $M_\odot$/yr, corresponding to total luminosities in the range ~ 1–10×10^{42} erg s^{-1} (assuming H_0=50 km/s/Mpc, Ω=0.3).

The SFR estimates based on equation (1) represent only a *lower limit* to the real SFR in these galaxies. To get an estimate of the extinction we use V–K colour indices (magnitudes taken from the ESO-EIS survey) and using STARBURST99, the evolutionary synthesis code of Leitherer et al. [10], for various star formation histories we calculated the range of intrinsic (extinction free) colours. By comparing the observed V–K colours to the model predictions we derive a median extinction A_V of 1.8 assuming a screen model for the dust distribution. This A_V value corresponds to a median correction factor for the SFR(H_α) of ~4.

Also the far-infrared luminosities can be used to infer SFR, especially since the ISOHDFS galaxies are dust enshrouded. For the same IMF as above, the SFR scales with the FIR luminosity as:

$$SFR\ (M_\odot/yr) = 2.6 \times 10^{-44} L_{FIR}\ (ergs^{-1}) \qquad (2)$$

The L(FIR) in equation (2) is calculated based on the method of Franceschini et al. [5] which uses the 15 μm flux and assumes a L_{FIR}/L_{MIR} ratio of ~10. We find that the SFR(FIR) estimates are on average a factor of 5–50 higher than the SFR estimates inferred from H_α uncorrected for extinction. The ratio SFR(FIR)/SFR(H_α) drops to 3 if we use extinction corrected H_α values, confirming that the extinction in these galaxies is much higher than inferred from UV or optical observations alone. Thus, ISOCAM galaxies are in fact actively star forming dust enshrouded galaxies. We finally note that the factor 3 inconsistency noted above for the FIR based SFR is consistent with the value obtained by Poggianti & Wu [12] for a sample of luminous IRAS starbursts.

5   Probing the Dynamical Stage of ISOCAM Galaxies

Having established that ISOCAM galaxies are actively starbursting systems, the next step was to investigate their dynamical stage. Spatially resolved H_α measurements allow us to probe the kinematics and estimate the stellar mass in the nuclear regions of these galaxies.

In August 2000 we carried out high resolution spectroscopy (R~5000) of selected targets for which we had obtained accurate spectroscopic redshifts from our previous work. The first of our targets ISOHDFS-27 was a spiral galaxy at a redshift z=0.58. Figure 3 shows the H_α line intensity profile. The shape of the profile implies that the emitting region is not centrally concentrated but most likely originates in a ring-like configuration. We have measured a peak-to-peak velocity of 550 km/s. After correcting for the inclination of the galaxy (i=49°) the actual rotational velocity of the system is 275 km/s, higher than that so far measured in most local spiral galaxies [17]. We estimate that the mass enclosed within the central 40 kpc is of the order 1×10^{12} $M_\odot$ significantly higher than that so far measured in most local spiral galaxies [6] or more distant systems [19].

During the same observing run we obtained spatially resolved H_α measurements for two other galaxies. One was found to be a double system consisting of two counter-rotating galaxies with equal masses (of the order of 2×10^{11} $M_\odot$) at a redshift of 0.58 while the other is a single massive (M~1×10^{12} $M_\odot$) system at a redshift of 1.27 [16]. Since ISOCAM galaxies are presumed to contribute significantly to the infrared background our findings suggest that the IR background is largely contributed by massive galaxies with substantial ongoing star formation activity.

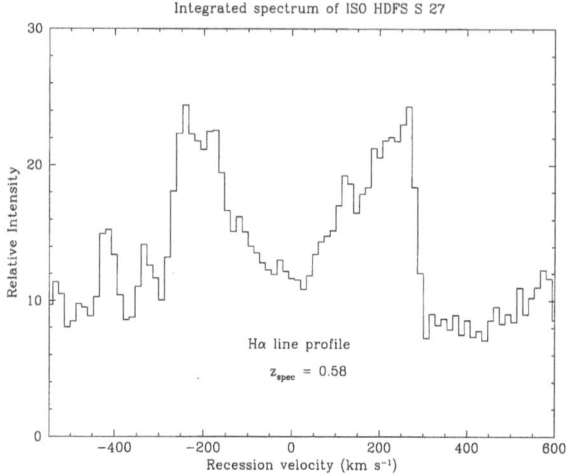

Fig. 3. H_α line profile of ISOHDFS-27 at a redshift of 0.58. The peak-to-peak velocity is 550 km/s.

References

1. H. Aussel et al.: in preparation (2001)
2. A. Barger et al.: AJ **117**, 102 (1999)
3. D. Elbaz et al.: A&A **351**, L37 (1999)
4. H. Flores et al.: ApJ **517**, 148 (1999)
5. A. Franceschini et al.: in preparation (2001)
6. R. Giovanelli et al.: **301** L7 (1986)
7. M. Hauser et al.: ApJ **481**, 49 (1998)
8. M. Kessler et al.: A&A **305**, L27 (1996)
9. R.C. Kennicutt: ApJ **388**, 310 (1992)
10. K. Leitherer et al.: ApJS **123**, 3 (1999)
11. S. Oliver et al.: MNRAS **316**, 749 (2000)
12. B. Poggianti & H. Wu: ApJ **529**, 157 (2000)
13. J-L. Puget et al.: A&A **308**, L5 (1996)
14. D. Rigopoulou et al.: A&A **305**, 747 (1996)
15. D. Rigopoulou et al.: ApJLett **537**, L85 (2000)
16. D. Rigopoulou et al.: in preparation (2001)
17. V. Rubin, D. Burstein, W.K. Ford, & N. Thonnard: ApJ **289** 81 (1985)
18. A. Sternberg: ApJ **506**, 721, (1998)
19. N.P. Vogt, et al.: ApJ **465**, 15 (1996)

Hard X-rays from Starburst Galaxies Near and Far

Edward C. Moran

University of California, Berkeley, CA 94720, USA

1 Introduction

Compared to other types of extragalactic objects, such as AGNs or galaxy clusters, starburst galaxies might appear to be fairly unspectacular X-ray sources. Their broadband X-ray luminosities range from a few times 10^{38} to at most 10^{42} ergs s^{-1}, and X-ray emission represents only one one-hundredth of a percent of their bolometric output. X-ray emission is, however, a direct consequence of star formation, and at energies above $\sim 2\,\mathrm{keV}$, X-ray observations offer a view of the starburst phenomenon that is relatively free of the effects of extinction and reprocessing—something which is not available in many parts of the electromagnetic spectrum.

Unfortunately, it is still somewhat uncertain exactly *how* hard X-rays in starbursts are produced, and much of this report focuses on that issue. Independent of the mechanism, however, starburst galaxies as a class are hard X-ray emitters, and I will also discuss our efforts to determine their integrated contribution to the sky brightness in the hard X-ray band.

2 Hard X-rays from Starburst Galaxies Near ...

Observations with the *ASCA* satellite by myself and others have demonstrated that the broadband X-ray spectra of starburst galaxies are very complex (see [8], and references therein). The hard X-ray fluxes appear to originate in the nuclear regions of the galaxies, where the bulk of the star formation is taking place. Although the hard X-ray spectral components are often modeled as a single power law, it has been recognized that several different types of sources, both thermal and nonthermal, could be contributing to the luminosities of starburst galaxies in this band.

In the Milky Way and Magellanic Clouds, the brightest hard X-ray sources we see associated with star formation are high-mass X-ray binary star systems (HMXBs), which consist of an early-type star and a compact stellar remnant. For this reason, it has been generally presumed that HMXBs must produce most of the hard X-ray luminosities of starburst galaxies. Unfortunately, little work has been done to relate the observed hard X-ray properties of starbursts directly to the star formation occurring within them, so the role of their HMXB populations remains unclear. For example, although various authors adopt very different values of the typical individual binary luminosity, ranging from 10^{34}–10^{35} ergs s^{-1} (e.g., [1]) to 10^{37}–10^{38} ergs s^{-1} (e.g., [15]), they generally conclude that the implied number of binaries required to explain a starburst's hard X-ray luminosity ($\sim 10^3$–10^6) is reasonable. D. Helfand and I have been trying to obtain a better quantitative handle on the binary contribution to the X-ray luminosities of starbursts, using the following reasoning:

In a region where the star formation rate is roughly constant, the number of O stars present will reach equilibrium in about 10^7 years. The X-ray emitting phase of a high-mass binary is much shorter than the lifetime of even the most massive stars, so the number of X-ray binaries present will reach equilibrium on the same time scale. Thus, the number of O stars in a star-forming region traces the number of HMXBs present, independent of the star-formation rate. Of course, what we're really after is not the *number* of binaries, but their contribution to the X-ray energy budget of starbursts. So we consider instead their *combined luminosity* relative to the number of O stars. If we could measure this quantity accurately in nearby star-forming regions, such as the Milky Way or Magellanic Clouds, we would then only need to estimate the number of O stars in distant starbursts to assess the role of binaries in their hard X-ray production.

Thus, we have made careful estimates of the specific X-ray luminosity per O star for HMXBs in the Local Group. Full details are reported in [5]; the results are summarized in Table 1. For the solar neighborhood ($R \leq 3\,\mathrm{kpc}$), our measurements are based on (1) time-averaged luminosities of ~ 60 HMXBs from light curves obtained with the *RXTE* satellite and (2) direct counts of nearby O stars. Less direct methods were used to obtain estimates for the Milky Way as a whole and other Local Group galaxies.

Table 1. Specific X-ray Luminosity per O Star for HMXBs.

Galaxy	$[L_{\mathrm{x}}/N(\mathrm{O})]_{\mathrm{HMXB}}$ $(\times 10^{34}$ ergs s^{-1} star$^{-1})$
Solar Neighborhood	1.6
Milky Way	3.5
LMC	5.4
SMC	20
M33	9–20
M31	20

The total 2–10 keV X-ray luminosity of high-mass binaries in a star-forming galaxy can be expressed as $L_{\mathrm{x}} = [L_{\mathrm{x}}/N(\mathrm{O})]_{\mathrm{HMXB}} \times N(\mathrm{O})$, where $[L_{\mathrm{x}}/N(\mathrm{O})]_{\mathrm{HMXB}}$ is an adopted value of the specific X-ray luminosity per O star for HMXBs, and $N(\mathrm{O})$ is the actual number of O stars present. Assuming an IMF slope of 2.35, an upper mass cutoff of $100\,\mathrm{M}_\odot$, and solar metallicity, the models of [6] predict that a region producing stars at a constant rate of $1\,\mathrm{M}_\odot$ yr^{-1} for a few times 10^7 yr will have 2.5×10^4 O stars and an associated bolometric luminosity of

$\sim 5.3 \times 10^{43}$ ergs s^{-1}. Provided that the young stellar population dominates the host galaxy's bolometric luminosity—which is approximately equal to its total infrared luminosity L_{IR} — the number of O stars can be scaled for a system of arbitrary star-formation rate: $N(O) = 2.5 \times 10^4 \, (L_{IR}/5.3 \times 10^{43})$. The binary luminosity expression then becomes $L_X = 4.7 \times 10^{-40} \, [L_X/N(O)]_{HMXB} \, L_{IR}$, or in terms of fluxes, $F_X = 4.7 \times 10^{-40} \, [L_X/N(O)]_{HMXB} \, F_{IR}$.

Using the latter equation, we have computed total HMXB X-ray fluxes expected at a given IR flux for the range of Local Group values of $[L_X/N(O)]_{HMXB}$. These are represented by the shaded region in Fig. 1. Also plotted are the locations in the $F_X - F_{IR}$ plane of several nearby starburst galaxies that have been

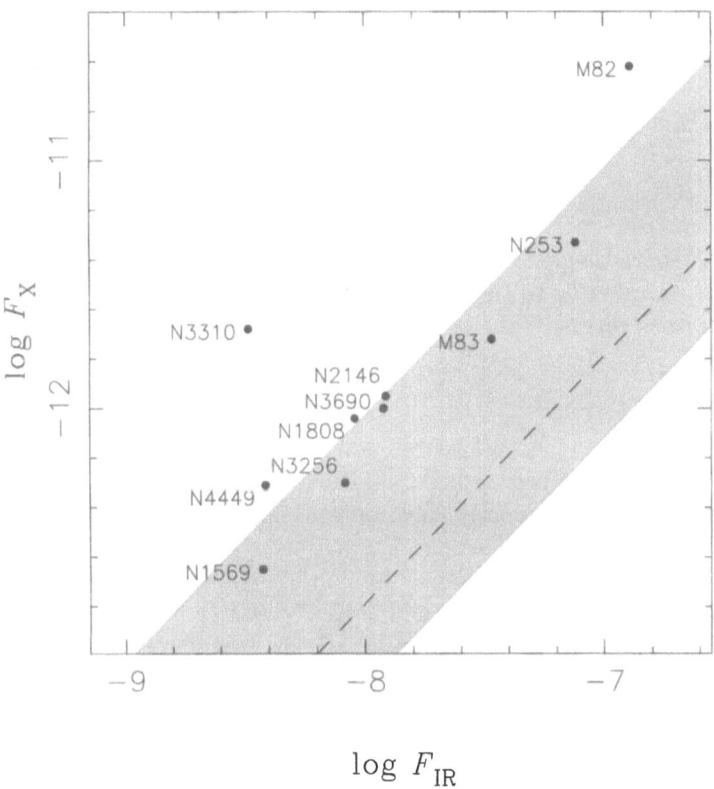

$$\log F_{IR}$$

Fig. 1. X-ray flux in the 2–10 keV band versus 8–1000 μm IR flux for ten starburst galaxies. The IR fluxes were computed using the highest reported *IRAS* flux densities and the formula of [11]. The X-ray fluxes are from published *ASCA* studies. The shaded region shows the range of X-ray fluxes expected from HMXBs at a given IR flux for the different determinations of the specific X-ray luminosity per O star in Local Group galaxies. The region is bounded on the lower-right by the value of 1.6×10^{34} ergs s^{-1} star^{-1} found for the solar neighborhood, and on the upper-left by the SMC value of 20×10^{34} ergs s^{-1} star^{-1}. The dashed line represents the upper limit obtained for the solar neighborhood.

studied with *ASCA*. Several important conclusions can be drawn from Fig. 1. First, there is a clear tendency for the 2–10 keV X-ray fluxes of starburst galaxies to increase with F_{IR}, indicating that their hard X-ray luminosities are largely governed by sources whose contributions are proportional to the star-formation rate. Second, in order for HMXBs to account for *all* of the hard X-rays produced in starbursts, their typical output per O star must be significantly greater than that observed in the Milky Way or the LMC. And finally, two objects, M82 and NGC 3310, deviate significantly from the $F_x - F_{IR}$ trend exhibited by the other starburst galaxies[1]. The X-ray fluxes of these two objects are clearly inconsistent with the level of emission expected from an HMXB population, even one similar to that of the SMC, suggesting that each galaxy possesses an extra component of hard X-ray luminosity that is weak (or absent) in the other starbursts. Indeed, *Chandra* images of NGC 253 [13] and M82 [4], which have similar IR luminosities but 2–10 keV X-ray luminosities that differ by a factor of 5, reveal strikingly different hard X-ray morphologies. The hard X-ray flux of NGC 253 is produced almost entirely by discrete sources, whereas in the more luminous M82, about half of the hard X-ray emission arises from a diffuse component coincident with the most active region of star formation. It has been suggested [10,12,8] that inverse-Compton scattered emission, resulting from the interaction of IR photons with supernova-generated relativistic electrons, may in some circumstances contribute appreciably to the hard X-ray fluxes of starburst galaxies. M82 and NGC 3310 thus represent the best sites for the investigation of this possibility.

3   ... and Far

Starburst galaxies are prevalent in the Universe and are sources of hard X-ray emission — how much do they contribute to the X-ray sky brightness? Unfortunately, the question is difficult to answer, even with current instrumentation, because of the low individual X-ray luminosities of starbursts. However, it has been known for some time that the faint radio source population is dominated by star-forming galaxies, and we have used this fact to estimate the integrated X-ray production of distant starburst galaxies.

First, we have to establish a relationship between the radio and X-ray properties of starbursts. In Fig. 2, the 5 keV flux densities of several nearby starbursts are plotted versus their nuclear 5 GHz flux densities, revealing a strong linear correlation. Thus, since the sub-mJy radio source population is dominated by star-forming galaxies, we can in principle use the 5 GHz number count–flux relation [2] to predict the integrated hard X-ray intensity of starbursts. The method is described in full by [8].

---

[1] M82 contains a variable hard X-ray source [7] that can, in outburst, dominate the 2–10 keV emission of the galaxy. However, preliminary *Chandra* results [4] indicate that there are many other sources of hard X-ray emission in M82 as well. The X-ray flux of M82 shown in Fig. 1 is the *lowest* flux measured by *ASCA*; it is likely to represent the integrated contribution of the "quiescent" sources of hard X-ray luminosity in the galaxy.

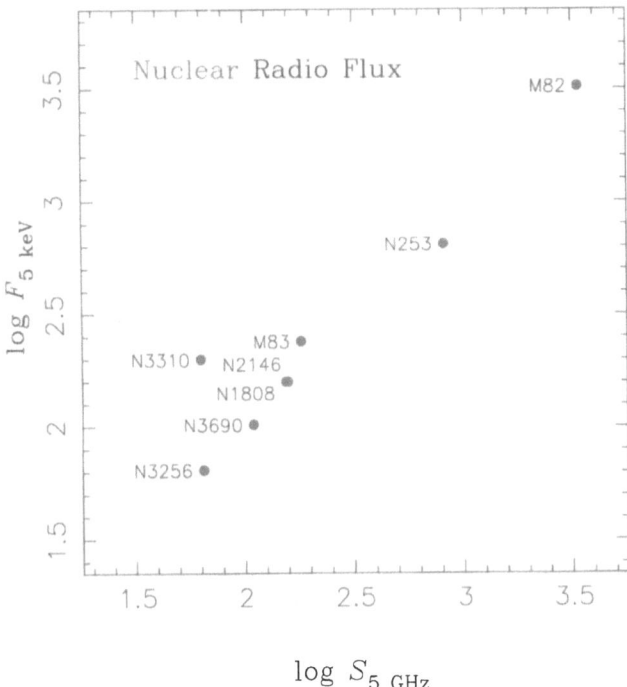

Fig. 2. Comparison of the 5 keV flux densities (in units of 10^{-15} ergs cm^{-2} s^{-1} keV^{-1}) and nuclear 5 GHz flux densities (in mJy) of nearby starburst galaxies. The strong linear correlation indicates a ratio of $F_{5\,\mathrm{keV}}/S_{5\,\mathrm{GHz}} = 10^{-18}$ ergs cm^{-2} s^{-1} keV^{-1} μJy^{-1}. The mean core-to-total 5 GHz flux ratio is 0.44 for these galaxies.

To obtain the 5 keV sky intensity due to starburst galaxies, we integrate $I_X = \int n(S)\, f_{SB}\, F_X(S)\, dS$, where $n(S)$ is the differential radio source counts measured by [2], f_{SB} is the fraction of faint radio sources that are starbursts, and $F_X(S)$ is an expression for the X-ray flux in terms of the radio flux, which is derived from Fig. 2. It is important to note that the X-ray and radio spectral indices of starbursts are, to first order, the same, so no K-correction is needed here. Recent deep VLA surveys suggest that at star-forming galaxies make up at least 60% of the μJy radio source population (e.g., [9]), so we adopt $f_{SB} \approx 0.6$. The upper limit of integration has little impact on the value of the integral, so we can treat it as infinity. For the lower limit of integration, the radio $\log N - \log S$ has only been measured down to a few μJy, so we need to extrapolate to fainter fluxes. Windhorst et al. (1993) have argued that radio source counts must flatten below about 20 nJy, or else radio sources would distort the microwave background spectrum at centimeter wavelengths. Thus, we conservatively adopt 0.1–1 μJy for the lower limit of integration. Putting everything into the integral, we find that the total intensity of starburst galaxies at 5 keV is 2–3 $\times 10^{-13}$ ergs cm^{-2} s^{-1} keV^{-1}, which is 10–15% of the 5 keV X-ray background intensity [3].

This empirical result is interesting for X-ray background research because the answer isn't negligible. In fact, even in the deepest *Chandra* images to date, some 20–30% of the X-ray background remains unresolved, much of which could be associated with emission from distant starbursts. The result also suggests that X-ray studies can be used to constrain models for the star-formation history of the Universe. For example, continued investigation of starburst galaxies and nearby star-forming regions, similar to that described above, will clarify the relationship between hard X-ray luminosity and star-formation rate. Combined with "Madau-Steidel" plots of star-formation history, such a relationship can be used to derive an independent estimate of the X-ray sky brightness associated with star formation, which can then be compared to the empirical value derived here and the fraction of the X-ray background that remains unaccounted for by active galactic nuclei.

References

1. H. Awaki, S. Ueno, K. Koyama, T. Tsuru, K. Iwasawa: PASJ **48**, 409 (1996)
2. E.B. Fomalont, R.A. Windhorst, J.A. Kristian, K.I. Kellerman: AJ **102**, 1258 (1991)
3. K.C. Gendreau, et al.: PASJ **47**, L5 (1995)
4. R.E. Griffiths, A. Ptak, E. Feigelson, G. Garmire, L. Townsley, W.N. Brandt, R. Sambruna, J.N. Bregman: Science **290**, 1325 (2000)
5. D.J. Helfand, E.C. Moran: ApJ, in press (2001)
6. C. Leitherer, T.M. Heckman: ApJS **96**, 9 (1995)
7. H. Matsumoto, T. Tsuru: PASJ **51**, 321 (1999)
8. E.C. Moran, M.D. Lehnert, D.J. Helfand: ApJ **526**, 649 (1999)
9. E.A. Richards: BAAS **30**, 1326 (1998)
10. G.H. Rieke, M.J. Lebofsky, R.I. Thompson, F.J. Low, A.T. Tokunaga: ApJ **238**, 24 (1980)
11. D.B. Sanders, I.F. Mirabel: ARA&A **34**, 749 (1996)
12. R. Schaaf, W. Pietsch, P.L. Biermann, P.P. Kronberg, T. Schmutzler: ApJ **336**, 722 (1989)
13. D. Strickland, T.M. Heckman, K.A. Weaver, M. Dahlem: AJ **120**, 2965 (2000)
14. R.A. Windhorst, E.B. Fomalont, R.B. Partridge, J.D. Lowenthal: ApJ **405**, 498 (1993)
15. A.L. Zezas, I. Georgantolpoulos, M.J. Ward: MNRAS **301**, 915 (1998)

Distribution of Star Formation
Between Starbursts and Normal Galaxies

Rodger I. Thompson

Steward Observatory, University of Arizona, Tucson, AZ 85721, USA

Abstract. The combination of optical and near infrared observations of a portion of the Hubble Deep Field provides a data set that can be used for photometric redshift and photometric extinction determination. The more than a factor of 5 wavelength coverage greatly mitigates the partial degeneracy between extinction and galaxy type encountered in optical only studies. Once the parameters of redshift, extinction and galaxy spectral energy distribution have been determined, the question of the distribution of star formation between starburst galaxies and normal galaxy star formation can be addressed.

1 Introduction

The history of star formation in the universe is the subject of intense study at this time. Related to this is the question of not only when but where did the majority of star formation occur. Are most stars formed in starburst galaxies or are more formed in the normal type of star formation that we observe in our own galaxy? Since we see both types of star formation going on currently it must be a mix of the two, but the percentage of each type is of great interest. A related question is whether most of star formation is hidden from the optical and near infrared view or whether those observations adequately account for the majority of star formation.

In 1998 a section of the Northern Hubble Deep Field [1] was imaged to very deep limits [2] with the NICMOS instrument on the Hubble Space Telescope. The combination of the optical data from the HST WFPC2 observations and the near infrared observations produced images in 6 photometric bands spanning the wavelength range between 0.3 and 1.6 μm. These observations provide a very accurate data set for the application of photometric redshift techniques. The wide wavelength coverage also provides an opportunity to determine photometric extinction levels. The extra wavelength range provided by the infrared observations helps break the partial degeneracy between extinction and spectral energy distribution that plagues optical only attempts to separate the effects of extinction from the intrinsic spectral energy distribution.

We will also utilize a third data set in this analysis, the NICMOS observations of the entire Northern Hubble Deep Field by a General Observer team led by Mark Dickinson. This data is now available in Hubble Space Telescope public archives. Although this data does not go as deep as the smaller area deep images, the area coverage is approximately seven times greater.

2   Star Formation History

In this work we utilize the results of an analysis of the star formation history in the deep infrared fields by [3]. The results of this study are shown in Fig. 1. In Fig. 1 and for the rest of this paper we adopt a cosmology of H_o equal to 65 and Ω_o equal to 0.3. The values in Fig. 1 from other work have been adjusted to this cosmology.

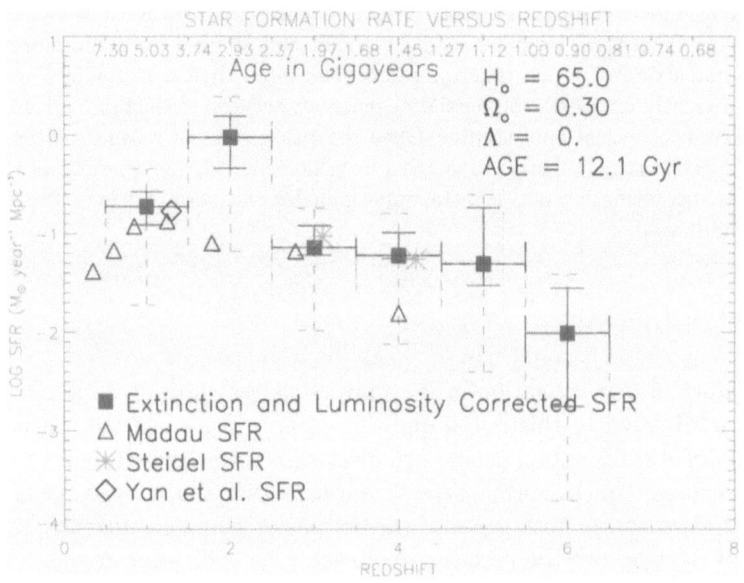

Fig. 1. History of star formation in the NICMOS Hubble Deep Field. The square black boxes show the results from the current study. Results from other work are displayed with alternate symbols. The solid error bars show the error from all sources except for large scale structure. The dashed error bars include the effect of large scale structure.

In this section we will briefly summarize the steps used to generate Fig. 1 that are relevant to the present discussion. These same steps are used to generate results from the full HDF optical and infrared images. The general method takes template galaxy spectral energy distributions (SEDs) both from observations and from calculations and then numerically redshifts and extincts them. We use the Calzetti obscuration law [4] derived from observed galaxies. The expected fluxes in the six photometric bands are then calculated. The values of the fluxes are then interpolated between SEDs to produce a finer grid. The observed fluxes for each galaxy are then matched to a model in the grid via a minimum chi squared technique. This produces a redshift, extinction, and intrinsic SED for each galaxy. The flux values along with the redshift then establish the luminosity of the galaxy. The extinction corrected star formation rate is then calculated via the relationship between the 1500 angstrom flux and the star formation rate [5] shown in Eq. 1.

$$UV_{1500} = 8.0 \times 10^{27} \cdot SFR(M_\odot/yr) \text{ ergs second}^{-1} \text{ Hz}^{-1} \qquad (1)$$

The star formation rate values in Fig. 1 have been corrected for missing luminosity from undetected galaxies and parts of galaxies via a new method discussed in [3] which utilizes the distribution function of star formation rate intensities. This method, however, does not apply to individual galaxies. Future discussion will only apply to galaxies and parts of galaxies that are actually detected. The images from the whole field HDF study have been analyzed by the same method as discussed above.

3   Identification of Starburst Galaxies

The analysis described above provides the star formation rate, the intrinsic luminosity, the redshift, and the extinction for each galaxy in the field. The extinction value indicates how much of the luminosity is removed by dust and reradiated as thermal infrared radiation. This information allows us to identify galaxies that have high star formation rates. For the purposes of this paper we will define a starburst galaxy as a galaxy with a star formation rate equal to or greater than 10 $M_\odot$ per year. For this analysis we use the data from the entire HDF as well as the deep image information, however, we will display the results separately for the entire HDF and the deep HDF image. All of the galaxies in the study that are determined to be starburst galaxies have luminosities greater than 10^{11} $L_\odot$ and more than half of their luminosity is reradiated in the infrared which categorizes them as LIRGs and ULIRGs (Luminous or Ultra Luminous Infra Red Galaxies). Not all of the LIRGS, however, are starburst galaxies by the definition of 10 $M_\odot$ per year. The breakdown of star formation rates by type is shown in Fig. 2

When the total star formation for all redshifts is integrated we find that the percentage of stars formed in starburst galaxies is 65% in the deep field and 50% in the whole HDF field. Although the star formation rate is roughly constant with redshift, the majority of stars were formed later than a redshift of 2 simply because most of the age of the universe is at redshifts less than 2. The results of this study assumes that we have seen most of the star formation through the combined optical and near infrared observations. If there is an extensive extremely extincted population of starburst galaxies then the total star formation could be tipped in favor of starbursts. The SCUBA sources may represent such a population.

4   The ULIRG Population

At a redshift of 2 the star formation rate in the deep field has a distinct maximum as shown in both Fig. 1 and Fig. 2. This spike in the star formation rate is due to

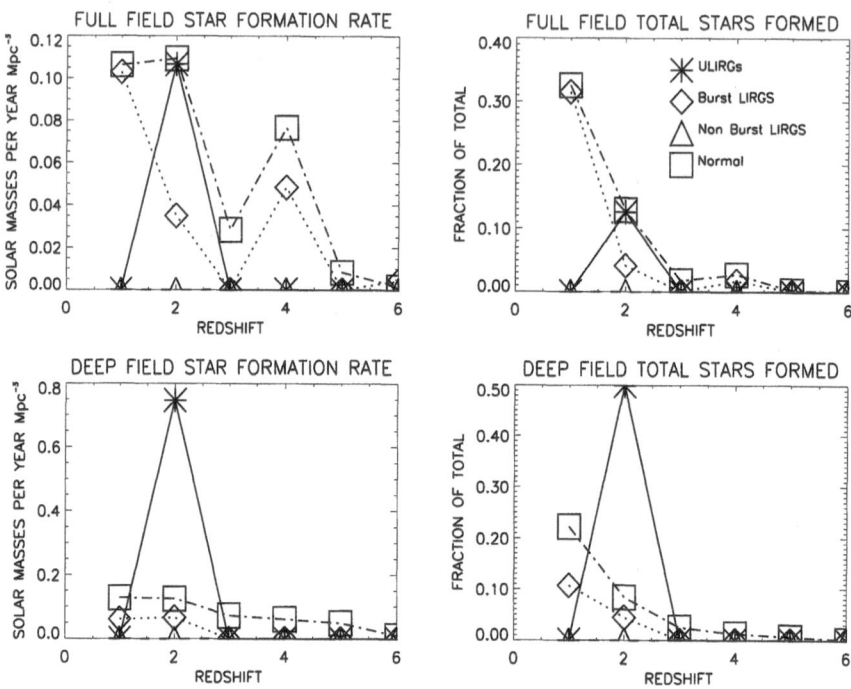

Fig. 2. Plots of the star formation rates and star formation fractions as a function of redshift for the full and deep HDF areas. The plots include the starburst galaxies denoted by ULIRGS and Burst LIRGS along with the non-starburst galaxies denoted by Non Burst LIRGS and Normal.

the presence of two ULIRGS in the deep field area. In the full field, which does not contain any additional ULIRGS, the effect of the two ULIRGS is diluted by the factor of 7 increased volume. At the present day density of the ULIRG population the probability of having a single ULIRG in the deep field is 10^{-4}. Some theories of the evolution of galaxies predict a thousand fold increase in the merger rate and hence starbursts near a redshift of two [6]. The presence of the two ULIRGs could be due to this effect. The possibility can not be dismissed that it was just chance that two ULIRGS appeared in the deep field. The ULIRG population is very probably clustered around large density fluctuations so the presence of two is not much more remarkable than the presence of one, particularly since they have very similar redshifts. If the presence of the two ULIRGs is not representative of the universe at large the star formation total by normal galaxies would be roughly twice that of star bursts. This illustrates the need for deep surveys over a much larger area than has been covered in this study. Another note of caution is that the dust obscuration law [4] is not the most appropriate for starburst galaxies.

5   SCUBA and ISO Sources

If the parameters we have determined from the above analysis are correct we can predict the flux present in any infrared band if we know the thermal infrared SED. In preliminary work we have used a starburst dust SED with PAH emission features to estimate the flux from each of our objects in the ISO and SCUBA photometric bands. To date this exercise has had reasonable success in identifying the observed sources, but the assumption of a single dust SED is clearly not accurate. We are in the process of adding additional dust SEDs in an attempt to account for the variations in the observed dust SEDs. Also for the ISO sources the photometric redshifts must be very accurate as the emission is dominated by the PAH features which vary rapidly with wavelength.

6   Summary

Using the results from a previous work to determine the star formation history in the northern Hubble Deep field we find that the total star formation is roughly split equally between starburst and other galaxies. This result is strongly influenced by the presence of two ULIRGs in the field which may be a chance result. If these ULIRGs are removed from the sample the star formation total in starbursts is then roughly half that of other galaxies. On the other hand a strong rise in the number of ULIRGs near a redshift of 2 is predicted in some theories. Finally the possibility that the SCUBA objects represent a hidden population of starbursts not observed in this sample remains a possibility. We are currently working on using various dust SEDs to see if we can match our observed sources with the known SCUBA sources in the HDF.

Acknowledgements: This work is partially supported by NASA grant NAG 5-3042 and utilizes data obtained from the NASA/ESA Hubble Space Telescope obtained at the Space Telescope Science Institute, which is operated by the Association of Universities for Research in Astronomy under NASA contract NAS5-26555. The author wishes to acknowledge the great contribution from his colleagues Ray J. Weymann and Lisa J Storrie Lombardi to this work.

References

1. R.E. Williams, B. Blacker, M. Dickinson, W.V.D. Dixon, II.C. Ferguson, A.S. Fruchter, M. Giavalisco, R.L. Gilliland, I. Heyer, R. Katsanis, Z. Levay, R.A. Lucas, C.B. McElroy, L. Petro, M. Postman: A.J. **112**, 1335 (1996)
2. R.I. Thompson, L.J. Storrie-Lombardi, R.J. Weymann, M.J. Rieke, G. Schnieder, E. Stobie, D. Lytle: A.J. **117**, 17, (1999)
3. R.I. Thompson, R.J. Weymann, L.J. Storrie-Lombardi: Ap.J. in press (2000)
4. D. Calzetti, A.L. Kinney, T. Storchi-Bergmann: Ap.J **429** 582 (1994)
5. P. Madau, L. Possetti, M. Dickinson: Ap.J **498**, 106 (1998)
6. A.W. Blain, A. Jameson, I. Smail, M.S. Longair, J.-P. Kneib: MNRAS **309**, 715 (1999)

How to Correct for Dust Absorption in Starbursts

Gerhardt R. Meurer and Mark Seibert

The Johns Hopkins University, Baltimore, MD 21218, USA

Abstract. We review new and published results to examine how well the bolometric flux of starbursts can be recovered from ultraviolet (UV) and optical observations. We show that the effective absorption of starbursts can be substantial, up to ~ 10 mag in the far UV, and ~ 5 mag in Hα, but apparently not as high as some claims in the literature (several tens to a thousand mag). The bolometric fluxes of an IUE sample of starbursts can be recovered to 0.14 dex accuracy using the UV flux and spectral slope. However, this relationship breaks down for Ultra Luminous Infrared Galaxies (ULIGs). The Hα flux combined with the Balmer decrement can be used to predict the bolometric flux to 0.5 dex accuracy for starbursts including most ULIGs. These results imply a foreground screen component to the dust distribution.

1 Introduction

Dust presents one of the biggest obstacles to interpreting observations of starburst galaxies in the optical and especially the ultraviolet (UV). The problem is difficult, because it depends not only on the amount of dust and its composition, but also the distribution of both dust and light sources. Faced with such complexity, the astronomical community's response includes assuming that star formation remains mostly unobscured by dust [12] to deriding those who even consider that UV and optical observations can be corrected for dust [26].

Here we use new and published observations to critically examine the importance of dust absorption in starburst galaxies, in order to answer the following: 1. Is dust important? 2. What is the dust geometry? 3. How does dust effect broad band colors and fluxes? 4. What does it do to emission line fluxes and line ratios? 5. *Can we recover the bolometric flux of a starburst from its UV or optical properties?*. The last question is a proxy for asking whether we can determine the star formation rate, but avoids distance uncertainties and assumptions about the lower end of the Initial Mass Function (IMF).

2 Samples and Tracers

We consider two samples of starburst galaxies, those observed in the UV by the *International Ultraviolet Explorer* (IUE), and starbursts found in the far-infrared (FIR) by the *InfraRed Astronomical Satellite* (IRAS). The samples are complementary. The IUE sample contains a lot of dwarfs as well as starbursts with $L_{\rm bol}$ as high as $\sim 10^{11.5}$ $L_\odot$. IUE starbursts are good templates for high-redshift

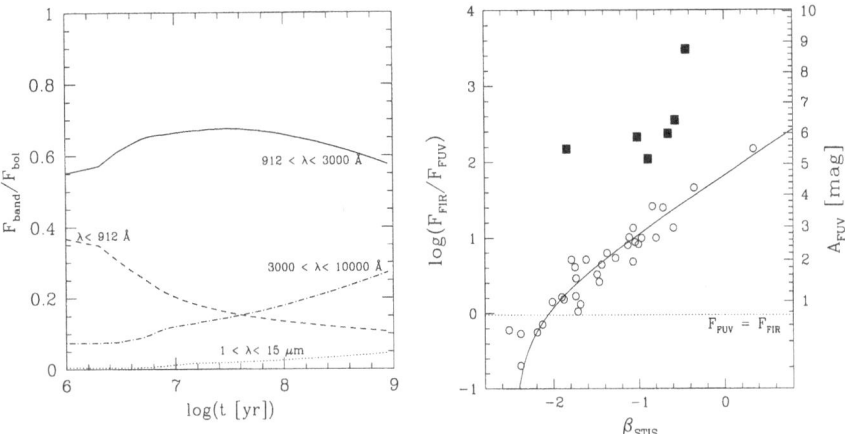

Fig. 1. (a)(*Left*): The fraction of the intrinsic bolometric luminosity emitted as ionizing radiation (dashed), UV (solid), optical (dot-dashed), and infrared (dotted) for a constant star formation rate stellar population with a Salpeter IMF (limits 1–100 $M_\odot$) and solar metallicity [11]. **(b)**(*Right*): Ratio of FIR to far UV (FUV) flux plotted against spectral slope $\beta_{\rm STIS}$ for IUE starbursts (open circles) and ULIGs (squares). Here the FUV flux, $F_{\rm FUV}$ and $\beta_{\rm STIS}$ are derived from the STIS bandpasses used for the ULIG observations [6]. The right axis shows the effective UV absorption. The solid line shows a linear fit of A_{FUV} to $\beta_{\rm STIS}$ to the IUE sample. The horizontal line shows where the bolometric corrected UV and FIR fluxes are equal.

Lyman Break Galaxies [15,16]. The IRAS sample includes the most luminous starbursts, the Ultra-Luminous Infrared Galaxies (ULIGs: $L_{\rm bol} \geq 10^{12}\ L_\odot$), but very few dwarfs. ULIGs make up < 6% of the FIR background [18], but may be good templates for high-redshift submm sources which could contribute significantly to the star formation rate density at $z \gtrsim 2$ [2].

We consider two tracers of star formation: the UV continuum, and Balmer emission lines. The UV continuum dominates the *intrinsic* (before dust) bolometric output of starbursts (Fig. 1a), and is sensitive to main sequence stars with $\mathcal{M}_* > 5$ $M_\odot$. The Balmer lines are among the strongest in the optical and provide a good measure of the ionizing flux, and hence to stars having $\mathcal{M}_* > 20$ $M_\odot$.

3   The IUE Starburst Sample

The FIR emission of a starburst represents the total luminosity absorbed by dust. For any star forming or young population, the dust heating is dominated by the UV, hence the FIR/UV flux ratio, or infrared excess (IRX) can be transformed directly into an "effective absorption". Figure 1b plots IRX versus UV spectral slope β ($f_\lambda \propto \lambda^\beta$) for both samples. Here F_{FUV} is a generalized flux λf_λ evaluated at rest wavelength $\lambda = 1515\text{Å}$, where f_λ is the flux density. Almost all galaxies emit more in the FIR than the UV, hence dust is clearly important for defining the spectral energy distribution. A strong correlation, the IRX-β relationship, is apparent for the IUE sample and readily fit by a linear relationship between effective UV absorption and β [14,16].

Figure 2a plots the absorption corrected UV flux, $F_{FUV,0}$ to F_{bol} ratio versus β. The absorption comes from the fit plotted in Fig. 1b. The mean logarithmic ratio of the IUE sample is $\langle \log(F_{FUV,0}/F_{bol}) \rangle = -0.13$ with a scatter of 0.14 dex and no residual correlation with β. For the IUE starbursts, the FUV flux and β are sufficient to recover the F_{bol} to 40% accuracy.

For this sample Balmer fluxes from IUE aperture matched spectra [13,20] are available. Figure 2b compares the ratio of dust corrected Hα flux $F_{H\alpha,0}$ and F_{bol} with $E(B-V)_g$, the intrinsic reddening of the ionized gas. Again, there is no correlation between the two quantities. Note that the dust correction, $A_{H\alpha} = 2.5E(B-V)_g$, assumes a standard Galactic extinction law. The mean logarithmic ratio is $\langle \log(F_{H\alpha,0}/F_{bol}) \rangle = -2.43$ with a dispersion of 0.28 dex. The Balmer fluxes can recover the F_{bol} to better than a factor of 2 in this sample. Calzetti et al. [4] find that $E(B-V)_g$ measured from IR Paβ and Brγ to Balmer line ratios agrees well with that measured from only the Balmer lines.

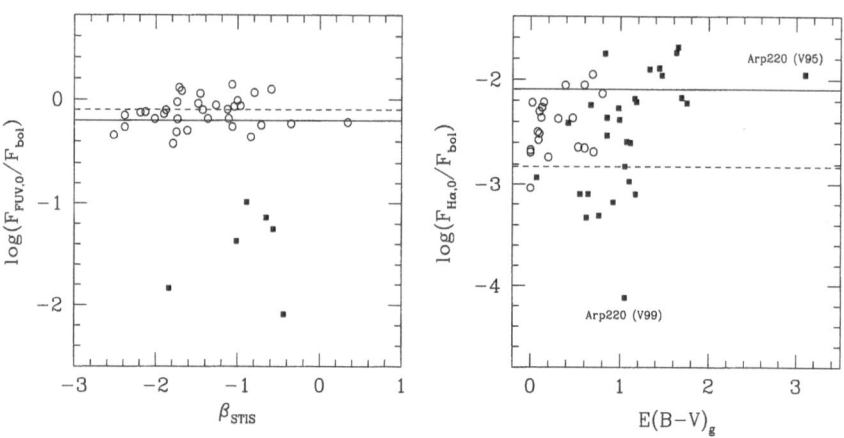

Fig. 2. Ratio of recovered flux (observed flux corrected for dust absorption as deduced from reddening) to bolometric flux, F_{bol}, plotted against reddening indicator. Symbols are as in Fig. 1b. For the IUE sample F_{bol} is a weighted sum of the observed UV and FIR fluxes, for the IRAS sample only the FIR flux is used. In **(a)**(*Left*), the numerator is the UV flux and the ratio is compared to the UV spectral slope β. In **(b)**(*Right*), the numerator is the Hα flux and the ratio is compared to the reddening $E(B-V)_g$ determined by the Hα/Hβ decrement. The horizontal lines show model predictions from Starburst99 [11] for a stellar population forming at a constant star formation rate for 100 Myr and having a Salpeter IMF with a lower mass limit of 1 M$_\odot$ and upper mass limits of 100 M$_\odot$ (solid line) and 30 M$_\odot$ (dashed line).

4   The FIR Selected Starburst Sample

Very few UV observations of ULIGs exist, perhaps because they were expected to be so dusty as to be invisible in the UV. However recent observations of ULIGs from the ground at $\lambda_c \sim 3400$ Å [22], and with HST at $\lambda_c \sim 2300, 1400$ Å [23] show that ULIGs do emit a small fraction of their bolometric luminosity in the UV.

We obtained *Space Telescope Imaging Spectrograph* UV images of seven galaxies with $L_{bol} \geq 10^{11.6} L_\odot$ [6]. In all cases the galaxies were detected in both the far UV (rest $\lambda_c \sim 1515$ Å) and the near UV (rest $\lambda_c \sim 2440$ Å) with some UV emission detected within a kpc of the nuclei [19]. However, in most cases the UV peak does not coincide well on the few hundred parcecs scale with the near IR emission. Figures 1b and 2a show that the IRX-β correlation underpredicts the bolometric flux of ULIGs by a factor ranging from ~ 7 to 90.

The situation is more optimistic with Balmer lines as shown in Fig. 2b which includes data on 28 IRAS galaxies with $L_{bol} > 10^{11.6} L_\odot$. The H$\alpha$ fluxes were derived from narrow band images [1], while spectra from a variety of published sources [24,25,27] were used to remove [NII] contamination and measure $E(B - V)_g$. After correcting for absorption, ULIGs have similar $\langle \log(F_{H\alpha,0}, F_{bol}) \rangle = -2.48$ to the IUE starbursts, with a somewhat larger dispersion: 0.51 dex.

One caveat is that these points represent total Hα fluxes corrected with *nuclear* $(R < 1$ kpc) line ratios. Gradients in $E(B - V)_g$ may mean that we overestimated $F_{H\alpha,0}$. Large spatial variations certainly exist in some galaxies as shown by the two data point for Arp220 in Fig. 2b. However, on average the gradients are shallow with typically $\Delta E(B - V)_g \approx 0.4$ mag out to $R = 8$ kpc (where the contribution to the total Hα flux is small) compared to $\langle E(B - V)_g \rangle \approx 1.1$ mag in the center [9]. Clearly, spectroscopy over the entire face of ULIGs is required to properly determine $F_{H\alpha,0}$. There is precious little of this available in the literature. While we can not yet rule out a fortuitous coincidence, Fig. 2b indicates that integrated Balmer line fluxes can be used to predict the bolometric flux of starbursts to a factor of about three accuracy, even in most ULIGs.

5   Discussion

UV color and/or Balmer line ratios can be used to crudely estimate the F_{bol} of starbursts, even ULIGs. When looking at large samples an accuracy of ~ 0.5 dex should be sufficient for measuring integrated star formation rate densities. The effective absorption implied by Figs. 1b and 2b is $\lesssim 10$ mag in the far UV, and $\lesssim 5$ mag in Hα. Five to ten magnitudes of dust absorption is large (factor of 10^2 to 10^4 in attenuation), but not overwhelming. These results contradict the notion that star formation is essentially buried behind unmeasurabley large absorption in the UV and optical. Why is this?

First of all, our results do not rule out some very buried star formation. Perhaps the completely buried phase is of short duration, before stars migrate from their birth site or the surrounding dust and gas is cleared away by the supernovae [10]. This scenario may also explain the higher dust column density affecting emission lines compared to continuum radiation [5]. However, we have not found cases where all the star formation is buried behind several tens of magnitudes of absorption in both the UV and Hα. Some appropriately reddened emission usually gets out.

Secondly, some claims for missing star formation are very model dependent. For example Poggianti [17] mentioned a short fall of a factor of three in Hα derived star formation rates compared to FIR estimates. However this is relative to a stellar population model of constant star formation rate with an assumed IMF upper mass limit $\mathcal{M}_u = 100$ M$_\odot$. A deficit of 0.5 dex compared to this model is completely consistent with our results (Fig. 2b) which suggests $\mathcal{M}_u \approx 50$ M$_\odot$ may be more appropriate. Aperture effects may be behind other claims of high extinction. For example Sturm et al. [21] use flux ratios of weak near to mid IR recombination and fine structure lines to infer a V band dust absorption of 30–80 mag for Arp220. However, the aperture size they use increases with wavelength, which can induce a spuriously large absorption since this source fills these apertures in Hα [1].

The correlation of effective absorption with β (IUE sample) and Balmer decrement (both samples) strongly suggests a foreground screen dust geometry [3,4]. While there is some hostility to such a model (e.g. [26]), we have yet to see these correlations well modeled without a screen contribution. However, this screen is not likely to be a thin uniform sheet encompassing all star formation tracers, otherwise the ULIGs would fall on the same IRX-β relationship as the IUE sample. The Charlot and Fall model [5] is a hybrid containing both foreground screen dust shells and mixed gas and dust. It works well for the IUE galaxies and perhaps can be adapted to fit the IRAS sample as well.

It should be no surprise that a foreground screen component is required, since a starburst can naturally produce such a screen. Its stellar winds and supernovae will evacuate a cavity around the starburst and power a galactic wind [7]. Most of the dust opacity will arise in the walls of this cavity. Any molecular clouds that wander into the cavity will be compressed by the high pressure within the starburst and hence have a low covering factor. Direct evidence for this scenario is given by Heckman et al. [8] who show that the metal content in the wind is directly related to the reddening. In particular, the depth of the blue shifted NaI absorption line in starbursts correlates well with both the optical continuum color and the Hα/Hβ ratio.

6   Conclusions

We conclude by answering the questions we posed at the start: 1. Yes, dust is important in most starbursts. 2. The dust geometry includes a strong foreground screen contribution, probably arising in a galactic wind. 3. Dust reddens the UV colors as the flux is diminished in the IUE starbursts, but this relationship breaks down for ULIGs. 4. Optical emission line flux ratios redden with increasing dust absorption for all types of starbursts. 5. The bolometric flux of starbursts can be recovered from their UV color (except ULIGs) or more crudely, from Balmer line flux ratios (all starbursts). These results bode well for estimating the star formation rate density locally and out to high redshift from UV and optical surveys.

Acknowledgements: We thank the conference organizers for the opportunity to research this topic. This contribution is a progression from what was shown at the meeting (including more points in Fig. 2b) where GRM lead the discussion on this topic. We thank Kurt Adelberger, Tim Heckman, Bianca Poggianti and Sylvain Veilleux for stimulating conversations that helped refine our case. We also thank our collaborators on the STIS ULIG project, Jeff Goldader, Tim Heckman, Daniela Calzetti, Dave Sanders, and Chuck Steidel.

References

1. L. Armus, T.M. Heckman, & G.K. Miley: ApJ **364**, 471 (1990)
2. A.J. Barger, L.L. Cowie, D.B. Sanders, E. Fulton, Y. Taniguichi, Y. Sato, K. Kawara, & H. Okuda: Nature **394**, 248 (1998)
3. D. Calzetti, A.L. Kinney, & T. Storchi-Bergmann: ApJ **429**, 582 (1994)
4. D. Calzetti, A.L. Kinney, & T. Storchi-Bergmann: ApJ **458**, 132 (1996)
5. S. Charlot, & S.M. Fall: ApJ **539**, 718 (2000)
6. J.D. Goldader, G.R. Meurer, T.M. Heckman, M. Seibert, D.B. Sanders, D. Calzetti, & C.C. Steidel: in preparation (2001)
7. T.M. Heckman, L. Armus, & G.K. Miley: ApJS **74**, 833 (1990)
8. T.M. Heckman, M.D. Lehnert, D.K. Strickland, L. Armus: ApJS **129**, 493 (2000)
9. D.-C. Kim, S. Veilleux, & D.B. Sanders: ApJ **508**, 627 (1998)
10. H.A. Kobulnicky, & K.E. Johnson: these proceedings [astro-ph/0011409] (2000)
11. C. Leitherer et al.: ApJS **123**, 3 (1999)
12. P. Madau, H.C. Ferguson, M.E. Dickinson, M. Giavalisco, C.C. Steidel, & A. Fruchter: MNRAS **283**, 1388 (1996)
13. K. McQuade, D. Calzetti, & A.L. Kinney: ApJS **97**, 331 (1995)
14. G.R. Meurer, T.M. Heckman, C. Leitherer, A. Kinney, C. Robert, & D.R. Garnett: AJ **110**, 2665 (1995)
15. G.R. Meurer, T.M. Heckman, M.D. Lehnert, C. Leitherer, & J. Lowenthal: AJ **114**, 54 (1997)
16. G.R. Meurer, T.M. Heckman, & D. Calzetti: ApJ **521**, 64 (1999)
17. B.M. Poggianti: these proceedings [astro-ph/0011430] (2000)
18. D.B. Sanders, & I.F. Mirabel: ARAA **34**, 749 (1996)
19. N.Z. Scoville et al.: AJ **119**, 991 (2000)
20. T. Storchi-Bergmann, A.L. Kinney, & P. Challis: ApJ **98**, 103 (1995)
21. E. Sturm et al.: A&A **315**, L133 (1996)
22. J.A. Surace, & D.B. Sanders: AJ **120**, 604 (2000)
23. N. Trentham, J. Kormendy, & D.B. Sanders: AJ **117**, 2152 (1999)
24. S. Veilleux, D.-C. Kim, D.B. Sanders, J.M. Mazzarella, & B.T. Soifer: ApJS **98**, 171 (1995) [V95]
25. S. Veilleux, D.-C. Kim, & D.B. Sanders: ApJS **522**, 113 (1999) [V99]
26. A.N. Witt, H.A. Thronson, & J.M. Capuano: ApJ **393**, 611 (1992)
27. H. Wu, Z.L. Zou, X.Y. Xia, & Z.G. Deng: A&AS **132**, 181 (1998)

HST Ultraviolet Spectra of Nearby Starbursts: Benchmarks for High Redshift Galaxies

Christy Tremonti[1,2], Claus Leitherer[2], Daniela Calzetti[2], and Tim Heckman[1]

[1] Department of Physics and Astronomy, The Johns Hopkins University, 3400 North Charles Street, Baltimore, MD 21218, USA
[2] Space Telescope Science Institute, 3700 San Martin Drive, Baltimore, MD 21218, USA

Abstract. Abundance determination has traditionally been a challenge for high redshift galaxies whose optical metallicity indicators are redshifted into the near-infrared. In the present study we develop an empirical method of metallicity estimation based upon restframe ultraviolet spectral features. We have obtained a sample of nearby starburst galaxies with high quality FOS or GHRS spectra available in the HST archives and optically derived metallicity estimates available in the literature. From this sample we construct six template spectra with metallicities spanning the range 1/10 to 2 times solar. We find a high degree of correlation between the strength and shape of the stellar wind line profiles and the metallicity of the gas. Our template spectra provide useful benchmarks for comparison with rest frame UV spectra of high redshift galaxies.

1 Introduction

The vacuum ultraviolet (UV) is an excellent regime in which to study star forming galaxies because it contains the direct spectroscopic signatures of the hot young massive stars which dominate the bolometric luminosity, as well as an abundance of interstellar features which provide valuable tracers of the gas. In the present study we attempt to exploit the diagnostic power of this rich spectral region to obtain empirical estimates of metallicity.

The effect of metallicity on the vacuum-UV spectra of individual stars has been well established. In particular, the strength of the O-star P-Cygni features (N V $\lambda1240$, Si IV $\lambda1400$, C IV $\lambda1550$) is known to depend upon the mass loss rate, which is a sensitive function of metallicity [5]. This dependence is evident in comparisons of stars of the same spectral type and luminosity class in the Milky Way and the Magellanic Clouds [1,7,8]. However, the situation is less clear for composite stellar populations, such as those found in star clusters or galaxies. In a spatially integrated starburst spectrum, the strength of the P-Cygni features is a function of both the metallicity and the number of O-stars present, which depends upon the burst age and IMF [3]. The question we seek to answer in this study is whether metallicity leaves a strong enough imprint on the UV spectral morphology of an entire galaxy for the vacuum-UV to provide a useful tracer of metallicity independent of the galaxy's star formation history. We begin by examining HST spectra of nearby starburst galaxies with well known metallicities.

2   The Sample

Our sample is composed of 12 starburst galaxies with high quality spectra in the range 1200–1800 Å available in the Hubble Space Telescope archives and optically derived metallicity estimates in the literature. Six of our spectra were obtained with the Faint Object Spectrograph (FOS) using the H13 and H19 gratings which provided a dispersion of $\sim$1 Å per diode in the 1140–1600 Å regime and $\sim$1.5 Å per diode from 1590–2300 Å. The other six spectra were observed with the Goddard High Resolution Spectrograph (GHRS) using the G140L grating which provided a dispersion of $\sim$0.57 Å over a 285 Å variable bandpass in the 1100–1900 Å range. We combined the calibrated spectra for each object into a single spectrum, including multiple pointings within a galaxy. The 12 spectra that we include in our final analysis were selected to have S/N of at least 10 in the continuum.

Table 1. Starbursts observed with HST.

Galaxy	Instr.	Dist. (Mpc)	M_B (mag)	12+[O/H]	Morphology
IRAS 08339+6517	GHRS	85.4	-20.8	8.53	Pec HII
He 2-10	GHRS	10.3	-18.0	8.93	I0 pec
NGC 1068	GHRS	16.0	-21.5	9.26	(R)SA(rs)b, Sey2
NGC 1705	GHRS	5.0	-15.7	8.36	SAO- pec, HII
NGC 1741	GHRS	52.5	-20.5	8.31	Group
NGC 2363	FOS	4.6	-17.0	7.85	HII Region in IB(s)m
NGC 3690	FOS	49.3	-21.5	8.80	Pair
NGC 4214	FOS	4.1	-17.9	8.15	IAB(s)m, HII
NGC 4670	FOS	15.6	-17.9	8.16	SB(s)0/a pec, BCD
NGC 5253	FOS	3.3	-16.6	8.16	Im pec, HII
NGC 7552	FOS	21.2	-20.5	9.20	SA(s)c pec, LINER
NGC 7714	GHRS	40.6	-20.2	8.53	SB(s)b: pec, LINER

The sample galaxies and some of their global properties are listed in Table 1. While our sample is small and not rigorously selected, our starburst galaxies nevertheless span a fairly broad swath of parameter space. That is, they include a wide range of morphological types (irregulars, nuclear starbursts, blue compact dwarfs) and activity classes (HII galaxies, LINERS, Seyfert 2's) and encompass a broad range of global galaxy parameters such as metallicity and absolute blue magnitude. In the future we hope to expand this sample with the addition of STIS data.

A systematic study of the vacuum-UV properties of a sample of 45 starburst galaxies observed with IUE was previously undertaken by Heckman et al. [2]. While this study laid some important ground work, the superior capabilities of HST offer some distinct advantages over IUE, namely in terms of spectral resolution and S/N. The modest spectral resolution of IUE (6 Å $\sim$1200 km s^{-1}) was barely sufficient to distinguish between the broad stellar wind lines and the much narrower interstellar features. With HST it is possible to go beyond measuring integrated line strengths to studying the actual line profiles, such as those shown in Figure 4. However, it should be kept in mind that the projected aperture size of our HST observations is a factor of ∼100 smaller than that of IUE. While the 10" × 20" IUE aperture provided a good match to the circumnuclear sizes of starbursts, the HST apertures employed in this sample (typically the 1.74" square Large Science Aperture for GHRS and the 0.86" circular aperture for the FOS) often encompass only a single bright star cluster. However, since one or a few bright clusters can account for a substantial portion of the UV light (40%, on average, for this sample), we assume that our conclusions would be valid if the entire galaxy were observed, as is the case for high redshift galaxies. We note, however, that age-metallicity degeneracy issues affecting the P-Cygni featrues are more of a concern in cluster spectra which are likely to be relatively coeval populations.

3   The Effects of Metallicity

The Heckman et al. [2] study of starburst galaxies observed with IUE established that, apart from the effects of dust, a starburst's metallicity is the single most important parameter in determining its UV spectral morphology. This conclusion is certainly borne out by our data. Figure 1 shows a two dimensional representation of the trend in spectral morphology with metallicity, where color is used to represent the depth of the lines. This figure has been created by simply ordering our spectra by metallicity and interpolating between them. The relatively smooth trend evident in our sample galaxies attests to the importance of metallicity over other properties such as galaxy morphological type, velocity dispersion, and bolometric luminosity, although many of these also correlate to some extent with the metallicity.

The trend of increasing line strength with metallicity is most prominent in the high ionization lines of Si IV λ1440 and C IV λ1550. Figure 2 shows a fairly tight correlation between the measured equivalent width of the absorption component (excluding the P-cygni emission bump) and the metallicity as parameterized by

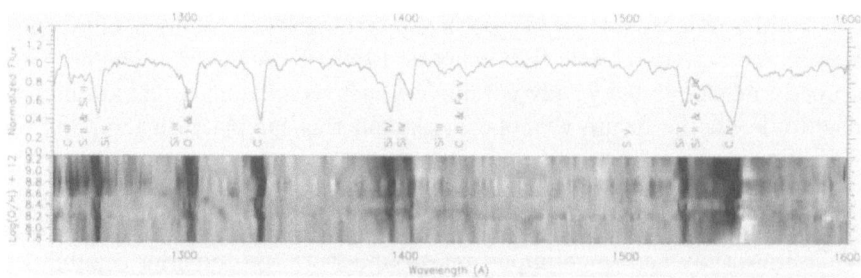

Fig. 1. Spectral morphology as a function of metallicity.

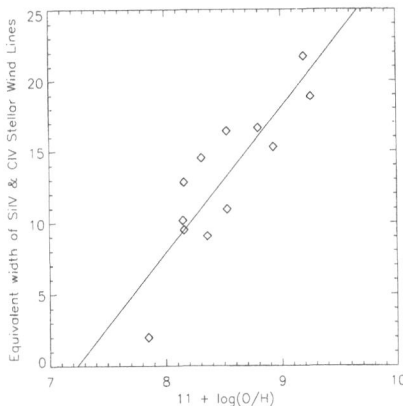

Fig. 2. Equivalent width of the stellar wind lines as a function of metallicity.

$12 + \log(O/H)$. This Z-dependence is not surprising given that these lines have a strong contribution from radiatively driven stellar winds. The strengths of the low ionization interstellar lines (Si II λ1260, O I λ1302 + Si II λ1304, and C II λ1335) also show a weak Z-dependence. This is consistent with a picture where the strong lines are highly optically thick, and thus only weakly dependent on ionic column density.

We have created six template spectra which reflect the natural metallicity groupings of our data in the range $12 + \log(O/H) = 7.85$–9.20. These are shown in Fig. 3. By combining multiple spectra into a single template, we improve the S/N and wash-out the effects of other factors such as the age of the stellar population.

The template spectra show pronounced changes in their spectral morphology with metallicity. At the low end, spectra have nearly featureless continua, weak interstellar absorption, and, most notably, weak stellar wind absorption troughs comparable in depth to the height of the P-Cygni emission component. As [O/H] increases, the galaxies develop deeper and broader interstellar lines and stellar wind absorption troughs, and continua which show broad depressions due to the collective effects of weak photospheric features. Figure 4 compares a subset of the metallicity template spectra in the region of the Si IVλ1400 and C IVλ1550 stellar wind features. Interestingly, the P-Cygni emission part of the C IV line is seen to decrease with increasing metallicity, which we interpret as an effect of increased line blanketing by photospheric iron lines. This effect provides a valuable diagnostic, as it cannot be mimicked by changing the O-star content of the population.

4   Local Starbursts as Benchmarks

The metallicity templates we have created can be utilized as benchmarks for comparison with restframe ultraviolet spectra of high redshift galaxies. Abundance

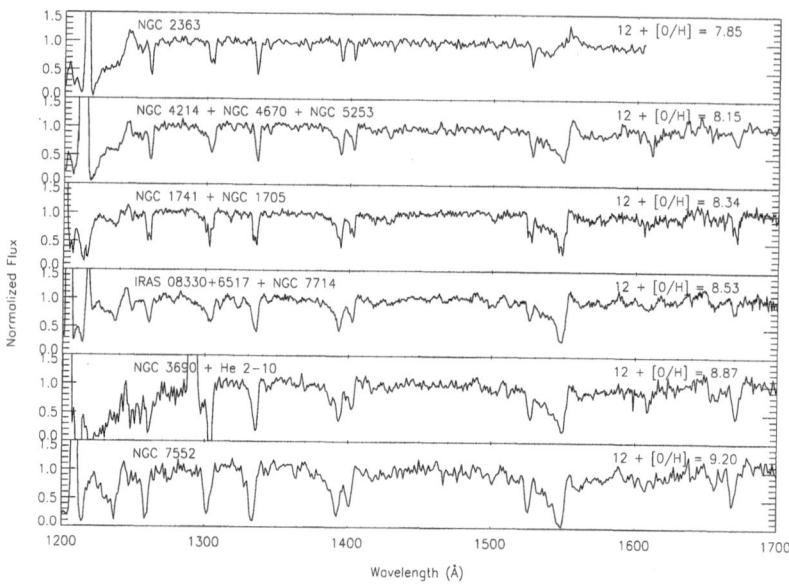

Fig. 3. Metallicity template spectra.

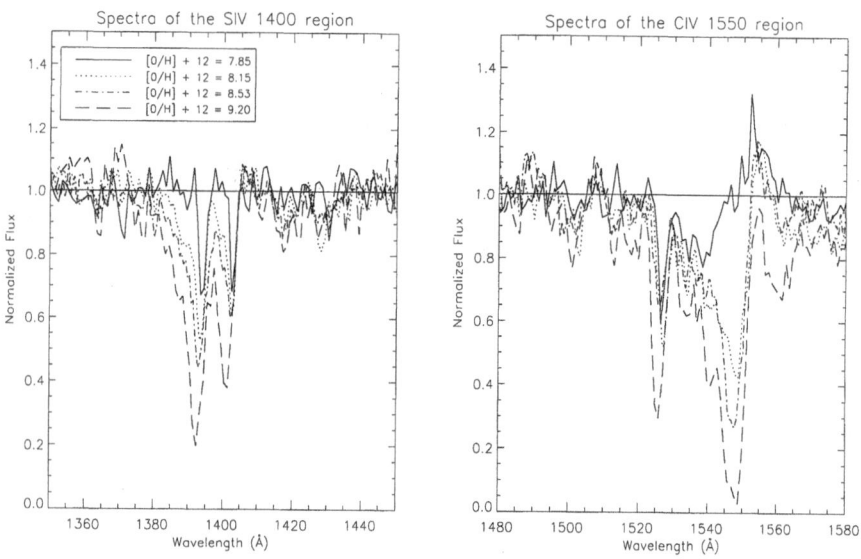

Fig. 4. Template spectra of the stellar wind lines Si IV and C IV.

determination using the traditional optical metallicity indicators is difficult for galaxies at redshifts > 2 where the restframe optical spectrum is shifted into the near-infrared. As a test of the diagnostic power of our templates, we computed the χ^2 of the difference between each of our templates and the high signal-to-noise spectrum of the $z = 2.73$ galaxy cB58 [4]. In this instance we have excluded the interstellar lines from the fit, since their extreme strength in the cB58 spectrum is a consequence of the galaxy's high star formation rate and the larger covering factor of the gas. We find the minimum χ^2 with the $12 + [O/H] = 8.34$ template, shown overplotted on the cB58 spectrum in Fig. 5. The metallicity of cB58 as measured from the optical lines using NIRSPEC on Keck II is 8.39 [6], in good agreement with the metallicity derived from our templates.

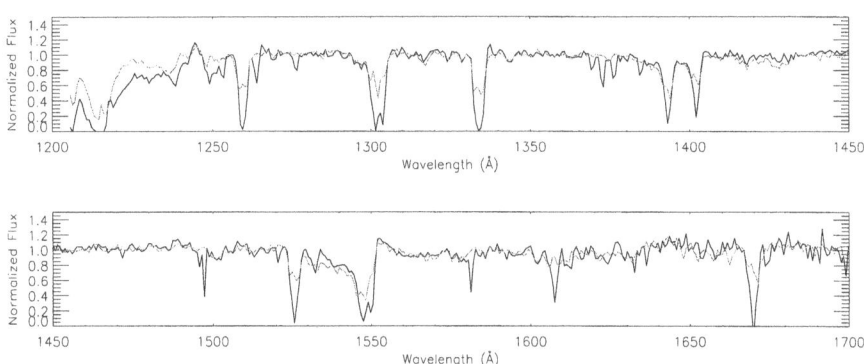

Fig. 5. Spectrum of cB58 compared to a template with $12 + \log(O/H) = 8.34$.

5   Summary and Future Work

In conclusion, the restframe ultraviolet spectral morphology of a starburst galaxy correlates strongly with metallicity. Our template spectra can be used for empirical abundance determination of high redshift objects for which metallicities are otherwise difficult to obtain. In the future we plan to improve our template spectra with the addition of more galaxies. We will also explore correlations of various restframe UV properties with other physical galaxy parameters such as mass.

References

1. C. D. Garmany, P. S. Conti: ApJ **293**, 407 (1985)
2. T. M. Heckman, C. Robert, C. Leitherer, D. R. Garnett, F. van der Rydt: ApJ **503**, 646 (1998)
3. C. Leitherer, et al.: ApJS **123**, 3 (1999)
4. M. Pettini, C. C. Steidel, K. L. Adelberger, M. Dickinson, M. Giavalisco: ApJ **528**, 96 (2000)
5. J. Puls, et al.: A&A **305**, 171 (1996)
6. H. I. Teplitz, et al.: ApJ **533L**, 65 (2000)
7. N. R. Walborn, D. J. Lennon, S. M. Haser, R. Kudritzki, S. A. Voels: PASP **107**, 104 (1995)
8. N. R. Walborn, et al.: PASP **112**, 1243 (2000)

Part VI

Cosmic Star Formation History

Deep NICMOS/HST and Radio 1.4 GHz Observations of Galaxies at High Redshifts

Lin Yan

The Carnegie Observatories, 813 Santa Barbara St., Pasadena, CA 91101, & SIRTF Science Center, Caltech, Pasadena, CA 91125, USA

Abstract. An increasing number of observations have shown that dusty starburst galaxies are probably much more numerous at high redshifts than today, and optical surveys of the distant universe suffer from large extinction corrections. In this talk, I present the quantitative estimate on how much dust extinction correction needed to apply to the rest-frame UV luminosity density at z ~ 1.3. In addition, I will discuss the recent deep 1.4GHz radio observations of extremely red galaxies (EROs), covering an area of $26' \times 26'$ with complimentary deep optical and near-IR images. We found that the fraction of bright EROs (H < 20) detected in deep 1.4GHz images is small, only 8–17%. The implication is that a large fraction of bright EROs are probably old ellipticals or systems with a small amount of star forming activities at z ~ 1 − 2. We found that ~ 20% of well detected micro-Jansky radio sources are very faint or even not detected in the optical and near-IR. These faint micro-Jansky sources have H > 20.5 and K > 19.5 − 20. The interpretation of these optically faint, micro-Jansky radio sources is that they are potentially candidates of dust enshrouded starburst galaxies beyond a redshift of 1. Our 1.4GHz detection threshold of 40μJy (5σ) sets the minimum limit of star formation activities of ~ 200$M_\odot$/yr at $z \sim 1.5$, which can be probed by our deep VLA data. An increasing number of recent observations [2,4] support our speculation that optically faint, micro-Jansky radio sources are potential dusty starbursts at high redshift. This may be the most efficient way of detecting a large sample of sub mm sources with the current available instruments.

1 Introduction

Optical surveys of high redshift galaxies preferentially sample star forming systems with a little or no dust [1]. The existence of dusty starburst galaxies at high redshifts has been dramatically revealed by the observations from the SCUBA bolometer array at sub-mm wavelengths [14,12]. Star formation rates measured from the galaxies detected at 850μm are several orders of magnitude higher than that of the rest-frame optical-UV selected galaxies at similar redshifts. Although the fraction of dusty starbursts relative to the total galaxy population at z ~ 3 is still largely unknown, the estimated total luminosity density from dusty systems at z ~ 3 could be higher or at least comparable to what has been measured from the optical surveys. At near infrared wavelengths, several recent observations have presented quantitative results on the amount of dust extinction in high redshift galaxies. In this talk, I will first discuss the results from the NICMOS parallel observations, which provide the quantitative estimate of dust extinction in galaxies at $z \sim 1-2$. In particular, I will present the deep 1.4GHz radio observation of a wide field with complimentary deep optical and near-IR images. This combined multi-wavelength dataset shows that faint micro-Jansky radio sources could be sign-posts of dusty star forming galaxies at high redshift.

2 H_α Emission Line Galaxies at $z \sim 0.7 - 1.9$

The NICMOS parallel imaging and grism observations were both made with Camera 3 with a field of view $\sim 52'' \times 52''$. The imaging data were taken with broad band filters F110W and F160W at 1.1μ (J band) and 1.6μ (H band). The grism data has a spectral resolution of 200 per pixel and covers wavelength regions from 0.8μ to 1.2μ (G096) and 1.1μ to 1.9μ (G141). We have reduced and analysed the NIC3 parallel grism G141 data, covering ~ 65 sq. arcminutes. The NICMOS data at the near-IR offers the unique advantage of accessing the redshift range between $1-2$ where little is known about the properties of normal galaxies by using redshifted H_α emission lines. In addition, the measurements of star formation rates based on near-IR spectra suffer less uncertainty from dust extinction. A more detailed description of the survey and data can be found in [17,18], and [19].

We found a total of 33 H_α emission line galaxies over an effective co-moving volume of $10^5 \, h_{50}^{-3}$ Mpc^3 for $q_0 = 0.5$. The implied co-moving number density of emission line galaxies in the range $0.75 < z < 1.9$ is $3.3 \times 10^{-4} \, h_{50}^3$ Mpc^{-3}, very similar to that of the bright Lyman break objects at $z \sim 3$. These objects have a median H_α luminosity of 2.7×10^{42} erg sec^{-1}. Most, if not all, of the emission lines detected are either H_α or an unresolved blend of H_α+[NII]6583/6548. This identification is mostly based on H-band apparent magnitudes (median value of 20.5), the emission line equivalent widths and the lack of other detected lines within the G141 bandpass. The redshifts of 6 galaxies in our sample have been confirmed by detection of [OII]3727 emission in the optical spectra using LRIS on the Keck 10m telescope. The spectra of these 33 galaxies can be found in [18].

We compute the H_α luminosity function LF based on this sample of emission-line galaxies. All of the detailed results are in [18]. Figure 1 shows our derived H_α LF at $z = 1.3$ and the local H_α LF as measured by [6]. This plot shows strong evolution in the H_α luminosity density from $z \sim 0$ to $z \sim 1.3$. This is no surprise given the evolution in the ultraviolet luminosity density, but our result provides an independent measure of evolution for H_α emission alone. The LF is well fit by a Schechter function over the range $6 \times 10^{41} < L(H\alpha) < 2 \times 10^{43}$ erg sec^{-1} with $L^* = 7 \times 10^{42}$ erg sec^{-1} and $\phi^* = 1.7 \times 10^{-3}$ Mpc^{-3} for $H_0 = 50$ km s^{-1} Mpc^{-1} and $q_0 = 0.5$. The integrated H_α luminosity density at $z \sim 1.3$ (our median z) is $1.64 \times 10^{40} \, h_{50}$ erg s^{-1} Mpc^{-3}, ~ 14 times greater than the local value reported by [6].

We compute the star formation rates from the integrated H_α and the UV continuum luminosity densities using the relations from [11]. In Fig. 2, we plot uncorrected measurements of the star formation rate at various epochs. The clear

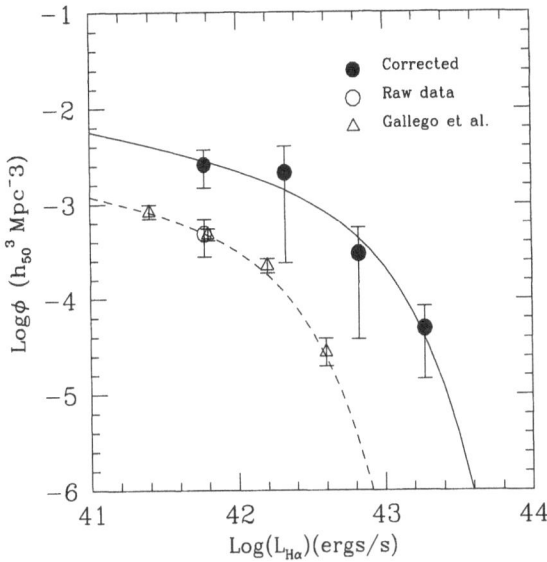

Fig. 1. Hα luminosity function at $0.7 < z < 1.9$. The open circles represent our raw data and the filled circles are our points corrected for incompleteness. The open triangles show the local Hα luminosity function by [6]. The solid and dashed lines are the best fits to the data at $z \sim 1.3$ and $z \sim 0$ respectively.

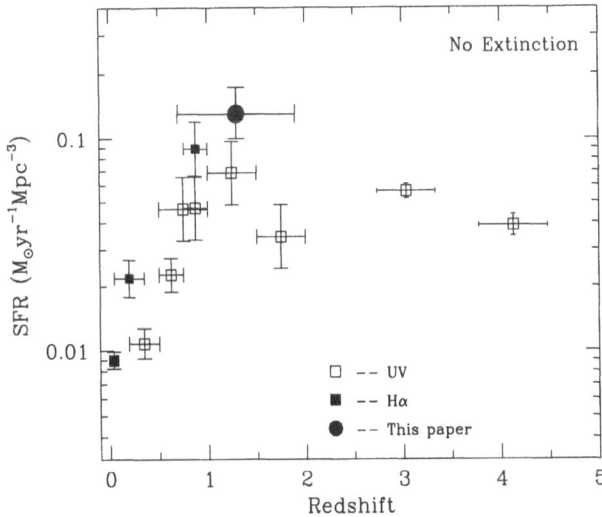

Fig. 2. The global star formation rate vs. z without extinction correction. The open squares represent the measurements at 2800Å or 1500Å by [12,5,15], whereas the filled squares are the measurements using Hα 6563Å by [6,16,7]. Our result is shown in the filled circle.

trend for the longer wavelength determinations of the star formation rate to exceed those based on UV continua is one of the pieces of evidence for significant extinction at intermediate and high redshifts. The amplitude of the extinction correction is quite uncertain. Our measurement spans $0.7 < z < 1.9$, overlapping with the [5] photometric redshift sample and allowing a direct comparison between the observed 2800Å luminosity density and that inferred from Hα. Our Hα-based star formation rate is three times larger than the average of the three redshift bins measured by [5]. If we attribute the entire difference to reddening, the total extinction corrections at 2800Å and Hα are large and model-dependent. The dust extinction at 2800Å is 2.1 magnitude with a MW reddening curve and $1 - 2$ magnitude higher if we use the local starburst reddening law.

3   High Redshift Dusty Star Forming Galaxies Probed by Deep 1.4 GHz Observations

Several deep surveys at 850μm have revealed a population of dust enshrouded starburst galaxies at redshift beyond 1 [12,14,2]. However, many outstanding questions regarding the high redshift universe at sub-mm to mm wavelengths remain unclear. This is primarily due to the fact that only a very small number of 850μm sources have reliable identifications in the optical, near-IR and radio wavelengths. Most deep sub-mm surveys were done without pre-selection of the targets. For these surveys, the poor spatial resolution in the submillimeter made the source identifications at the optical and near-IR virtually impossible. Furthermore, it is very difficult to achieve high completeness in optical spectroscopic surveys of submm galaxies due to the extreme faintness of the candidate opticals counterparts (e.g. [10,2]).

Because of the tight far-IR/radio correlation [9], deep 1.4 GHz radio maps reaching down to micro-Jansky levels could be used to pinpoint the locations of star forming activities at high redshifts. Several observational results support the above speculation. At the sub-mJy level, the previous spectroscopic work has shown that these radio sources are primarily star forming galaxies rather than AGNs at $z < 1$ [3,8]. The recent 1.4GHz radio maps of the HDF by [13] suggests that a large fraction of micro-Jansky radio sources are star forming galaxies as well. The most direct supportive evidence came from the SCUBA observation by [2]. They have observed 15 targets selected to be both micro-Jansky 1.4GHz radio sources and optically/near-IR faint. They detected 5 of them at 850μm.

To fully understand how micro-Jansky radio observations could be used as an efficient tool, in combination with deep optical and near-IR images, for selecting potential dusty starbursts at high redshifts, we obtained very deep 1.4 GHz radio map (VLA B array) of a field where we also have deep optical and near-IR images. The field of view of all datasets (V, H and 1.4GHz) covers $26' \times 26'$; and the sensitivities reach V of 27mag, H of 21mag and 1σ RMS noise of 8.8μJy at 1.4 GHz. We detected 280 radio sources over 400 square arcminutes above 5σ 40μJy. Figure 3 shows the 1.4GHz flux histogram of the detected sources. As shown, the majority of the sources have fluxes less 200μJy. The surface density

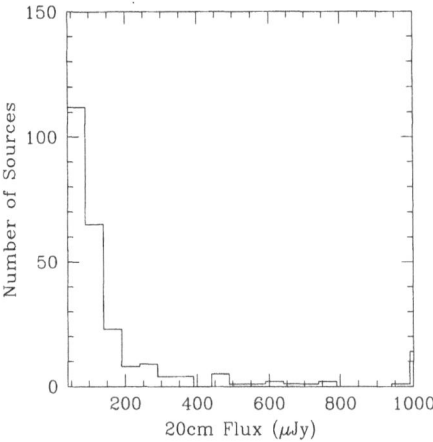

Fig. 3. 1.4GHz radio flux histogram of the sources detected over 400 square arcminutes. As seen, the majority of the sources are faint micro-Jansky sources, with fluxes between 40 to 200μJy.

of sources with 40μJy < S(1.4GHz) < 200μJy is $\sim$ 0.6/sq.arcmin., consistent with previous work [13].

Figure 4 shows the color-magnitude diagram of galaxies detected in our data, with high-lighted solid points indicating the sources also detected in 1.4GHz. It is clear from this plot that *the fraction of red galaxies with V - H > 5.5 detected in the radio is small, 8 − 17% for H < 20.* This implies that systems like ERO HR 10 are relatively rare. Perhaps a large fraction of bright EROs (H < 20, K $\sim$ 19) is indeed old ellipticals or systems with a small amount of star formation activities, but they are below our 1.4GHz detection threshold.

In addition, $\sim$ 20% of radio sources, well detected in 1.4GHz, are found to be very faint or not even detected in the optical and near-IR. Their 1.4GHz fluxes are mostly at the micro-Jansky level, with a few exceptions having submilli-Jansky fluxes. Our speculation is that these optically faint, micro-Jansky sources could be fundamentally different in their nature from the H-band bright red galaxies which are also detected in the 1.4GHz map. These faint micro-Jansky radio sources generally have H > 20.5 or K > 19.5 − 20.0. Their low level of optical/near-IR fluxes could be interpreted as strong dust extinction in distant starburst systems. The recent observation by [2] fully supports this scenario, although their sample is fairly small. Our radio observations show that optical-IR faint micro-Jansky radio sources are perhaps the best candidates for the sub-mm detections. From the radio/far-infrared correlation, we can estimate the minimum star formation rate set by our 1.4GHz detection threshold, with the assumption of the source redshift. For 40μJy limit and with $z \sim$ 1.5, the minimum SFR is about 200M$_\odot$/yr. Although this type of calculation has large uncertainties, the estimated SFR does provide a rough indicator of the type of starburst population probed by deep radio maps from the VLA.

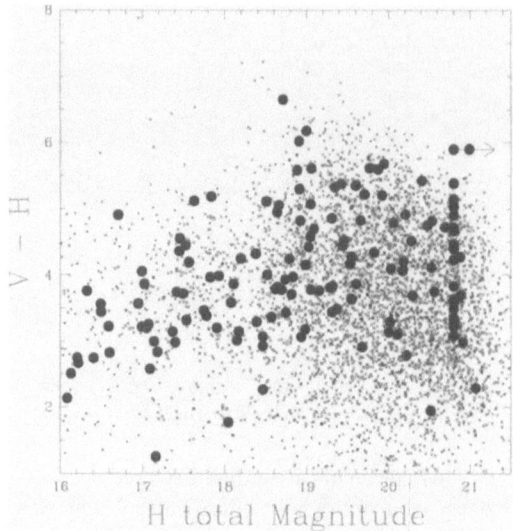

Fig. 4. Color V - H versus H magnitude diagram for all of the galaxies detected within a field of view of $26' \times 26'$. Larger solid dots represent the systems detected in deep 1.4GHz radio map, as well as in V and H. Arrows indicate the lower limits either in V or H band. A large number of sources detected in the radio are too faint in the optical V and/or the near-IR H band. These sources do not show up in this plot.

References

1. K. Adelberger, C. C. Steidel: ApJ submitted (2000)
2. A. Barger, L. Cowie, E. Richards: AJ **119**, 2092 (2000)
3. C. R. Benn, M. Rowan-Robinson, R. G. McMahon, et al.: MNRAS **263**, 98 (1993)
4. S. Chapman, et al.: ApJ submitted (2000)
5. A. Connolly, A. S. Szalay, M. Dickinson, et al.: ApJ **487**, L13 (1997)
6. J. Gallego, J. Zamorano, A. Aragon-Salamanca, M. Rego: ApJL **459**, 1 (1995)
7. K. Glazebrook, C. Blake, F. Economou, S. Lilly, M. Colless: MNRAS **306**, 843 (1999)
8. F. Hammer, D. Crampton, S. J. Lilly, et al.: MNRAS **286**, 470 (1995)
9. G. Helou: ApJ **311**, 33L (1986)
10. R. Ivison et al.: ApJ **542**, 271 (2000)
11. R. Kennicutt: ARAA **36** (1999)
12. S. Lilly, et al.: ApJ **518**, 641L (1999)
13. E. A. Richards: ApJ **533**, 611 (2000)
14. I. Smail, et al.: ApJ **490**, 5 (1997)
15. C. C. Steidel, K. Adelberger, M. Giavalisco, M. Dickinson, M. Pettini: ApJ **519**, 1 (1999)
16. L. Tresse, S. J. Maddox: ApJ **495**, 691 1998
17. L. Yan, P. McCarthy, et al.: ApJ **503**, 19L (1998)
18. L. Yan, P. McCarthy, et al.: ApJL **519**, 47 (1999)
19. L. Yan, P. McCarthy, et al.: AJ **120**, 575 (2000)

Star Forming Galaxies at z ≥ 2

Alan Moorwood[1], Paul P. van der Werf[2], Jean Gabriel Cuby[3], and Tino Oliva[4]

[1] European Southern Observatory, Karl-Schwarzschild-Str. 2, 85748 Garching, Germany
[2] Leiden Observatory, P.O. Box 9513, 2300 RA Leiden, The Netherlands
[3] European Southern Observatory, Alonso de Cordova 3107, Santiago, Chile
[4] Osservatorio Astrofisico di Arcetri, Largo E. Fermi 5, 50125 Firenze, Italy

1 Introduction

The availability of a new generation of infrared spectrometers on 8–10 m class telescopes has recently opened up the exciting prospect of studying the rest frame optical properties of galaxies at z ≥ 2. In particular, the familiar nebular emission lines [OII]($\lambda\lambda3727$), Hβ,[OIII]($\lambda\lambda4959,5007$), Hα and [NII]($\lambda\lambda6548,6584$) can be exploited to determine star formation rates, extinction, metal abundances and masses for comparison with those of nearby galaxies measured in the same way. The detection limits now achievable should also make it feasible to obtain infrared spectra of relatively large samples of high z galaxies and hence of pursuing statistically significant studies of the abundance and velocity versus magnitude, and other relationships, which may provide important clues to the process of galaxy formation.

The one practical problem that remains is the selection of suitable galaxies with redshifts within the range of interest, and such that the target lines of interest fall both within the atmospheric windows and between the forest of strong OH sky lines which plague near infrared spectroscopy from the ground. The situation is most favourable at z ≃ 2.5–3.5 where redshifts are now available for several hundred galaxies detected by the Lyman Break technique [16]. Pioneering infrared spectroscopy [13] of a few of these using UKIRT first illustrated the potential for the type of nebular line studies mentioned above. The fact that integration times approaching a night per line were required also illustrated the need for more sensitive instrumentation in order to make further progress. This is now the case and relatively high quality spectra have recently been obtained [14] for a sample of 15 z ≃ 3 redshift selected Lyman Break galaxies using NIRSPEC [9] at Keck and ISAAC [11] at the VLT.

At redshifts of z ≃ 2 the Lyman Break technique is not very efficient. Photometric redshifts are becoming available for an increasing number in this range (see e.g. [2]) but spectroscopic redshifts are difficult to obtain due to the absence of suitable lines in the visible. Based on the first attempt to construct the cosmic star formation history from the UV fluxes of galaxies at z ≤ 1 and z ≥ 3 [7], however, it appeared that this epoch might correspond to the peak of star formation activity. This stimulated our own attempt to assemble a sample of galaxies at z ≃ 2 by conducting an infrared narrow band imaging survey for Hα emission redshifted to around 2.1 μm. By selection of the survey filter wavelengths this technique also satisfies the requirement for follow-up spectroscopy that at least the target line is clear of strong OH sky lines. The technique is also not new. Narrow band surveys in the K band to various depths have been made previously [8,17,20,21]. These have produced $\simeq 30$ candidate line emitting galaxies but mostly in fields containing quasars or radio galaxies at the target redshift. To my knowledge, only one of these has been confirmed spectroscopically [1].

I will briefly summarize here the main results of our survey for Hα emission at $z = 2.2$ which reached the deepest limits to date, covered 100 sq. arcmin including the HDFS and has already been followed up with near infrared spectroscopy of the most promising candidates. More details can be found in reference [12].

2   Candidate Hα Emitting Galaxies at z = 2.2

The survey part of this programme was conducted with SOFI [10] at the ESO NTT and covered several individual fields of $\simeq 20$ sq. arcmin centred on the WFPC2 and STIS HDFS fields plus a randomly selected one about 30$°$ away. Each of the HDFS fields were observed in two narrow band filters (1% at 2.09 μm and 1.3% at 2.12 μm) as well as the broad Ks filter, yielding line flux limits of 5–10×10^{-17} erg cm^{-2} s^{-1}, a total effective survey area of $\simeq 100$ sq. arcmin and a total co-moving volume of ~ 8600 Mpc^3 for Hα emitters at $z = 2.2$ (adopting $H_0 = 50$ km s^{-1} Mpc^{-1} and $q_0 = 0.5$). About 10 candidate line emitting objects were detected of which 6 have so far been spectroscopically confirmed as described below. Their volume density is therefore $\sim 10^{-3}$ Mpc^{-3}. The density of objects found by narrow band infrared imaging thus appears to be comparable to the $\sim 7 \times 10^{-4}$ Mpc^{-3} estimated for Lyman Break galaxies from their mean surface density of $\simeq 0.5$ per sq. arcmin in the redshift range $z \simeq 2.5$–3.5 [16].

Line emission has been confirmed spectroscopically with ISAAC at the VLT in 6 of the 10 candidate objects. Re-inspection of the survey data is consistent with one of the others being spurious and another two probably being fainter than the spectrosopic sensitivity reached. Although small, it is worth noting that this is the first and hence largest sample of spectroscopically confirmed candidates found in a narrow band infrared imaging survey. Unfortunately, the spectra exhibit only a single line whose identification remains somewhat uncertain. Hα is considered the most plausible and the absence of the [NII] companion lines normally seen in lower z galaxies is reconcilable with sub-solar abundance, high excitation and/or simply the low s/n ratios. The same argument could be used to support its identification with [OIII]($\lambda\lambda$5007) except that its factor of 3 fainter companion at $\lambda\lambda$4959 is not clearly detected. One of the striking recent results from infrared spectroscopy of the Lyman Break galaxies, however, is that [OIII]($\lambda\lambda$5007) is always brighter, by up to a factor 2, than Hα (or as inferred from Hβ) [5,14,18,19]. This implies that, contrary to earlier assumptions, narrow band surveys in the K band are almost as sensitive to [OIII]($\lambda\lambda$5007) at $z = 3.2$ as Hα at $z = 2.2$ from galaxies with the same star formation rate. A follow up programme to clarify this issue is in progress.

2.1 Star Formation Rates and the SFRD

Star formation rates derived for individual galaxies from their Hα fluxes are in the range 20–35 $M_\odot$ yr^{-1} (using the Kennicutt [4] relationship appropriate for continuous star formation and a Salpeter IMF extending from 0.1–100 $M_\odot$). These are 0.7–3.7 times those derived alternatively from the rest frame UV continuum fluxes of the same galaxies deduced from their WFPC2 I bands magnitudes and using the corresponding relationship from Kennicutt. This difference has been attributed mostly to the lower extinction at the wavelength of Hα – expected on the basis of the intrinsic wavelength dependence of dust extinction. The actual situation, however, is more complicated. Firstly, Hα traces primarily the youngest stars whose local extinction within their parent clouds acts in the opposite direction by decreasing the intrinsic Hα/UV continuum ratio. Secondly, the Hα/UV continuum ratio also decreases during the decaying starburst phase when the Kennicutt relationships for continuous star formation no longer apply. It is therefore interesting that the same comparison made recently on the larger sample of Lyman Break galaxies yields variations of a factor of $\simeq 2$ between the SFRs but no systematic trend for those derived from the Balmer lines to be higher [14]. At least in UV and Hα selected galaxies therefore it cannot, contrary to the initial expectations, be claimed that the Balmer lines yield significantly more reliable estimates of the SFRs than the UV continuum fluxes. The situation in more highly obscured, ULIRG type galaxies appears to be quite different but is beyond the scope of this paper.

Hα estimates of the high z star formation rate density (SFRD) are still of interest because they provide the only direct comparison with that of the local universe estimated using the same star formation tracer.

Using the $1/V_{max}$ estimator, a crude Hα luminosity function has been constructed for our small sample with just three bins in the range log $L_{H\alpha}$(erg s^{-1}) = 42.2–43. Of interest first is the fact that it is well represented by the same Schechter function fitted by Yan et al. [22] to their Hα luminosity function at $z \sim 1.3$ derived from NICMOS grism survey data. Integrating over this function yields a SFRD of $\sim 0.14 \, M_\odot$ yr^{-1}, which is about a factor of 15 larger than that derived by Gallego et al. [3] from their Hα survey of nearby galaxies, and in agreement with the steep rise up to $z \simeq 1$ deduced from the UV continuum fluxes of CFRS galaxies by Lilly et al. [6]. It is also equal to the latest values derived from extinction corrected UV continuum fluxes of Lyman break galxies at $z = 3$–4.5 by Steidel et al. [15]. The star formation rate history traced from $z = 0$–2.2 in Hα therefore appears fully consistent with that deduced from UV continuum fluxes.

2.2 Galaxy Dynamics

Perhaps the most interesting and unexpected result concerns the dynamics of these galaxies and the potential for future work in this area. For this, the spectra obtained so far are not ideal due to a combination of the relatively short integration times of $\simeq 1$ hr and poor seeing of typically 1–1.5″ leading to the use of a 2″ slit in most cases. Nevertheless, the line is generally resolved in velocity with observed FWHM of 150–479 km s^{-1} and derived velocity dispersions of 50–175 km s^{-1} which are comparable to those measured in the Lyman Break galaxies [14]. The simplest interpretation is that this implies masses of $\sim 10^{10}$ M$_\odot$ within just the central few kpc of these galaxies – a conclusion of extreme significance within the context of hierarchical galaxy formation models. Unfortunately, there remains the possibility that the emission line velocity dispersions are dominated by outflows or other gas motions unrelated to the gravitational mass. Of most interest, therefore, is one of the galaxies which was observed with 0.6″ seeing and whose spatially resolved line emission is consistent with ordered rotation. Interpreting it as a rotation curve implies a terminal velocity of ≥ 140 km s^{-1} at $\simeq 3$ kpc which falls within the range observed for nearby spirals. The WFPC2 I band image also shows a single, elongated object (by chance aligned with the slit) thus making this probably the best candidate to date for a relatively massive ($\geq 10^{10}$ M$_\odot$) disk formed at z ≥ 2. Of interest also is that, even with its relatively large velocity width, the absolute B magnitude of this galaxy derived from its observed H band magnitude still lies 2–3 mag above the Tully-Fisher relationship for nearby spirals.

Acknowledgements: I wish to thank the organisers of this extremely interesting and enjoyable meeting.

References

1. S.V.W. Beckwith, D. Thompson, F. Mannucci, S.G. Djorgovski: ApJ **504**, 107 (1998)
2. A. Fontana, S. D'Odorico, F. Poli, et al.: AJ **120**, 2206 (2000)
3. J. Gallego, J. Zamorano, A. Aragón-Salamanca, M. Rego: ApJ **455**, L1 (1995)
4. R.C. Kennicutt: ARA&A **36**, 189 (1998)
5. H.A. Kobulnicky, D. Koo: ApJ, in press (2000)
6. S.J. Lilly, O. Le Fèvre, F. Hammer, D. Crampton: ApJ **460**, L1 (1996)
7. P. Madau, H.C. Ferguson, M.E. Dickinson, et al.: MNRAS **283**, 1388 (1996)
8. F. Mannucci, D. Thompson, S.V.W. Beckwith, G.M. Williger: ApJ **501**, L11 (1998)
9. I.S. McLean, J.R. Graham, E.E. Becklin, et al.: SPIE **4008**, 1048 (2000)
10. A.F.M. Moorwood, J.G. Cuby, C. Lidman: The Messenger **91**, 9 (1998)
11. A.F.M. Moorwood, J.G. Cuby, P. Ballester, et al.: The Messenger **95**, 1 (1999)
12. A.F.M. Moorwood, P.P. van der Werf, J.G. Cuby, E.E. Oliva: A&A **362**, 9 (2000)
13. M. Pettini, M. Kellog, C.C. Steidel, et al.: ApJ **508**, 539 (1998)
14. M. Pettini et al.: ApJ, submitted, and astro-ph/0101254 (2000)
15. C.C. Steidel, et al.: ApJ **519**, 1 (1999)
16. C.C. Steidel: SPIE **4005**, 22 (2000)
17. H.I. Teplitz, M.A. Malkan, I.S. McLean: ApJ **506**, 519 (1998)
18. H.I. Teplitz, M.A. Malkan, C.C. Steidel, et al.: ApJ **542**, 18 (2000a)
19. H.I. Teplitz, I.S. McLean, E.E. Becklin, et al.: ApJ **533**, L65 (2000b)
20. D. Thompson, F. Mannucci, S.V.W. Beckwith: AJ **112**, 1794 (1996)
21. P.P. van der Werf, A.F.M. Moorwood, M.N. Bremer: A&A **362**, 509 (2000)
22. L. Yan, P.J. McCarthy, W. Freudling, et al.: ApJ **519**, L47 (1999)

Identification of ISOPHOT Deep Field Sources

David B. Sanders[1,2], Sylvain Veilleux[3], Min Yun[4], Lennox L. Cowie[1],
Kimiaki Kawara[5], Yoshiaki Taniguchi[6], Shinki Oyabu[5], Takashi Murayama[6]
and Haruyuki Okuda[7]

[1] Institute for Astronomy, Univ. of Hawaii, 2680 Woodlawn Dr., Honolulu, HI 96822,
USA
[2] Max-Planck-Institut für extraterrestrische Physik, Postfach 1312, 85741 Garching,
Germany
[3] Department of Astronomy, University of Maryland, College Park, MD,20742, USA
[4] Department of Astronomy, University of Massachusetts, Amherst, MA 01003, USA
[5] Institute of Astronomy, U. Tokyo, 2-21-1 Osawa, Mitaka, Tokyo 181-8588, Japan
[6] Astronomical Institute, Tohoku University, Aoba, Sendai 980-8578, Japan
[7] Gunma Astronomical Observatory, 6860-86 Nakayama, Takayama, Agatsuma,
Gunma 371-0702, Japan

Abstract. We have completed an analysis of the deepest far-infrared ($90\,\mu$m,$175\,\mu$m)
survey made using the ISOPHOT photometer aboard the *Infrared Space Observatory*
(ISO). The cumulative number counts for two $44' \times 44'$ regions in the Lockman Hole appear to be ~ 10 times higher than expected from a no-evolution model. Optical counterparts for the brightest $175\,\mu$m sources, ($S_{175} > 150\,$mJy), have recently been identified
using I-band and K-band imaging plus VLA-B 20cm continuum maps. Keck spectroscopy of a subsample of these sources shows that they are dusty emission line galax-
ies at redshifts $z < 1.5$, with infrared luminosities[1] in the range $L_{\mathrm{ir}} \sim 10^{10} - 10^{13}\,L_\odot$.
These ISOPHOT sources plausibly represent a link between the low-z luminous in-
frared galaxies (LIGs) detected by IRAS and the more distant submillimeter sources
that have recently been detected in deep $850\,\mu$m surveys with SCUBA.

1   Introduction

One of the main goals of the Japan/UH cosmology program using ISAS guaranteed time on ISO (see review [19]) was to carry out a deep far-infrared survey of
the Lockman Hole (LH) at $90\,\mu$m and $170\,\mu$m using the ISOPHOT photometer
[15]. These observations were motivated partly by a desire to test whether the
hint of a strongly evolving population of ultraluminous infrared galaxies (ULIGs)
as seen at low redshifts ($z < 0.25$) by IRAS (e.g. [13]) continued to higher redshifts, as well as to better gauge the relative contribution of "dust-enshrouded"
far-infrared sources to the total energy budget of the universe. These ISOPHOT
observations have proved all the more exciting due to the post-mission discovery of what appears to be a substantial population of ULIGs at high redshifts
($z > 1.5$), using the submillimeter array SCUBA on the James Clerk Maxwell
Telescope (JCMT) (e.g. [26,11,2,16]). The ISOPHOT data offer the opportunity
to "bridge the redshift gap" between the IRAS and SCUBA sources (see Fig. 1).

---

[1] $L_{ir} \equiv L\,(8 - 1000\,\mu m)$ in the object rest frame. $H_{\mathrm{o}} = 75\,\mathrm{km\,s}^{-1}\mathrm{Mpc}^{-1}$, $q_{\mathrm{o}} = 0$.

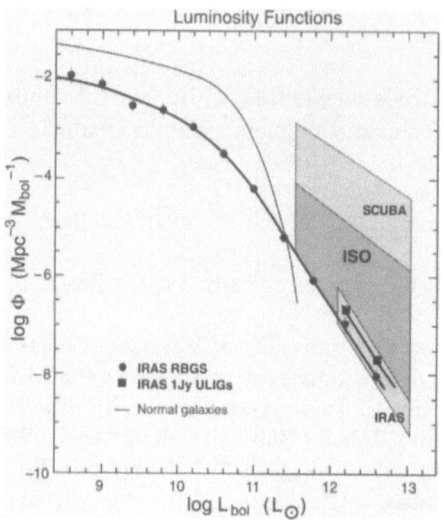

Fig. 1. Luminosity functions (LFs) of infrared selected galaxies (thick lines with data points: IRAS RBGS [22]; IRAS 1-Jy ULIGs [14]. Shown for comparison is the luminosity function of nearby optically selected galaxies [24]. Shaded regions represent the plausible range of values for the LF of LIGs derived from the current measured and "best-guess" redshift distributions for sources detected in the ISO and SCUBA deep fields.

1.1 The Lockman-Hole Deep Fields

Two $44' \times 44'$ deep fields were mapped with the ISOPHOT detector, both of which were located in the Lockman Hole [17]. LH_EX was centered on the Rosat X-ray deep field [9] and LH_NW, approximately one degree WNW of the Rosat field, was centered on the true minimum in the HI column density maps. Both fields were mapped with two filters: C_90 (centered at $90\,\mu$m) and C_160 (centered at $170\,\mu$m). The maps were taken with half-beam pixels ($23''$/pixel at $90\,\mu$m and $46''$/pixel at $170\,\mu$m), and reduced with the PHT Interactive Analysis Package. Final flux scaling was based on the IRAS fluxes of the brightest source F10507+5723 (UGC 06009). A complete description of the data reduction can be found in [12] and [18].

2 Source Counts

Analysis of the cleaned and calibrated ISOPHOT maps yields cumulative source counts, N (sources/sr), of 1.4×10^5 at $S_{90} = 150\,$mJy and 5.4×10^4 at $S_{170} = 250\,$mJy. These values are 3–10 times higher than expected from a simple extrapolation of the IRAS cumulative $100\,\mu$m counts to lower flux levels (assuming similar far-infrared rest frame colors for ISOPHOT and IRAS galaxies), and indicate strong evolution in the far-infrared galaxy population, even above that expected from the strongly evolving model-"E" of [7]. An even stronger increase in the source counts is found when pushing down to flux levels of 70 mJy and 100 mJy at 90 μm and 170 μm respectively, consistent with the results reported by [21,5] for ISOPHOT deep surveys of the ELAIS and Marano deep fields. Matsuhara et al. [18] also give constraints on the number counts down to $S_{90} = 35$ mJy and $S_{170} = 60$ mJy, which continue to indicate the existence of strong evolution, as well as yielding an integrated sky brightness of 0.09 - 0.30 MJy/sr at 90 μm and 0.053 - 0.15 MJy/sr at 170 μm. Compared to the COBE measurements of the cosmic far-infrared background [20,6,10] our new results suggest that 15–40 % of this background can now be attributed to the integrated light from discrete far-infrared sources.

3   Identification of Optical Counterparts

The reduced ISOPHOT deep field maps were overlaid on I-band and K'-band images in an attempt to identify optical counterparts of the far-infrared sources. Figure 2 is a "finder chart" for source #9 in the LH_NW field, illustrating the fact that several optical sources are typically seen within the 170 μm beam. Addition of the 90 μm beam greatly aids in source identification, but it is not until much higher resolution radio continuum data is added (in this case, VLA-B 20cm) that the source of the far-infrared flux is pinned down. The R-K color of the identified object is consistent with that observed for ULIGs (e.g. [13]), as is its radio/far-infrared flux ratio (e.g. [4]). The case of LH_NW9 is typical of many of the finder charts in that the true source is often not well centered in the ISOPHOT 170 μm beam, most likely due to the fact that sources are riding on top of a fluctuating background composed of unresolved galactic cirrus.

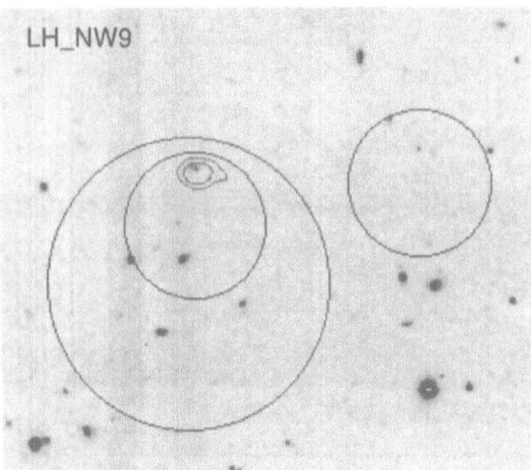

Fig. 2. ISOPHOT deep field source LH_NW9 ($S_{170} = 252$ mJy, $S_{90} = 139$ mJy) at $z = 1.107$. Large and small circles represent 170 μm and 90 μm sources respectively [12]. Contours are VLA-B 20cm emission [29]. The I-band image was obtained with the 8K×8Kcamera on the UH2.2m telescope.

4   Redshifts and Infrared Luminosities

We have used I-band, K′-band and VLA 20cm continuum maps to construct finder charts for all (~ 80) sources with $S_{170} > 100\,\mathrm{mJy}$ in both LH fields. The ESI and NIRSPEC spectrographs on KeckII were then used to obtain redshifts of the likely optical/NIR counterparts. ESI observations of the brightest sources ($S_{170} > 200\,\mathrm{mJy}$), taken during 2000 Mar, yielded reliable redshifts for all 18 sources observed. Spectra for the 9 lowest redshift sources are shown in Fig. 3. All objects show emission features typical of starburst and/or narrow-line AGN (c.f. [28]) Many of the spectra also show evidence of heavy dust obscuration, both from a large Balmer decrement as well as red (B-I) colors.

The redshifts of the 18 sources ranged from 0.028 to 1.107 with resultant infrared luminosities in the range $L_{\mathrm{ir}} \sim 10^{10} - 10^{13}\,L_{\odot}$. These results are shown in Fig. 4. Four sources qualify as ULIGs ($L_{\mathrm{ir}} > 10^{12}\,L_{\odot}$) one of which is a hyperluminous infrared galaxy (HyLIG: $L_{\mathrm{ir}} > 10^{13}\,L_{\odot}$). The significance of finding one HyLIG in a region as small as $\sim 0.25\,\mathrm{deg}^2$ is that it dramatically illustrates the existence of strong evolution in the comoving space density of LIGs. No-evolution models calibrated on local IRAS data ($z < 0.15$) would predict only one such source in $\sim 20\,\mathrm{deg}^2$ at a level of $S_{170} > 200\,\mathrm{mJy}$.

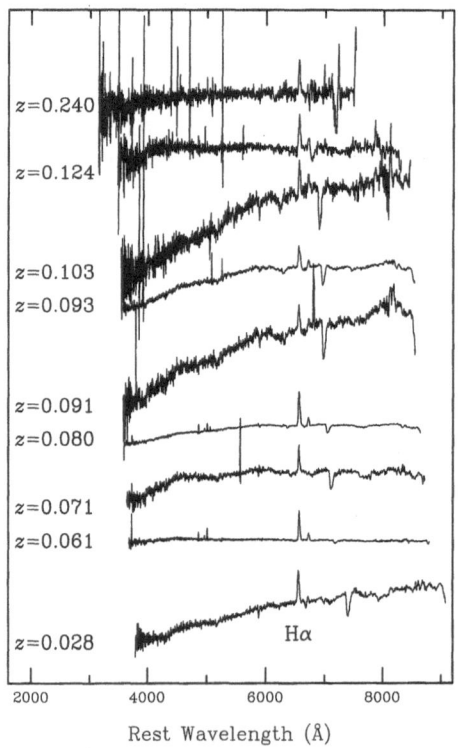

Fig. 3. Optical spectra of a sample of the brightest ISOPHOT 170 μm sources ($S_{170} > 200\,\mathrm{mJy}$), obtained with the KeckII telescope.

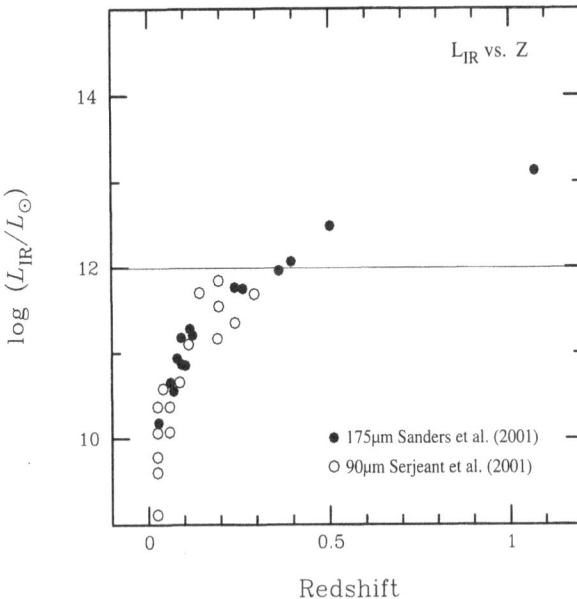

Fig. 4. Infrared luminosity, L_{ir} (8-1000 μm), versus optical redshift for bright ISOPHOT 170μm sources ($S_{170} > 200$ mJy) with secure redshifts. Shown for comparison are the luminosities and redshifts for the few bright ISOPHOT 90 μm sources ($S_{90} > 100$ mJy) in the ELAIS S1 region [5] with previously known redshifts as reported by [25].

Recently, [25] have reported 90 μm sources identified in their ELAIS deep fields, and have listed redshifts for 14 of these sources based on previous measured redshifts of optical counterparts. The redshifts and computed far-infrared luminosities of these 90 μm sources have also been plotted in Fig. 4 for comparison with our 170 μm sources. Figure 4 suggests that both surveys may be sampling the same source population, with the 170 μm observations perhaps probing to larger redshift due to the fact that this wavelength is closer to the redshifted far-infrared peak of the galaxy spectral energy distribution (SED).

5   Conclusions

Our recent spectroscopic survey of the brightest 170 μm sources that were detected in two ISOPHOT deep fields in the Lockman Hole confirm that these ISO sources are likely to be dusty LIGs at $z < 1.5$. The space density of these objects confirms the existence of a strongly evolving population of LIGs whose presence was suggested based on number counts at the faintest IRAS flux levels. These ISOPHOT sources are plausibly the low redshift tail of a an even more numerous high-z population of ULIGs that have recently been detected in submillimeter surveys with SCUBA.

Acknowledgments: DBS is grateful for support from a senior award from the Alexander von Humboldt Stiftung and from NASA JPL Contract 961566.

References

1. H. Aussel, C.J. Cesarsky, D. Elbaz, J.L. Starck: A&A **342**, 313 (1999)
2. A.J. Barger, L.L. Cowie, D.B. Sanders, et al.: Nature **394**, 248 (1998)
3. A.W. Blain, J.-P. Kneib, R.J. Ivison, I. Smail: AJ **119**, 2092 (1999)
4. C. Carilli, M.S. Yun: ApJL **513**, L13 (1999)
5. A. Efstathiou, S. Oliver, M. Rowan-Robinson, et al.: MNRAS **319**, 1169 (2000)
6. D.J. Fixsen, E. Dwek, J.C. Mather, et al.: ApJ **508**, 123 (1998)
7. B. Guiderdoni, F.R. Bouchet, J.-L. Puget, et al.: Nature **390**, 257 (1997)
8. P. Hacking, J.R. Houck: ApJS **63**, 311 (1987)
9. G. Hasinger: A&AS **120**, 607 (1996)
10. M.G. Hauser, R.G. Arendt, T. Kelsall, et al.: ApJ **508**, 25 (1998)
11. D. Hughes, S. Serjeant, J. Dunlop, et al.: Nature **394**, 241 (1998)
12. K. Kawara, Y. Sato, H. Matsuhara, et al.: A&A **336**, L9 (1998)
13. D.-C. Kim: Ph.D. Thesis, University of Hawaii (1996)
14. D.-C. Kim, D.B. Sanders: ApJS **119**, 41 (1998)
15. D. Lemke, U. Klaas, J. Abolins, et al.: A&A **315**, L64 (1996)
16. S.J. Lilly, S.A. Eales, W.K.P. Gear, et al.: ApJ **518**, 641 (1999)
17. F.J. Lockman, K. Jahoda, D. McCammon: ApJ **302**, 432 (1986)
18. H. Matsuhara, K. Kawara, Y. Sato, et al.: A&A **361**, 407 (2000)
19. H. Okuda: in 'ISO Surveys of a Dusty Universe', Lecture Notes in Physics V. 548, eds. D. Lemke, M. Stickel, K. Wilke (New York: Springer-Verlag, 2000), p. 40
20. J.-L. Puget, A. Abergel, J.-P. Bernard, et al.: A&A **308**, L5 (1996)
21. J.-L. Puget, G. Lagache, D.L. Clements, et al.: A&A **345**, 29 (1999)
22. D.B. Sanders, J.N. Mazzarella, D.-C. Kim, et al.: ApJS, submitted (2001)
23. D.B. Sanders, I.F. Mirabel: ARA&A **36**, 749 (1996)
24. P. Schechter: ApJ **203**, 297 (1976)
25. S. Serjeant, A. Efstathiou, S. Oliver, et al.: MNRAS, in press (2001) (astro-ph/0010025)
26. I. Smail, R.J. Ivison, A.W. Blain: ApJ **490**, L5 (1997)
27. B.T. Soifer, D.B. Sanders, B.F. Madore, et al.: ApJ **320**, 238 (1987)
28. S. Veilleux, D.E. Osterbrock: ApJS **63**, 295 (1987)
29. M. Yun, D.B. Sanders, K. Kawara, et al.: in 'Deep Millimeter Surveys: Implications for Galaxy Formation and Evolution', eds. J. Lowenthal, D. Hughes (World Scientific, 2001), in press

The History of Starburst Galaxies

Andrew W. Blain

Institute of Astronomy, Madingley Road, Cambridge CB3 0HA, UK

Abstract. Dusty galaxies with luminosities in excess of $10^{11}\,\mathrm{L}_\odot$ have been detected out to redshifts $z \sim 1$ by the *Infrared Space Observatory (ISO)*, and to higher redshifts using millimetre(mm)- and submm-wave cameras on ground-based telescopes. The integrated properties of these more distant galaxies are also constrained by measurements of the intensity of the submm-wave background radiation. While it is generally unclear whether their energy is released by gravitational accretion or by star formation, circumstantial evidence favours star formation. Unless these high-redshift galaxies are extremely massive, which is not expected from standard models of galaxy evolution, this luminosity cannot be sustained for more than a fraction of a Hubble time, and so they are undergoing some sort of 'bursting' behaviour. The interpretation and analysis of this population is discussed, and the key observations for deriving a robust history of their evolution, which is likely to be the history of starburst activity, are highlighted.

1 The Evolution of Luminous Dusty Galaxies

From observations of low-redshift dusty galaxies using the *IRAS* satellite [28] close to the peak of their restframe spectral energy distributions (SEDs), it is known that a similar amount of energy in the local Universe is produced by stars in dust-enshrouded and dust-free environments. The comoving luminosity density of dusty galaxies is also known to evolve strongly, from the slope of the faint counts of *IRAS* galaxies at $60\,\mu\mathrm{m}$ [3], which provide information to $z \simeq 0.2$. *ISO* observations at both shorter [13] and longer [19] wavelengths confirm that strong evolution continues to $z \sim 1$.

At longer wavelengths, the redshifted emission from very luminous, high-redshift dusty galaxies can be detected in the mm and submm wavebands. Independent surveys made using the 450/850-µm SCUBA camera at the JCMT [27] have determined the counts of high-redshift dusty galaxies. 1.2-mm surveys using the MAMBO detector array at the IRAM 30-m telescope [8] have detected a similar population of galaxies. In three cases, the detection of CO emission from gas located at the position and redshift of a suspected optical identification (at $z = 1.06$, 2.55 and 2.80) [15] provides absolute confirmation of the identification. Extremely deep VLA radio images of the survey fields can be used to impose constraints on the redshifts and SEDs of the detected galaxies [10,26]. It is likely that the detected galaxies are at $\bar{z} \simeq 2 - 3$, and there are very few plausible low-redshift ($z \leq 1$) counterparts. The counts and redshift distributions of these distant dusty galaxies can be used to constrain models of galaxy evolution at high redshifts.

In addition to the detection of individual submm-selected galaxies, the background radiation intensity in the mm and submm wavebands [14] traces the integrated emission from the entire population of dusty high-redshift galaxies. The mm/submm-wave background spectrum has the form $I_\nu \propto \nu^{2.64}$, and there is no clear spectral break down to 2000 μm. The lack of a break to a steeper slope supports the idea that high-redshift galaxies with redshifted SEDs peaking at about 1000 μm are still contributing to the background intensity, indicating a high maximum redshift of the population of about 10. If the shape of the SED is assumed not to evolve significantly, then the luminosity density must evolve as $\rho_L \propto (1+z)^{\simeq -1.1}$ [5,7] for $z \gg 1$ in order to generate this background spectrum, a result which is independent of cosmology. Such a gently declining high-redshift luminosity density is naturally consistent with the incomplete redshift distributions of galaxies detected in mm/submm-wave surveys.

Note that any model of the evolution of dusty galaxies must predict the redshift distribution of submm-detected galaxies correctly. Submm-wave surveys are very sensitive to high-redshift galaxies [4], and it is easy to propose models that fit both the observed submm-wave counts and background radiation spectrum at the expense of a redshift distribution that is biased far too high. The verification of predicted redshift distributions is therefore a crucial test of such models.

In this paper, the forms of evolution of galaxies that were previously derived in the context of far-infrared(IR) and submm-wave data are updated to take account of the much greater amount of information that has become available, particularly from *ISO* surveys. The results are similar to, but less uncertain than, those derived earlier [6,7]. The results have also been updated to include the currently favoured cosmological parameters $\Omega_0 = 0.3$, $\Omega_\Lambda = 0.7$ and $H_0 = 65 \, \mathrm{km \, s^{-1} \, Mpc^{-1}}$ are assumed.

2   Constraining the Evolution of Dusty Galaxies

A Baseline at Low Redshifts. The luminosity function of *IRAS* galaxies is best constrained at 60 μm [24]; and information about the same population of galaxies is also available at 100 μm [28]. These wavelengths are close to the peak of the SED of a nearby ($z \simeq 0$) dusty galaxy for any reasonable dust temperature. The ratio of the bright counts at 60 and 100 μm imply a luminosity-averaged dust temperature $T \simeq 35$–45 K. 850-μm observations of galaxies detected by *IRAS* [12], with a long wavelength baseline to provide an excellent probe of the SED, indicate that $T = 36 \pm 5 \, \mathrm{K}$ and the Rayleigh–Jeans spectral index is 3.3 ± 0.2. The population of low-redshift dusty galaxies can be divided into relatively short-lived warm interacting/starbursting galaxies and long-lived cooler quiescent galaxies [2,7]; however, the details of this distinction are relatively unimportant for studies of high-redshift galaxy evolution. Any low-luminosity, low-temperature dusty galaxies missing from existing surveys do not contribute significantly to the luminosity density, even at low and moderate redshifts.

The form of evolution of the baseline low-redshift far-IR luminosity function $\Phi_0(L)$ must be dominated by pure luminosity evolution, that is $\Phi(L,z) \simeq$

$\Phi_0[L/g(z), 0]$, to ensure that the submm galaxy counts and the background radiation intensity are both predicted correctly. Number density evolution is certainly also likely to be involved, but must be dominated by luminosity evolution [6]. The evolution function $g(z)$ is determined by demanding that the background radiation intensity, counts and redshift distributions of dusty galaxies are all in agreement with observations. These observations are, in order of increasing redshift, the faintest counts and redshift distributions of 60-μm *IRAS* galaxies, deep 90- and 170-μm counts from *ISO*, the spectrum of background radiation from *COBE*, and the faint counts and limited redshift information of distant galaxies detected using SCUBA at 450 and 850 μm and MAMBO at 1.25 mm.

Several approaches can be taken to investigate the evolution. The simplest is to assume a form for Φ, which implicitly includes details of all the physical processes taking place in galaxies, but is fitted to the data without investigating the processes in detail [6]. This has the advantage of requiring few parameters to model the galaxy SED and the form of evolution, in fact fewer than the number of constraining pieces of data. Well-motivated additions of greater complexity can thus be introduced to the models as the observations improve, without invoking parameters too numerous to constrain reliably and uniquely. A more physically motivated approach connects the evolving mass function of galaxies to the associated luminosity function using a prescription for both star formation and the fueling of active galactic nuclei (AGN) [11,16,29]; however, care must be taken to avoid getting lost in the space of free parameters. Without an additional population of short-lived, very luminous galaxies, standard semi-analytical models, which include star formation in the gas that cools in galaxy disks, fail to account for the observed surface density of SCUBA and MAMBO sources [7].

Describing Evolution with a Simple Luminosity Function. The first results derived using this approach [6] followed rapidly behind the first results of SCUBA surveys [25]. A low-redshift 60-μm luminosity function was assumed [24], an SED was defined by a dust temperature T, and a form of low-redshift evolution $g(z) = (1 + z)^\gamma$ was included. T and γ were determined by requiring that the form of the 60-μm *IRAS* counts and the early results of deep 175-μm *ISO* surveys were reproduced; $T = 38 \pm 4 \, \text{K}$ and $\gamma = 3.9 \pm 0.2$ were required [6]. As discussed above, this temperature is consistent with subsequent SCUBA measurements of dust temperatures for *IRAS* galaxies [12], while γ matched the value inferred from optical surveys [21], rather than the value of $\gamma = 3$ that is often assumed to describe the evolution of galaxies in the far-IR waveband. Note that $\gamma \simeq 4.5$ is derived from 15-μm *ISO* surveys, taking into account the complex restframe SED of a dusty galaxy between 5 and 10 μm [30].

It is now possible to use the much more extensive data from 90- and 175-μm *ISO* counts [19], and a more popular non-zero-Λ cosmology to revisit the results. There are no substantial changes; formally $T = 37 \pm 3 \, \text{K}$ and $\gamma = 4.05 \pm 0.15$ are the latest results, if a Rayleigh–Jeans spectral index of 3.5 is assumed.

The form of evolution at higher redshifts, too distant for *ISO* observations, is constrained by the background radiation intensity [14] and the counts of

SCUBA galaxies. From 2002, mid-IR *SIRTF* observations should make a major impact in this area. Constraints from the background and SCUBA data are rather degenerate, although even the first SCUBA data in 1997 [25] provided the tighter constraint: see [6], in which various models of high-redshift evolution were considered. The so-called Gaussian model, in which $g(z)$ is represented by a Gaussian in cosmic epoch, provided the best description of the redshift distribution of SCUBA galaxies. The more accurate 175- and 850-μm counts now available provide some additional information, and are useful for updating the results. Progress has also been made in developing a more appropriate form of $g(z)$, which is fully compatible with models of cosmic chemical evolution, and naturally includes a peak in the evolution function [17,22]:

$$g(z) = (1 + z)^{3/2} \text{sech}^2[b\ln(1 + z) - c]\cosh^2 c. \tag{1}$$

At low redshifts $\gamma \simeq (3/2) + 2b\sqrt{1 - \text{sech}^2 c}$. Using all available observational data, the results $b = 2.2 \pm 0.1$ and $c = 1.84 \pm 0.1$ are obtained; see the solid line in Fig. 1. These results are similar to those derived earlier [6], but take account of revised cosmological parameters and tighter error bars on some of the constraining data. The model described above is consistent with all the observed background radiation, counts and redshift distributions of galaxies at wavelengths longer than 60 μm. The dominant source of energy in the Universe remains the restframe far-IR radiation of starlight and AGN emission reprocessed by dust.

Describing Evolution Using a Model of Merging Galaxies. In an alternative investigation, we took a simple form of the evolution of the merger rate of dark-matter halos [5,7], which adequately reproduces the results of recent N-body simulations [18]. We assumed that a certain redshift-dependent fraction $x(z)$ of the total mass of dark and baryonic matter involved in mergers is converted into energy by nucleosynthesis in high-mass stars with an efficiency $0.007c^2$. Note that the same formalism is appropriate for describing the evolution of AGN fueling events at the epochs of mergers [7]. The form of evolution and normalization of $x(z)$ can be determined by a joint comparison with the background radiation intensity and low-redshift *IRAS* counts. Using an appropriate form of $x(z) = g(z)/(1 + z)^{3/2}$ [17,22], $b = 1.95 \pm 0.1$, $c = 1.6 \pm 0.1$ and $x(0) = 1.35 \times 10^{-4}$ are required in the standard cosmology, assuming the galaxy SED discussed above. The resulting history of galaxy evolution is shown by the thick dashed line in Fig. 1.

In addition, the counts of luminous galaxies associated with merger-induced bursts of activity can be determined if a fraction F of mergers are assumed to generate luminous bursts of duration σ. The product $F\sigma$, which could depend on redshift, is the function that controls the results. Using information derived from the counts of both low- and high-redshift galaxies, the form of $F\sigma(z)$ required to account for the observations can be determined. If the form $F\sigma(z) = F\sigma(0)\exp(az + bz^2 + cz^3)$ is chosen, then values of $F\sigma(0) = 2.4\,\text{Gyr}$, $a = -4.14$, $b = -0.56$ and $c = 0.46$ are required.

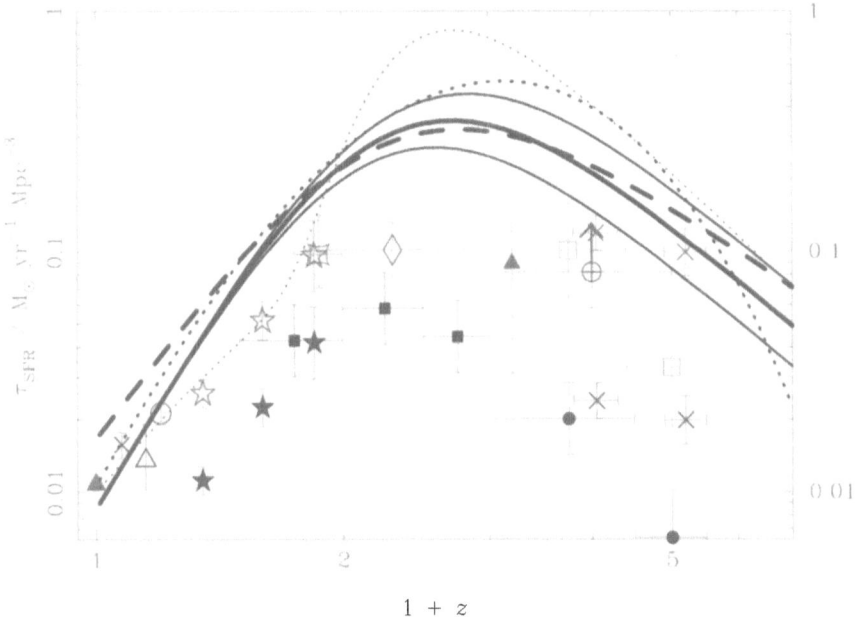

Fig. 1. The history of star formation inferred using both methods discussed in the text. The thick solid and dashed lines represent the simple luminosity evolution model and the hierarchical model respectively. The thinner solid lines show the approximate envelope of 68% uncertainty in the results of the simple model. The data points are taken from a variety of sources: references can be found elsewhere [7,20]. The thin and thick dotted lines represent the best-fitting results obtained in previous derivations, the modified Gaussian [6,1] and 35-K hierarchical models [7] respectively. The absolute normalisation of the curves depends on the assumed stellar initial mass function and the fraction of the dust-enshrouded luminosity of galaxies that is generated by AGN.

Faint Radio and 15-μm Mid-IR Counts. Both the simple and hierarchical models account for all the current observations in these wavebands. If the standard form of the far-IR–radio correlation is assumed, with a radio spectral index of -0.65, then the 8.4-GHz counts brighter than $10\,\mu$Jy predicted in the two models are 1.05 and $0.98\,\mathrm{arcmin}^{-2}$, with slopes of -1.4 and -1.3 respectively, matching the observed count $N(\geq S) = (1.01 \pm 0.14)(S/10\mu\mathrm{Jy})^{-1.25\pm0.2}$ [23]. The faintest 1.4-GHz counts [9] are also reproduced accurately. If the mid-IR SED is described by a power-law $f_\nu \propto \nu^\alpha$ with $\alpha = -1.95$ at wavelengths shorter than the peak of the SED, then the normalization and general features of the deep counts of galaxies determined using *ISO* at $15\,\mu$m [13] are also reproduced, including the marked change of slope at flux densities between 0.5 and $1\,$mJy. The presence or absence of a PAH emission feature in the SED has little effect on the results. In the hierarchical model, the slope of the predicted 15-μm counts at flux densities between 1 and $10\,$mJy is steeper as compared with that in the simple model, in better agreement with the observations.

3   Conclusions

These models, which involve a minimum number of free parameters, provide a reasonable description of all the counts and redshift distributions of dusty galaxies at both high and low redshifts. As additional data, especially more complete redshift distributions and very deep *SIRTF* mid-IR counts, become available, more details can be incorporated into the models to reveal further information about the properties of evolving distant dusty starbursts/AGN.

Acknowledgements: The author, Raymond and Beverly Sackler Foundation Research Fellow at the IoA, thanks the Foundation for generous financial support, ESO for support at the meeting, and Vicki Barnard and Kate Quirk for helpful comments on the manuscript. These results are updated from SCUBA Lens Survey work with Ian Smail, Rob Ivison and Jean-Paul Kneib.

References

1. A.J. Barger, L.L. Cowie, I. Smail, R.J. Ivison et al.: AJ **117**, 2656 (1999)
2. V.E. Barnard, A.W. Blain: MNRAS submitted (2001)
3. E. Bertin, M. Dennefeld, M. Moshir: A&A **323** 685 (1997)
4. A.W. Blain, M.S. Longair: MNRAS **264**, 509 (1993)
5. A.W. Blain, M.S. Longair: MNRAS **265**, L21 (1993)
6. A.W. Blain, I. Smail, R.J. Ivison, J.-P. Kneib: MNRAS **302**, 632 (1999)
7. A.W. Blain, A. Jameson, I. Smail, M.S. Longair et al.: MNRAS **309**, 715 (1999)
8. C.L. Carilli, F. Bertoldi, A. Bertarini et al.: preprint (astro-ph/0009298)
9. C.L. Carilli: this volume, astro-ph/0011199 (2001)
10. C.L. Carilli, M.S. Yun: ApJ **530**, 618 (2000)
11. S.M. Cole, A. Aragón-Salamanca, C.S. Frenk et al.: MNRAS **271**, 781 (1994)
12. L. Dunne, S.A. Eales, M. Edmunds, R. Ivison et al.: MNRAS **315**, 115 (2000)
13. D. Elbaz, C.J. Cesarsky, D. Fadda, H. Aussel et al.: A&A **351**, 37 (1999)
14. D.J. Fixsen, E. Dwek, J.C. Mather, C.L. Bennett et al.: ApJ **508**, 123 (1998)
15. D.T. Frayer, R.J. Ivison, N.Z. Scoville et al.: ApJ **514**, L13 (1999)
16. B. Guiderdoni, E. Hivon, F.R. Bouchet et al.: MNRAS **295**, 877 (1998)
17. A. Jameson: PhD thesis, University of Cambridge (1999)
18. A. Jenkins, C.S. Frenk, S.D.M. White et al.: MNRAS in press (astro-ph/0005260)
19. M. Juvela, K. Mattila, D. Lemke: A&A **360**, 813 (2000)
20. A.F.M. Moorwood, P.P. van der Werf et al.: A&A in press (astro-ph/0009010)
21. S.J. Lilly, O. Le Fèvre, F. Hammer, D. Crampton: ApJ **460**, L1 (1996)
22. M.S. Longair: in AIP Conf. Proc. No. 516, pp. 3 (AIP, New York, 2000)
23. R.B. Partridge, E.A. Richards, E.A. Fomalont et al.: ApJ **483**, 38 (1997)
24. W. Saunders, M. Rowan-Robinson, A. Lawrence et al.: MNRAS **242**, 318 (1992)
25. I. Smail, R.J. Ivison, A.W. Blain: ApJ **490**, L5 (1997)
26. I. Smail, R.J. Ivison, F.N. Owen et al.: ApJ **528**, 612 (2000)
27. I. Smail, R.J. Ivison et al.: MNRAS submitted (2001)
28. B.T Soifer, G. Neugebauer: AJ **101**, 354 (1990)
29. R.S. Somerville, J.R. Primack, S.M. Faber: MNRAS in press (astro-ph/0006364)
30. C. Xu: ApJ **541**, 134 (2000)

Radio Observations of High Redshift
Star Forming Galaxies

Chris L. Carilli

NRAO, Socorro, NM 87801, USA

Abstract. I summarize recent results from radio observations of high redshift star forming galaxies, discuss radio continuum emission as a measure of star formation rate, and consider future capabilities at cm to IR wavelengths.

1 Radio Surveys to μJy Sensitivity

Source counts based on low frequency surveys of the sky with Jy sensitivity showed a significant departure from a Euclidean, non-evolving source population, indicating, for the first time, cosmic evolution in a source population. The source population entailed luminous radio galaxies, with spectral luminosities at 178 MHz: $P_{178} > 10^{32}$ erg s^{-1} Hz^{-1} [1,2]. In these sources the synchrotron radio emission is from high energy electrons accelerated in a relativistic jet emanating from the active galactic nucleus (AGN) [3].

Subsequent observations with sub-mJy sensitivity, starting with the WSRT and continuing with the VLA [5], revealed flattening of the source counts below 5 mJy (Fig. 1). Windhorst et al. hypothesized that this flattening was due to a new population of sources, namely star forming galaxies with $P_{178} < 10^{31}$ erg s^{-1} Hz^{-1}. The radio emission in these sources is from relativistic electrons accelerated in supernovae remnant shocks [4].

There has been a recent revival of deep radio surveys, motivated in large part as follow-up to deep optical, infrared, and (sub)mm surveys [6,7,9,10]. The frequency of choice is 1.4 GHz for these deep surveys, allowing for μJy sensitivity with arcsecond resolution and a wide field of view (FWHM = 30'). In the coming years the Expanded VLA (EVLA) will push to the sub-μJy level, while in the coming decades the Square Kilometer Array (SKA) will potentially probe nJy sources. As pointed out in numerous papers in these proceedings (Bertoldi, Adelberger, Hughes, Sanders), radio observations play an important, complimentary role to observations at other wavelengths, in that:

- They are not plagued by extinction corrections.
- They provide a rough estimate of source redshifts, in combination with (sub)mm observations [11,12].
- Low order CO transitions redshift into the cm bands, revealing large reservoirs of less dense, cooler gas [13,14]. This topic will not be discussed herein.
- They provide arcsecond astrometry and imaging, thereby avoiding the confusion problems inherent in deep searches for optical counterparts of (sub)mm sources discovered in low resolution single dish bolometer array surveys.

For example, at the optical limits of the HDF ($I_{AB} < 29$) one expects, by chance, three faint galaxies within the $6''$ error circle of a typical SCUBA or MAMBO source [15]. At 1.4 GHz the source counts between $40\,\mu$Jy and $1\,$mJy obey: $N(> S_{1.4}) = 2.2 \times 10^{-6}\, S_{1.4}^{-1.4}$ arcsec^{-2}, with $S_{1.4}$ in mJy [6]. The number of spurious $S_{1.4} \geq 22\,\mu$Jy sources within the error circle is only 0.02.

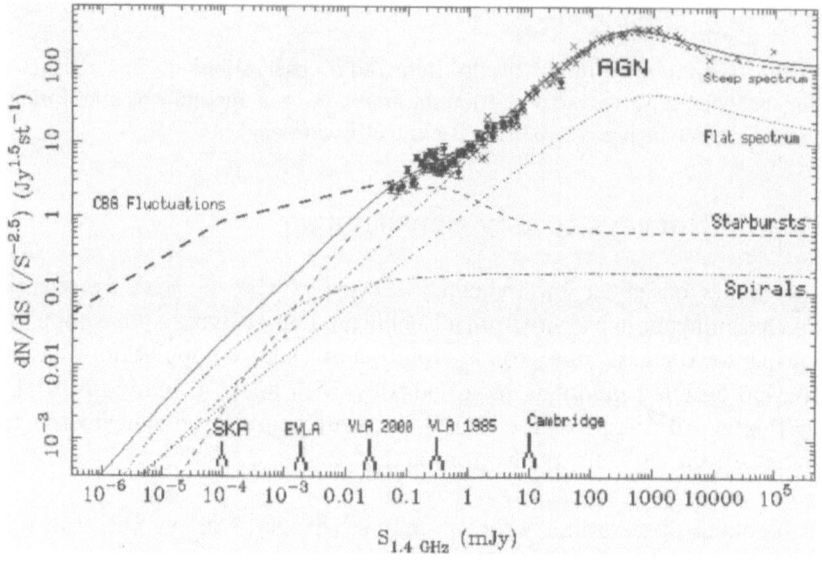

Fig. 1. Radio source counts at 1.4 GHz (adopted from [8]).

Since deep radio images are usually follow-up observations of deep fields at other wavebands, there has been very rapid progress in determining the nature of the parent galaxies of the μJy radio source population [6,16]. This short summary will focus on recent results on the μJy radio source population, emphasizing the unique information about high redshift star forming galaxies coming from radio observations. I will also discuss the radio-to-far infrared (FIR) correlation, re-deriving star formation rates based on this correlation, and conclude with a short discussion of future instrumentation.

Richards [6] finds that 75% of the $S_{1.4} \geq 40\,\mu$Jy radio sources are identified to $I < 24$, with a median of $I = 22.1$. Interestingly, he also finds that 25% of the sources are unidentified to $I > 25$. And perhaps more interestingly, a number of groups [6,17–19] find that (sub)mm observations of these μJy radio sources with faint (or absent) optical counterparts results in a $> 40\%$ detection rate at mJy levels. This suggests that the optically faint, μJy radio sources are equivalent to the mJy (sub)mm source population.

Haarsma et al. [16] have used an extensive spectroscopic and photometric redshift analysis to determine the redshift distribution of the μJy radio source population. They find that (roughly):

- 50% of the sources are spirals, or irregular galaxies, at $z < 1$.
- 25% are ellipticals, presumably low luminosity AGN, at $0.3 < z < 1.5$.
- 25% are optically faint (or absent), and red.

They propose that these later sources are likely to be high redshift $(1.5 < z < 4)$, dust obscured starbursts. This idea is consistent with the SCUBA and MAMBO results discussed above [6,17–19].

The angular size distribution of the μJy radio source population remains a point of debate. WSRT observations of the HDF at 1.4 GHz, $15''$ resolution, to 8μJy rms detect a number of sources not detected in the VLA survey at similar sensitivity but with $1''$ resolution [9]. This would suggest a significant population of sources larger than $1''$. However, confusion is a serious issue at this sensitivity level at $15''$ resolution. Combined MERLIN+VLA observations of the HDF suggest that most of the sources have angular sizes between $0.7''$ and $2''$, with a median of $1.4''$ (Muxlow et al. in prep). On the other hand, VLA observations of the cluster A2125 suggest that most of the sources are unresolved, with upper limits of typically $1''$ [10]. If the sources are indeed $1''$ in size, this presents a significant problem for the SKA, since the sky will become naturally confusion limited at the few nJy level, independent of the resolution of the instrument [7], well above the sensitivity of the SKA.

Richards [6] finds that the mean spectral index for a 1.4 GHz selected sample of sources is –0.8, typical of star forming galaxies. Not surprisingly, an 8 GHz selected sample shows a flatter mean spectral index of –0.4.

Barger et al. [17] find that only 15% of the radio sources with $S_{1.4} \geq 25\,\mu$Jy in the Hawaii Deep Field are X-ray sources with 2 to 10 keV fluxes of: $I_x \geq 1 \times 10^{-15}$ erg s^{-1} cm^{-2}. This result is consistent with the idea that the majority of μJy radio sources are star forming galaxies.

One problem with deep radio surveys is the limited area covered, such that cosmic variance can lead to substantial differences between counts derived for different fields. For instance, the counts at the 100 μJy level in the Phoenix Deep Field [7] are a factor two higher than those in the Hubble Deep Field [6]. This variance can be seen as the increased scatter below 1 mJy in Fig. 1.

2   The FIR-Radio Correlation: Deriving Star Formation Rates from Radio Observations and the Importance of Inverse Compton Losses

Most recent derivations of star formation rates based on radio observations [16,20,21] use equs. 21 and 23 in Condon [4] to relate the star formation rate (SFR) to the 1.4 GHz spectral luminosity. This relationship was derived from the supernova rate and the integrated radio luminosity of the Milky Way.

An alternate method for deriving the relationship between radio luminosity and SFR relies on the tight correlation between FIR luminosity and radio luminosity in star forming galaxies [4]. This correlation is remarkable in the small scatter observed over at least three orders of magnitude in FIR luminosity. Further, it holds for optical and IR selected samples [22]. This alternate method uses spectral synthesis models for star forming galaxies [23], assuming a fraction, f_c, of the bolometric luminosity is absorbed by dust and re-emitted in the infrared, and then uses the FIR-radio correlation to relate the radio luminosity to the FIR luminosity. For a 10^8 yr continuous starburst, solar abundances, a Salpeter IMF from 0.1 to 100 $M_\odot$, and $f_c = 1$, this calculation leads to:

$$\text{SFR} = 5.1 \times 10^{-22} \ L_{1.4} \ M_\odot \ \text{yr}^{-1} \qquad (1)$$

where $L_{1.4}$ is the 1.4 GHz spectral luminosity in W Hz^{-1}. Scaling to the same IMF limits, this equation implies a factor 2.5 lower SFR than the equations in Condon [4].

There are a number of uncertainties in both calculations. In the case of Condon's calculation the Galactic supernova rate and radio continuum luminosity are both uncertain by at least 50%. For the stellar synthesis model calculation, the scatter in the FIR-radio correlation leads to a 50% uncertainty, while changing parameters in the starburst model changes the predicted SFR, eg. decreasing the starburst age to 10^7 years increases the SFR by 50%, and there is the uncertain f_c. It is currently not clear which, if either, calculation is correct. Indeed, the relative agreement is encouraging, given the very different methods. And given the different conditions in different galaxies (e.g. the age of the starburst, the dust covering factor, the IMF, ...), it is clear that there will be no globally correct relationship, only a statistically most likely one.

The FIR-radio correlation has a very simple heuristic explanation: both the FIR and radio emission relate to massive star formation, with the FIR coming from dust heated by the interstellar radiation field, and the relativistic electrons being accelerated in supernova remnant shocks. However, given the number of processes and parameters involved, it remains remarkable, and as yet unexplained, as to why the correlation is so tight [4,24]. For instance, it has long been known that the total radio luminosity of galaxies, both normal disks and nuclear starbursts, is an order of magnitude larger than that expected from the sum of the supernovae [4,25], although see [27]. This requires that the relativistic electrons be stored in the ISM of galaxies for a timescale, $t \geq \frac{U_{B,SNR}}{U_{B,ISM}} \times t_{SNR} \sim 10^7$yr, where U_B is the magnetic energy density.

This leads to the question of the importance of inverse Compton losses off the microwave background for relativistic electrons in the ISM of high redshift galaxies. The energy density in the microwave background increases as: $U_{MWBG} = 4.0 \times 10^{-13}(1 + z)^4$ erg cm^{-3}. This is shown in Fig. 2, along with the energy density in the magnetic field in a typical spiral arm [26]. The energy density in the magnetic field is larger than that in the microwave background to $z \sim 1$. Beyond this redshift, inverse Compton cooling will dominate over synchrotron radiation, limiting electron lifetimes, $t_{1.4}$, and leading to a departure from the radio-FIR relation.

This is not true, however, for compact nuclear starburst galaxies, ie. systems with SFRs > 100 $M_\odot$ year^{-1} in regions smaller than a few hundred parsecs [28]. In these systems the energy density in the magnetic field is thought to be almost three orders of magnitude larger than in the disk [29], in which case

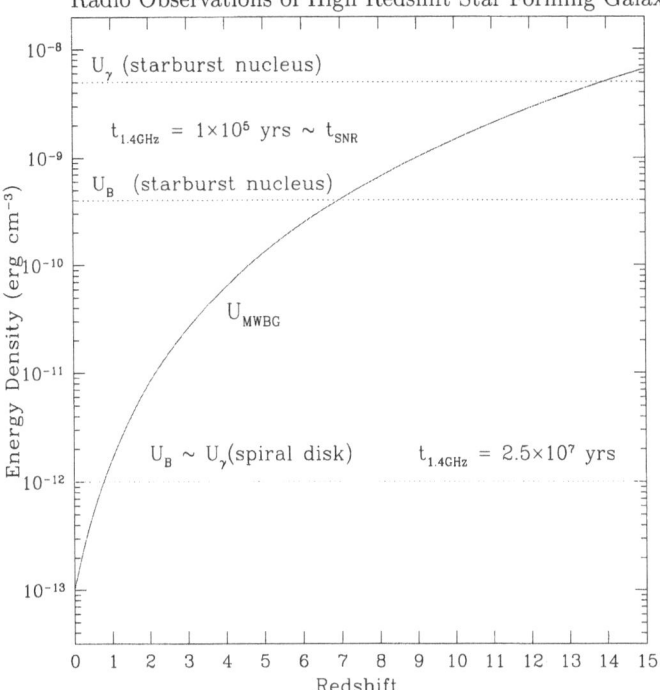

Fig. 2. The energy density of the microwave background, U_{MWBG}, vs. redshift. Also shown is the typical magnetic energy density, U_B, in the arm of a spiral galaxy, and U_B in compact nuclear starbursts, and the energy density in the radiation field, U_γ, in starbursts. $t_{1.4}$ is the radiative lifetime of an electron emitting at 1.4 GHz.

U_{MWBG} only becomes relevant at $z > 7$. But compact nuclear starbursts raise a different, related problem: the energy density in the IR radiation field from the starburst itself is larger still than that in the magnetic field. This means that inverse Compton cooling should remove the synchrotron emitting electrons on fairly short timescales ($\leq 10^5$ yr), and accentuates the question: why do nuclear starbursts follow the radio-FIR correlation? There is a large body of literature on this issue [4], but as yet no closure. Hence, we trade one problem, inverse Compton losses off the microwave background at high z, for a second, inverse Compton losses off the starburst IR radiation field at all z.

Until proven otherwise, we adopt the radio-FIR correlation as a given, and see how it can be used in the study of high z star forming galaxies.

3   Using the Radio-FIR Correlation to Study High z Star Forming Galaxies, and what the Future Holds

Haarsma et al. [16] have derived the cosmic star formation rate density (SFRD) vs. z based on μJy radio samples. They find a steep rise in the density from $z = 0$ to $z = 1$, as has been found in optical and IR studies. But they also find a systematically higher SFRD at all redshifts by a factor 3 relative to reddening corrected optical studies. This suggests that even larger dust corrections are needed in optical studies, or that optical studies miss a large population of dust-obscured galaxies. However, they use the equations in Condon [4] to derive SFRs from radio continuum luminosities. If we use equ. 1 above instead, the radio derived values agree well with the optical values.

A second area in which the radio-FIR correlation has been used in the study of high z star forming galaxies is as a redshift indicator [11,12,30–33,17]. The impetus in this case is the very faint optical counterparts being found for most faint (sub)mm sources, thereby precluding follow-up optical spectroscopy of a large sample of sources. The radio-FIR method relies on the opposing slopes of the synchrotron and thermal dust emission in star forming galaxies. Our most recent models use the average observed SED for nearby starbursts to relate redshift to the observed spectral index between 350 GHz and 1.4 GHz [12]. Figure 3 shows the model, along with a few sources with known redshifts, including sources with AGN and starburst optical spectra. The method is admittedly imprecise, especially at high redshift, and there are a few degeneracies in the solutions, such as the addition of cold dust or the presence of a radio loud AGN [30], but it is the only viable alternative for deriving redshifts for the large majority (90%) of the faint (sub)mm sources. The redshift distribution for the faint (sub)mm source population as derived using the curves from [12] can found in the contribution by Bertoldi in this volume.

An important point concerning radio follow-up observations of faint (sub)mm sources is the relative sensitivities. Comparing MAMBO images with deep radio images shows that 70% (10 of 14) of the ≥ 3.5 mJy (5σ) sources at 250 GHz have radio counterparts with $S_{1.4} \geq 22\,\mu$Jy within $3''$ (Bertoldi, this volume). We expect only 0.2 chance coincidences. This result lends confidence to the reality of the MAMBO detections. It also implies that the current sensitivity of deep VLA fields is well matched to that obtained with mm bolometer arrays, and in particular, that there is not a dominant population of very high z sources ($z > 4$), or of low z, cold dust sources.

Figure 4 shows the expected sensitivity of future cm to IR instruments compared to the expected flux density of Arp 220 at various redshifts. Overall, the next generation instruments are well matched to the expected flux density of Arp 220 out to $z \sim 2$ to 8. Clearly, the ALMA is by far the most sensitive telescope relative to the dust spectrum, and will detect low luminosity galaxies to very high z. However, this applies to pointed observations, which have a limited field-of-view (FWHM $\sim 20''$). For surveys of fields larger than about $15' \times 15'$, the next generation bolometer array cameras operating on large single dish telescopes will be competitive with ALMA, as will the EVLA and SIRTF.

Figure 5 shows the flux density of Arp 220 and M82 vs. z at various frequencies, relative to the sensitivities of future telescopes. The important point in this diagram is the interesting source selection function at (sub)mm wavelengths: typical (sub)mm surveys detect luminous star forming galaxies at essentially all redshifts, but they miss completely the low z, low luminosity galaxies. Hence (sub)mm surveys result in a very clean, but totally biased, sample of sources.

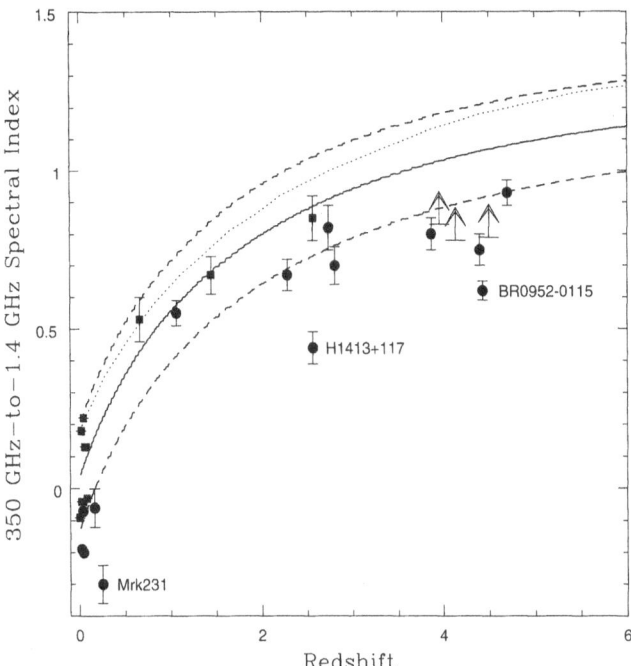

Fig. 3. The relationship between redshift and observed 1.4-to-350 GHz spectral index for an active star forming galaxy, derived from the observed SEDs of 17 low z galaxies [12]. The dash line shows the [31] model. Squares are observed values for galaxies with starburst spectra. Circles are values for sources with AGN spectra.

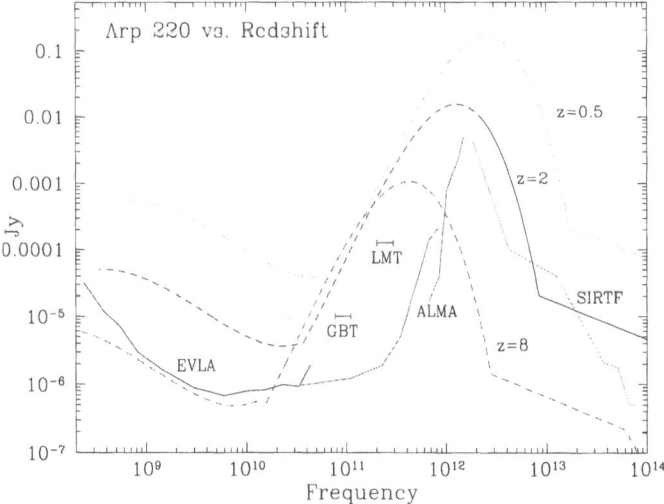

Fig. 4. The observed spectrum of Arp 220 at various redshifts, compared to the sensitivities for various existing and future telescopes.

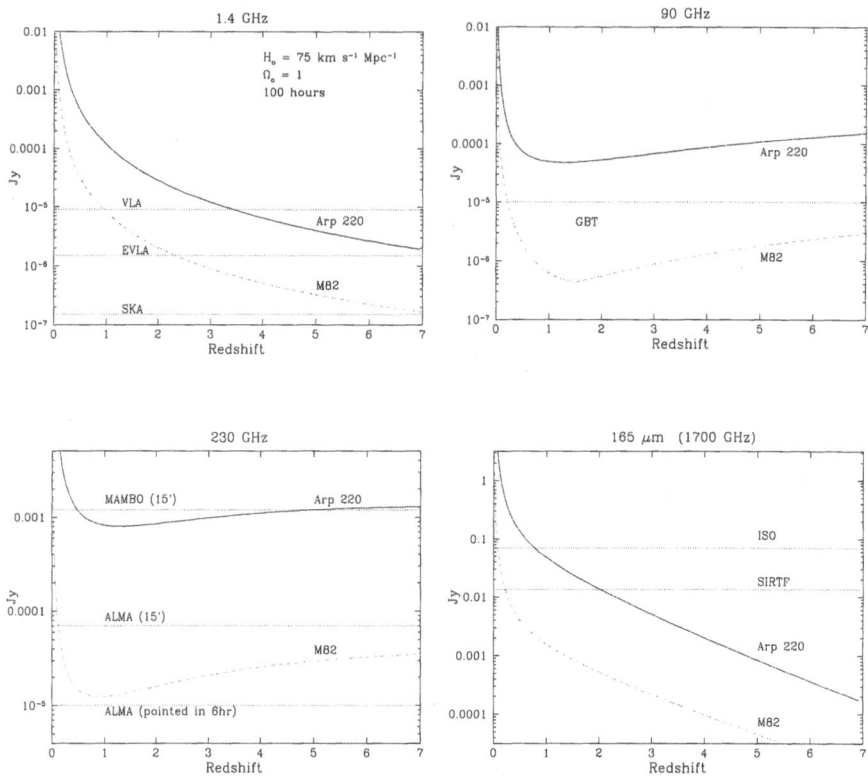

Fig. 5. The flux densities of Arp 220 and M82 vs. z at various observing frequencies. Also shown are the sensitivities of existing and future telescopes.

Radio observations result in a mixture of low z, low luminosity, and high z, high luminosity star forming galaxies, as well as radio loud AGN.

The closing debate at this workshop contrasted the IR vs. submm vs. optical views of high z galaxies and galaxy formation. Each side argued that they detect the dominant contribution to cosmic star formation at a specific epoch. Yet, each side has a very specific galaxy selection function, and the overlap between the populations apparently is small, $\sim 10\%$. Indeed, one might argue that μJy radio surveys are the least biased means of detecting all the source populations. But therein lies the fundamental problem: how to differentiate the source populations on a deep radio image? Overall, it is clear that all sides are currently seeing only limited, and perhaps orthogonal, aspects of galaxy formation. In order to address the general question of galaxy formation, or at least the formation of the stellar content of galaxies, requires wide field surveys using the EVLA, the next generation bolometers arrays on the LMT and GBT, and SIRTF, with very deep follow-up studies of selected samples of sources using ALMA and NGST.

The National Radio Astronomy Observatory is operated by Associated Univ. Inc., under contract with the National Science Foundation. I would like to thank A. Hopkins for allowing me to reproduce Fig. 1, and B. Poggianti, N. Miller, M. Yun, and F. Owen for useful discussions.

References

1. M. Ryle: ARAA **6**, 249 (1967)
2. D. Jauncey: ARAA **13**, 23 (1975)
3. C.L. Carilli, P.D. Barthel: A&A Reviews **7**, 1 (1996)
4. J.J. Condon: ARAA **30**, 575 (1992)
5. R. Windhorst, et al.: ApJ **289**, 494 (1985)
6. E. Richards: ApJ **533**, 611 (2000)
7. A. Hopkins, et al.: ApJ **519**, L59 (2000)
8. A. Hopkins, et al.: Exp. Ast. **10**, 419 (2000)
9. M.A. Garrett, et al.: A&A **361**, L41 (2000)
10. F. Owen, et al.: 2000, in preparation
11. C.L. Carilli, M.S. Yun: ApJ **513**, L13 (1999)
12. C.L. Carilli, M.S. Yun: ApJ **530**, 618 (2000)
13. P. Papadopoulos, et al.: 2000, Nature, submitted
14. C.L. Carilli, K.M. Menten, M.S. Yun: ApJ **521**, L25 (1999)
15. S.J. Lilly, et al.: ApJ **518**, 641 (1999)
16. D.B. Haarsma, et al.: 2000, ApJ, in press, astro-ph/0007315
17. A.J. Barger, L.L. Cowie, E.A. Richards: AJ **119**, 2092 (2000)
18. F. Bertoldi, et al.: 2000, in preparation
19. S.C. Chapman, et al.: 2000, ApJ, in press, astroph/0011066
20. L. Cram, et al: ApJ **507**, 155 (1998)
21. B. Mobasher, et al.: MNRAS **308**, 45 (1999)
22. J.J. Condon, Q.F. Yin: ApJ **357**, 97 (1990)
23. C. Leitherer, et al.: ApJS **123**, 3 (1999)
24. N. Duric, P. Crane (eds.) ASP Conf. Vol. 18 (San Francisco, ASP, 1991)
25. G.G. Pooley: MNRAS **144**, 101 (1969)
26. R. Beck, et al.: ARAA **34**, 155 (1996)
27. L. Colina, D.E. Perez-Olea: MNRAS **277**, 845 (1995)
28. D. Downes, P. Solomon: ApJ **507**, 615 (1998)
29. C.L. Carilli, G.B. Taylor: ApJ **532**, L95 (2000)
30. A. Blain: MNRAS **309**, 955 (1999)
31. L. Dunne, D. Clements, S. Eales: MNRAS, 2000, in press
32. D. Hines, F. Low: in: ASP Conf. Vol. 191, eds. Weymann et al., (San Francisco, ASP 1999), p. 165
33. D. Hughes, et al.: Nature **394**, 241 (1998)

The Cosmic Star-Formation History: The UV Finds Most

Kurt Adelberger

Harvard-Smithsonian Center for Astrophysics, 60 Garden St., Cambridge, MA 02138, USA

1 Introduction

The assigned subtitle of this talk was "The UV Takes it All!" The absurdity of that subtitle must have been obvious to many conference participants when they first read the program, and now, after four days of talks, I hope it is obvious to everyone here. The UV cannot possibly "take all" of the cosmic star-formation history. Data from low and high redshift alike overwhelmingly show that rapidly star-forming galaxies emit the bulk of their bolometric luminosities in the far-infrared, not the UV. Some extremely luminous high-redshift galaxies have not been detected in the UV at all (see, e.g., Hughes' and Blain's contributions to these proceedings); many others dominate relatively short $850\,\mu$m exposures but are barely detected in even the deepest (rest-frame) UV images. I was asked by the organizers to make a case for studying the distant universe in the rest-frame UV, and I will; but I cannot pretend that UV surveys are the best way to study all star-forming galaxies at high redshift. Some fraction of the stars in the universe formed in extremely dusty galaxies that are best studied in the infrared. Nevertheless the available data—even IR data!—suggest that *most* stars formed in objects that are easiest to detect in the rest-frame ultraviolet, and so I have modified the subtitle of this talk to a statement I am prepared to defend. UV observations may not detect most of the luminosity emitted by any single galaxy at high redshift, and may not detect some high redshift galaxies at all, but the average high redshift star-forming galaxy is much more easily detected through its UV radiation than its dust emission. In the first half of my talk I will lay out the arguments supporting that statement. In the second half I will discuss the attempts to estimate star-formation rates for UV-selected galaxies after they have been detected. It is here (as we shall see) that dust obscuration poses the greatest problems for UV surveys.

2 The UV Finds Most

Figure 1 helps illustrate the argument I'm going to make. The plot shows the level of dust obscuration observed among different types of star-forming galaxies in the local universe. The abscissa, $L_{\rm bol,dust} + L_{\rm UV} \equiv L_{\rm SFR}$, is the sum of the star-formation luminosity that is and that is not absorbed by dust; this provides a rough measure of total star-formation rate. The ordinate, $L_{\rm bol,dust}/L_{\rm UV}$, is the ratio of obscured to unobscured star-formation luminosity; it provides a rough measure of the level of dust obscuration. Here $L_{\mathrm{bol,dust}} \simeq 1.5\,L_{\mathrm{FIR}}$ is an estimate of a galaxy's bolometric dust luminosity and $L_{\mathrm{UV}} \equiv \lambda l_\lambda$ evaluated at $\lambda \sim 1600\,\text{Å}$. The well known trend of increasing dust obscuration with increasing star-formation rate is obvious. The expression $L_{\mathrm{bol,dust}}/L_{\mathrm{UV}} \propto L_{\mathrm{SFR}}$ (dashed line) provides a crude fit to the data. Some galaxies in the local universe are very heavily obscured ($L_{\mathrm{bol,dust}}/L_{\mathrm{UV}} > 100$), but these galaxies are rare and together they host only $\sim 5\%$ of the known local star-formation [15]. The vast majority of local star-formation occurs among less luminous and less obscured spiral and starburst galaxies.

Now imagine taking the local universe and placing it at redshift $z \sim 3$ (say) and observing it at $850\,\mu\mathrm{m}$ with SCUBA. The only galaxies bright enough to detect would be the heavily dust-obscured ultraluminous infrared galaxies (ULIRGs) with $100 < L_{\mathrm{bol,dust}}/L_{\mathrm{UV}} < 5000$. When analyzing this hypothetical $850\,\mu\mathrm{m}$ data it would be natural to conclude that all high-redshift star formation occurred in galaxies similar to the ones that were detected, in extremely dusty galaxies that would be difficult to detect in the rest-frame UV. But this conclusion would be wrong. It relies on the false assumption that the comparatively faint galaxies which host most of the star formation would be as dusty as the ultraluminous galaxies that were detected; it neglects the strong correlation of dust opacity and star-formation rate.

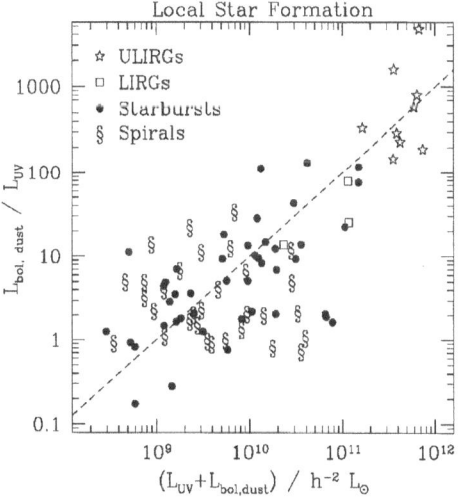

Fig. 1. Star formation in the local universe. The quantity on the abscissa is roughly proportional to star formation rate; the quantity on the ordinate is a measure of dust obscuration. Shown are ULIRGs from the samples of [18] and [12], starbursts from the sample of [14], and LIRGs and spirals from the sample described in [1]. The trend of decreasing obscuration with decreasing star-formation rate is crudely fit by $L_{\mathrm{bol,dust}}/L_{\mathrm{UV}} \propto L_{\mathrm{SFR}}$ (*dashed line*).

The real situation has obvious similarities to this hypothetical case. The typical galaxy detected by SCUBA appears to be heavily obscured by dust, orders of magnitude brighter in the (rest-frame) far-IR than the UV. But only ~25% of the $850\,\mu$m background is produced by objects brighter than SCUBA's 2 mJy detection limit [3], and only ~8% by objects bright enough to be included in the ~ 5 mJy samples of [4] and [7] — the only SCUBA samples with well established optical counterparts[1]. The 5 mJy sub-mm/radio samples have provided the strongest support for the popular belief that most high-redshift star formation occurred in extremely dusty galaxies that are not detected in UV surveys, but the fraction of high-redshift star formation that occurs in the extremely dusty objects of the 5 mJy sub-mm/radio samples is as small as the fraction of local star formation that occurs in LIRGs and ULIRGs. Should we believe that objects in the 5 mJy samples are any more representative of typical star-forming galaxies at high redshift than LIRGs and ULIRGs are of star-forming galaxies in the local universe?

The data suggest that they are not. In the remainder of this section I will argue that dust opacity and star-formation rate are as strongly correlated at high redshift as in the local universe, and that this implies that the fainter galaxies responsible for most of the $850\,\mu$m background are much less obscured in the UV than typical objects detected by SCUBA, sufficiently unobscured, in fact, that their UV emission is easier to detect than their dust emission.

The correlation of dust opacity and star-formation rate can be established at $z \sim 1$ by comparing galaxies selected at $850\,\mu$m with SCUBA to those selected at $15\,\mu$m with ISO. Owing to the strength of the 7.7 and $8.6\,\mu$m PAH emission features, $z \sim 1$ star-forming galaxies are comparatively bright at $15\,\mu$m and deep $15\,\mu$m ISO images detect dusty $z \sim 1$ galaxies many times fainter than those detectable with SCUBA. Figure 2a compares the star-formation rates and dust opacities of five $850\,\mu$m sources at $z \sim 1$ to those of the 17 ISO $15\,\mu$m sources in the HDF with $0.8 < z < 1.2$. The trend of decreasing dust opacity with decreasing star-formation rate is clear; the comparatively faint ISO sources are nowhere near as heavily dust obscured as the brighter ~ 5 mJy $850\,\mu$m sources. Since neither $15\,\mu$m nor $850\,\mu$m observations detect a very large fraction of $z \sim 1$ galaxies' bolometric dust luminosities, large multiplicative corrections were required when estimating $L_{\mathrm{bol,dust}}$ for this plot from the detected $15\,\mu$m or $850\,\mu$m fluxes. I adopted corrections that assume these $z \sim 1$ galaxies have dust SEDs similar to those of rapidly star-forming galaxies in the local universe. Details can be found in [1]; but the main result, the correlation of star-formation rate and dust opacity, is insensitive to the adopted corrections: altering $L_{\mathrm{bol,dust}}$ moves galaxies diagonally on the plot, parallel to the correlation.

At higher redshifts, galaxies' PAH features are redshifted well outside ISO's $15\,\mu$m window and another approach is required to measure the dust opacity of galaxies significantly less luminous than the 5 mJy SCUBA sources. One approach, adopted by [6] and [8], is to point SCUBA at many galaxies which

---

[1] See also [16] and [13], which discuss sub-mm samples selected in a different way but with the similarly bright typical $850\,\mu$m fluxes.

are too faint to be detected individually, and sum the measured fluxes. The 33 Lyman-break galaxies at $z \sim 3$ targeted by [6] and [8] were found to have a mean $850 \, \mu\text{m}$ flux of $0.72 \pm 0.31 \, \text{mJy}$, corresponding to $L_{\text{bol,dust}} \sim 2.7 \times 10^{11} h^{-2} L_{\odot}$ for the $\Omega_M = 0.3$, $\Omega_\Lambda = 0.7$ cosmology adopted throughout. The mean value of $L_{\text{bol,dust}}/L_{\text{UV}}$ for the galaxies in this sample was 13, significantly lower than the values observed among brighter SCUBA sources at similar redshifts (Fig. 2b). Since $0.7 \, \text{mJy}$ is close to the typical flux of the sources that dominate the $850 \, \mu\text{m}$ background [5,3], one might expect Lyman-break galaxies' moderate ratio of dust to UV luminosity $L_{\text{bol,dust}}/L_{\text{UV}} \sim 10$ to be more representative of average high-redshift star-forming galaxies than the extreme ratios $L_{\text{bol,dust}}/L_{\text{UV}} \gtrsim 200$ of the bright $850 \, \mu\text{m}$ samples. In any case, the $850 \, \mu\text{m}$ observations of Lyman-break galaxies provide further evidence that star-formation rate and dust opacity are correlated at high redshift.

If we accept that dust opacity and star-formation rate are correlated at high redshift in a way similar to what Figs. 2a and b suggest, we can proceed to estimate the characteristic level of dust obscuration among the comparatively faint galaxies that dominate the $850 \, \mu\text{m}$ background and see whether most of them are easier to detect in the sub-mm or the UV. Given a rest-frame UV detection limit of $f_\nu(1600 \, \text{Å} \times (1 + z)) > 0.2 \, \mu\text{Jy}$ ($m_{\text{AB}} = 25.5$), roughly appropriate for the Lyman-break survey of [17], and a sub-mm flux limit of $f_\nu(850 \mu\text{m}) = 2 \, \text{mJy}$, the confusion limit of SCUBA on the JCMT, a galaxy will be easier to detect in the rest-frame UV than the sub-mm if it has $L_{\text{bol,dust}}/L_{\text{UV}} \lesssim 500$ at $z \sim 1$ or $L_{\text{bol,dust}}/L_{\text{UV}} \lesssim 80$ at $z \sim 3$ [1]. Fig-

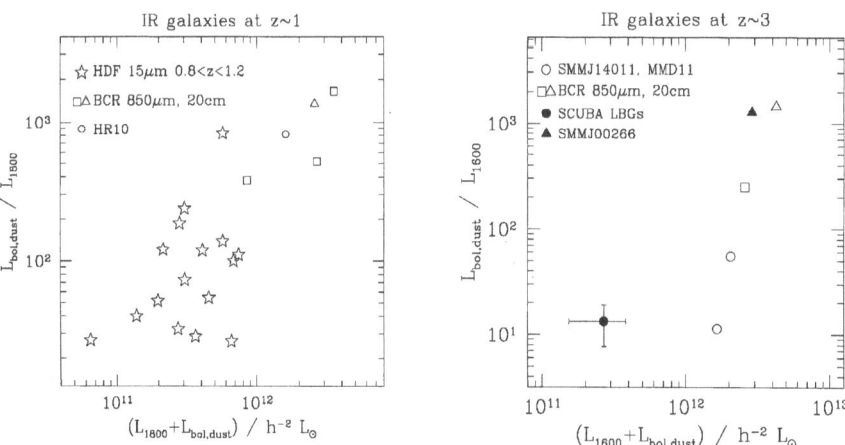

Fig. 2. The correlation of star-formation rate and dust opacity at high redshift. Left panel (**a**): $z \sim 1$. The bright $850 \, \mu\text{m}$ sources from [4] (BCR sub-sample with radio/sub-mm photometric redshift $1 < z_{\text{mm}} < 2$) and [9] (HR10) have larger ratios of dust to UV luminosity than the intrinsically fainter $15 \, \mu\text{m}$ ISO sources from [2] (see also [1]). Right panel (**b**): $z \sim 3$. Bright $850 \, \mu\text{m}$ sources with estimated/measured redshifts $z > 2$ (from [4,6,10,11]) have larger ratios of dust to UV luminosity than the (fainter) Lyman-break galaxies with similar redshifts that have been statistically detected at $850 \, \mu\text{m}$ by [8].

ure 2a suggests that $\langle L_{\mathrm{bol,dust}}/L_{\mathrm{UV}} \rangle = 500$ is characteristic of galaxies with $L_{\mathrm{SFR}} \simeq 10^{12} h^{-2} L_{\odot}$ at $z \sim 1$. Galaxies with lower star-formation luminosities will be less dust obscured on average and easier to detect in the UV than the sub-mm. $10^{12} h^{-2} L_{\odot}$ corresponds to $f_{\nu}(850\,\mu\mathrm{m}) \simeq 3.3\,\mathrm{mJy}$ for $\Omega_M = 0.3$, $\Omega_{\Lambda} = 0.7$, and the local dust SED shape assumed throughout; roughly 85% of the $850\,\mu\mathrm{m}$ background is produced by objects fainter than $3.3\,\mathrm{mJy}$ [3]. This suggests that the majority of star-formation at $z \sim 1$ occurs in galaxies that are easier to detect in the rest frame UV than the sub-mm.

A similar conclusion follows at $z \sim 3$. Figure 2b suggests that $\langle L_{\mathrm{bol,dust}}/L_{\mathrm{UV}} \rangle = 80$ is characteristic of galaxies with $L_{\mathrm{SFR}} \simeq 10^{12} h^{-2} L_{\odot}$ $(f_{\nu}(850\,\mu\mathrm{m}) \simeq 2.5\,\mathrm{mJy})$ at $z \sim 3$. 80% of the $850\,\mu\mathrm{m}$ background is produced by objects fainter and presumably less obscured.

This completes my simple argument: a correlation between star-formation rate and dust opacity is seen to exist at every redshift we can probe ($0 < z \lesssim 3$), and its slope is sufficient to imply that most of the $850\,\mu\mathrm{m}$ background was produced by galaxies that are easiest to detect in the rest-frame UV. It is an empirical argument based solely upon the observed ratios of dust to UV luminosity among galaxies whose dust and UV emissions have both been detected. It largely sidesteps the long and uncertain chain of reasoning – the incompleteness corrections, the luminosity-function extrapolations, the conversions of fluxes to bolometric luminosities to star-formation rates, and so on (see Carilli's contribution) – that makes attempts to address the same question through "Madau diagrams" so famously contradictory and unreliable.

3   Interpreting UV Surveys

The reason UV surveys detect a large fraction of high redshift star formation is that the UV luminosities of star-forming galaxies are largely independent of dust opacity. This is a consequence of the correlation of star-formation rate and dust opacity: the dustiest objects tend to be the most luminous, and the two effects – increased obscuration and increased luminosity – mostly cancel out at $\lambda \sim 1500\,\text{Å}$. As can be seen in Fig. 3, observed UV luminosities are similar for galaxies with dust obscurations $L_{\mathrm{bol,dust}}/L_{\mathrm{UV}}$ spanning four orders of magnitude. ULIRGs in the local universe are as bright in the UV as starburst dwarfs; high redshift galaxies that have been detected through their dust radiation are typically as bright in the UV as those that have been detected through their UV radiation.

This fact makes UV surveys of star-forming galaxies easy to construct but hard to interpret. Referring to Fig. 3, one can see that a UV luminosity-limited survey of the local universe would net a tremendously broad range of objects, an indiscriminate haul of ULIRGs and blue compact dwarfs and spirals like the Milky Way. The same appears to be true of luminosity-limited surveys at high redshift. But UV luminosities don't give us much to distinguish between these vastly different objects: if we were told only that a local galaxy had $L_{\mathrm{UV}} = 10^9 h^{-2} L_{\odot}$, for example, we would have no idea whether it was a ULIRG or a spiral or a starburst dwarf. How can we tell what sort of objects we have detected in a high redshift UV selected survey? How can we estimate any single object's star-formation rate to within even an order of magnitude? Measuring mid-IR, far-IR, sub-mm, or radio fluxes for the objects would be ideal, but most high-redshift star formation occurs in objects too faint to be detected at any of these longer wavelengths, and in any case the small fields of view of current instruments at mid-IR to mm wavelengths ($\sim 2' \times 2'$) are poorly matched to the large fields of view of modern optical instruments (up to $\sim 40' \times 40'$). Until the next generation of long wavelength instruments becomes available, much of our knowledge of galaxy formation at high redshift will have to be derived from UV data alone.

Meurer and collaborators have developed one technique for estimating dust luminosities (and hence star-formation rates) from UV observations (e.g. [14], hereafter MHC). Dustier local starbursts are not fainter in the UV on average, but they are redder, and MHC observed that a galaxies' redness in the UV was a surprisingly good predictor of its dustiness $L_{\mathrm{bol,dust}}/L_{\mathrm{UV}}$. Among the galaxies in MHC's sample, a measurement of the UV spectral slope β (in $l_\lambda \propto \lambda^\beta$ at $1200 \lesssim \lambda \lesssim 2000\,\text{Å}$) was sufficient to predict $L_{\mathrm{bol,dust}}/L_{\mathrm{UV}}$ to within a factor of 2. This β/far-IR correlation is the basis of many attempts to estimate the star-formation rates of UV-selected high-redshift galaxies.

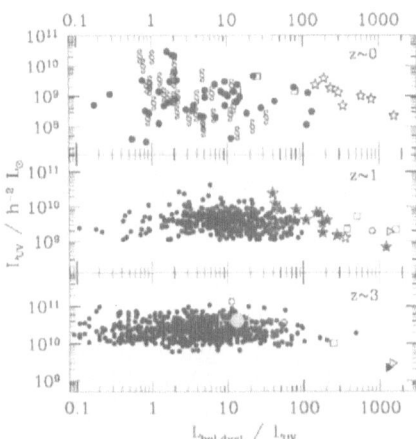

Fig. 3. Far-UV luminosity versus dust obscuration for star-forming galaxies at $z \sim$ 0 (top panel), $z \sim 1$ (middle panel), and $z \sim 3$ (bottom panel). The darker solid circles in the middle and bottom panel represent UV-selected star-forming galaxies with $L_{\mathrm{bol,dust}}$ estimated from the β/far-IR correlation; the lighter solid circle with error bars in the bottom panel shows the mean UV luminosity and dust obscuration ($\pm$ standard deviation of the mean) for Lyman-break galaxies observed in the sub-mm by [8]. Otherwise the symbols are as in Figs. 1 and 2. In general extremely dusty galaxies are no fainter in the far-UV than relatively dust-free galaxies.

Do high-redshift galaxies obey MHC's β/far-IR correlation? The evidence is ambiguous. In some cases the β/far-IR correlation appears to hold. One example is the lensed Lyman-break galaxy SMMJ14011+0252 at $z = 2.565$. Its observed 850 μm and 20 cm fluxes of 15 ± 2 mJy, 115 ± 30 μJy [13] agree well with the predictions of 23^{+14}_{-9} mJy, 150^{+95}_{-67} μJy that the β/far-IR correlation implies [1]. In this case the total star-formation rate can be deduced with reasonable accuracy from UV observations alone. The same appears to be true for the ten brightest Lyman-break galaxies in the HDF. According to MHC, the β/far-IR and far-IR/radio correlations predict a total 20 cm flux for these galaxies of 105 ± 24 μJy, nicely consistent with the total measured flux of 100 ± 33 μJy. But there are counter-examples as well. Baker and van den Werf discussed two in their talks, the galaxies MS1512+36-cB58 and MS1358+62-G1, where the β/far-IR correlation appears to overpredict the observed 850 μm flux by an order of magnitude or more. This suggests – worryingly – that failures of the β/far-IR correlation at high redshift are not restricted to the dustiest galaxies ($L_{\mathrm{bol,dust}}/L_{\mathrm{UV}} \gtrsim 100$) as they appear to be in the local universe (see Meurer's contribution to these proceedings).

Analysis of the largest available sample of UV-selected high-redshift galaxies with independent constraints on $L_{\mathrm{bol,dust}}$ suggests that most high-redshift galaxies obey the β/far-IR correlation better than MS1512+36-cB58 and MS1358+62-G1 but perhaps not as well as SMMJ14011+0252. Figure 4 shows the predicted and observed 20 cm fluxes of 69 "Balmer-break" galaxies in the Hubble Deep and Flanking Fields with spectroscopic redshifts $z = 1.00\pm0.11$ [1]. The objects that are predicted (via the β/far-IR and far-IR/radio correlations) to have larger 20 cm fluxes clearly do on average; the data do not support the view that MHC's β/far-IR correlation generally misestimates dust luminosities by more than an order of magnitude (cf. van den Werf's contribution to these proceedings). But nor do they obviously support the view that the β/far-IR correlation can predict far-IR fluxes as accurately at high redshift as in the local universe. The measured 20 cm fluxes of the galaxies in the brightest bin of predicted flux are roughly a factor of 3 times lower than expected, for example. Assessing the significance of this shortfall is difficult because of the large error bars in the predicted fluxes – a reasonable fraction of the objects in the brightest predicted bin were probably scattered out of the next brightest bin by various errors – and in any case a significant discrepancy between the observed and predicted fluxes might be due to a failure of the far-IR/20 cm rather than the β/far-IR correlation. But it is safe to say that the available data do not inspire absolute confidence in the validity of the β/far-IR correlation at high redshift.

An additional cause for concern is the fact that a red spectral slope β does not appear to be the most distinctive characteristic of the high-redshift galaxies known to have large dust luminosities. For example, these are the differences in mean color between $z \sim 1$ galaxies in the HDF that are and are not ISO 15 μm sources: $\Delta(U_n - G) = 0.11$, $\Delta(G - \mathcal{R}) = 0.45$, $\Delta(\mathcal{R} - I) = 0.17$. The SEDs of the dusty and luminous ISO sources differ from those of fainter and less obscured galaxies primarily in the range $2500 \lesssim \lambda_{\mathrm{rest}} \lesssim 3500$ Å, not $\lambda_{\mathrm{rest}} \lesssim 2000$ Å where

β is measured. This does not show that MHC's prescription for dust correction is wrong, only that it can be improved, but it adds to concerns that their prescription may not perform as well at high redshift as in the local universe.

Even if the β/far-IR correlation fails on some objects, however, there is reason to hope that it may perform reasonably well on an ensemble of high redshift galaxies. For example, the correlation implies a mean dust obscuration for $z \sim 3$ Lyman-break galaxies of $\langle L_{bol,dust}/L_{UV}\rangle \simeq 8$, a plausible value that is midway between the dust obscuration observed in local spirals ($\langle L_{bol,dust}/L_{UV}\rangle \simeq 5$) and local UV-selected starbursts ($\langle L_{bol,dust}/L_{UV}\rangle \simeq 15$) [1] and that is consistent with the mean obscuration $\langle L_{bol,dust}/L_{UV}\rangle \simeq 13 \pm 6$ derived from sub-mm observations of Lyman-break galaxies [8]. Moreover many of the major results to emerge from recent analyses of 850 μm data – the brightness of 850 μm background, the domination of the 850 μm background by ~ 1 mJy sources, the $\sim$ thousand-fold increase in number density of ULIRGs from $z \sim 0$ to $z \gtrsim 3$ – could have been predicted from UV observations alone by applying the β/far-IR correlation to UV-selected high-redshift galaxy populations [1]. Applying MHC's β/far-IR correlation to high-redshift galaxy populations at least leads to results that are reasonable, and so it seems sensible to continue using it cautiously until further long-wavelength observations can validate it or suggest better alternatives.

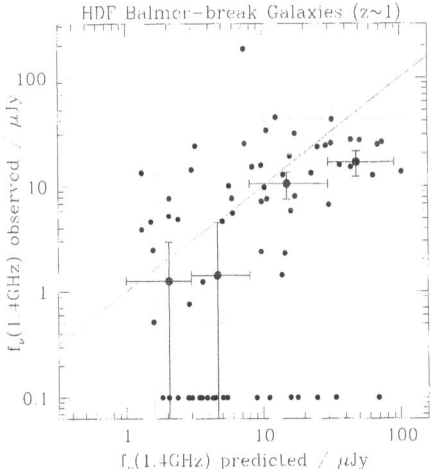

Fig. 4. Predicted and observed 20 cm fluxes of $z \sim 1$ galaxies in the HDF. Small solid circles represent individual galaxies. Objects with measured fluxes lower than 0.1 μJy have been drawn with measured flux equal to 0.1 μJy. Error bars for two representative galaxies are dotted. The large circles with solid error bars show the mean observed flux ($\pm$ standard deviation of the mean) for galaxies in different bins of predicted flux. The galaxy with $f_{pred} \sim 7$, $f_{obs} \sim 200$ likely has a significant AGN contribution to its 20 cm flux; it was the only outlier rejected before calculating the mean fluxes.

4   Conclusions

This conference concluded with a debate on the best way to explore high-redshift star formation. The best way? Understanding the history of galaxy formation is a large and difficult task, a shared enterprise that will benefit from observations at many wavelengths. I am not sure how much we stand to gain from arguing whose contributions are the most important.

Nevertheless I would have felt derelict in my assigned role as defender of the UV if I had not attacked one statement that was repeated several times at the meeting. This was the claim that UV and IR surveys detect "orthogonal" populations of galaxies, that a significant fraction of stars form in a "hidden" population of dusty galaxies that is not detected in UV surveys. I am not aware of any evidence supporting this claim. Where is the dusty hidden population in the local universe? A luminosity-limited UV survey deep enough to detect most spirals and blue starbursts would detect most LIRGs and ULIRGs as well (cf. Fig. 3). Where is it at $z \sim 1$? Existing UV-selected surveys easily detect most if not all of the $z \sim 1$ 15 μm sources detected by ISO. Where at higher redshift? A UV-selected survey would need to reach $\mathcal{R} \sim 26.5$ to detect the majority of the $\sim 1800\,\text{Å}$ luminosity density at $z \sim 3$ [17], and at this magnitude limit most if not all of the 850 μm sources in the samples of [4], [13][2], and [7] would be detected. It is easy to construct a "Madau diagram" in which galaxies detected at wavelength x contribute far more to the comoving star-formation density than galaxies detected at wavelength y. Data will confess to anything if you torture them enough. The proper approach is not to compare Madau diagrams – not to see who benefited most from the grotesquely compounded uncertainties – but to turn to the galaxies themselves and measure their luminosities at both wavelengths. When this is done (as we have seen) the answer is unequivocal: any UV survey at a given redshift $z \lesssim 3$ deep enough to detect majority of the UV luminosity density will detect the majority of IR-selected galaxies as well.

But there is one sense in which most star formation *is* hidden from UV surveys: only a small fraction of the energy emitted by massive stars emerges from typical galaxies in the UV. The majority is absorbed by dust and reradiated in the far-IR. Unless there is some way to estimate the total luminosity of a galaxy's massive stars from the (often) trace amount that is detected in the UV, it will be impossible to estimate star-formation rates for the detected galaxies and the usefulness of UV surveys will be diminished. Sect. 3 discussed one method for estimating far-IR fluxes from far-UV fluxes that is known to work in the local universe. As we saw, this method appears to work within a factor of ~ 3 at high redshift, but it is unclear if it works much better. The available data inspire more hope than confidence.

The strength of UV surveys is that they detect large numbers of high-redshift galaxies, even ones that are intrinsically very faint, in large and representative comoving volumes. The weakness is that star-formation rates are difficult to estimate for the detected galaxies. IR surveys complement them perfectly: star-

---

[2] Including correction for magnification by gravitational lensing.

formation rates can be estimated with reasonable confidence, but only small regions of the sky can be surveyed and only the most luminous sources can be detected. Harnessing the strengths of both types of surveys would lead to a rapid advance in our understanding of galaxy formation at high redshift. One obvious strategy would be to use IR observations of some UV-selected high-redshift galaxies to attempt to derive (or validate) a method for estimating dust luminosities from UV observations alone, a method that can then be applied to the large numbers of UV galaxies which have not been (and often cannot be) detected in the IR. This sort of multiwavelength cooperation will take us much farther than arguments about which wavelength is superior – despite the best efforts of wavelength partisans like the author of this screed.

Acknowledgements: I would like to thank the organizers for support at this enjoyable meeting. Special thanks are due to my collaborators S. Chapman, M. Dickinson, M. Giavalisco, M. Pettini, A. Shapley, and C. Steidel for their contributions to this work.

References

1. K.L. Adelberger, C.C. Steidel: ApJ **544**, 218 (2000)
2. H. Aussel, C.J. Cesarsky, D. Elbaz, J.L. Starck: A&A **342**, 313 (1999)
3. A.J. Barger, L.L. Cowie, D.B. Sanders: ApJ **518**, L5 (1999)
4. A.J. Barger, L.L. Cowie, E. Richards: AJ **119**, 2092 (2000)
5. A.W. Blain, J.-P. Kneib, R.J. Ivison, I. Smail: ApJ **512**, L87 (1999)
6. S.C. Chapman et al.: MNRAS **319**, 318 (2000)
7. S.C. Chapman, E. Richards, G. Lewis, G. Wilson, A. Barger: ApJ Let., submitted
8. S.C. Chapman et al. in preparation
9. A. Dey, J.R. Graham, R.J. Ivison, I. Smail, G.S. Wright, M.C. Liu: ApJ **519**, 610 (1999)
10. D.T. Frayer et al.: ApJ **514**, L13 (1999)
11. D.T. Frayer, I. Smail, R.J. Ivison, N.Z. Scoville: AJ **120**, 1668 (2000)
12. J.D. Goldader et al. in preparation
13. R.J. Ivison et al.: MNRAS **315**, 209 (2000)
14. G.R. Meurer, T.M. Heckman, D. Calzetti: ApJ **521**, 64 (1999)
15. D.B. Sanders, I.F. Mirabel: ARAA **34** 749 (1996)
16. I. Smail, R.J. Ivison, F.N. Owen, A.W. Blain, J.-P. Kneib: ApJ **528**, 612 (2000)
17. C.C. Steidel, K.L. Adelberger, M. Giavalisco, M. Dickinson, M. Pettini: ApJ **519**, 1 (1999)
18. N. Trentham, J. Kormendy, D.B. Sanders: AJ **117**, 2152 (1999)

The Cosmic Star-Formation History: The Far-IR and Sub-mm View

Alberto Franceschini

Padova University, Astronomy Department, Vicolo Osservatorio 5, 35122 Padova, Italy

Abstract. We briefly review recent discoveries on high-redshift star-formation obtained with observations at long-wavelengths by ground-based mm telescopes and space IR observatories. These observations indicate extremely high rates of cosmic evolution for IR galaxies and dramatic starburst activity at moderate to high redshifts. Our interpretative scheme considers a bimodal star formation (SF) in galaxies, including long-lived quiescent SF, and enhanced SF taking place during transient luminous events triggered by interactions and merging. An increase with z of the rate of interactions between galaxies (*density evolution*) and an increase of their IR luminosity due to the more abundant fuel available in the past (*luminosity evolution*) explain the strong observed evolution. Many of the phenomena revealed by these long-wavelength observations (e.g. the rate of cosmic evolution and luminosities of IR galaxies, and the IR background energetics) were largely unexpected based on UV-optical observations. history

1 Introduction

High-redshift galaxies and the generation of stars during the past cosmic history are most usually investigated by targetting the stellar integrated emissions in the UV/optical/near-IR. During the last few years, however, it has become more and more evident that observations of the high-z universe at longer wavelengths do not only provide complementary information, needed e.g. for appropriate bolometric corrections to the observed optical fluxes, but do in fact also reveal entirely new phenomena which are already significantly impacting on our view of galaxy formation and evolution.

2 Discovery of the Cosmic IR Background (CIRB)

The detection of the CIRB in the all-sky COBE maps was one of the most relevant achievements of observational cosmology during the last decade, bringing to the *first detection ever of the integrated emission of distant galaxies* in the form of an isotropic signal in the far-IR and sub-mm. The original detection of the CIRB by [18] has been later confirmed with independent analyses by various other groups using FIRAS on COBE [8], as well as data from the DIRBE experiment in three broad-band channels at $\lambda = 240$, 140 and 100 μm ([12], see Fig. 1).

In the wavelength interval $4\mu m < \lambda < 60\mu m$, where any cosmological flux is far dominated by the the Inter-Planetary Dust (IPD) emission, indirect constraints can be inferred from measurements of the cosmic opacity for $\gamma - \gamma$ interactions [23]: very high energy (TeV) photons emitted by low-redshift Blazars interact with photons of the CIRB producing electron-positron pairs. A measure of the high-energy opacity from TeV spectral observations allows to set limits on the IR background energy density. Stanev & Franceschini [22] have used TeV observations of MKN 501 (z=0.034) during an outburst in 1997 to derive the upper limits on CIRB (dotted histograms in Fig. 1).

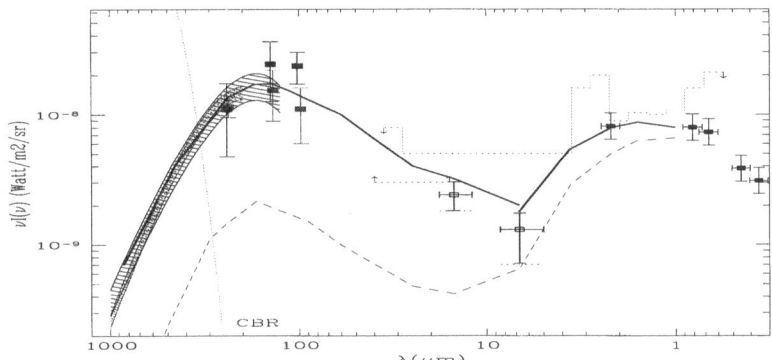

Fig. 1. The Cosmic Infrared Background (CIRB) spectrum as measured in the all-sky COBE maps compared with estimates of the optical background based on ultradeep HST surveys (see text). The lower dashed line marks the expectation based on the assumption that the IR emissivity of galaxies does not change with cosmic time. The thick line is the predicted CIRB intensity by the presently discussed model for IR galaxy evolution.

Altogether, the integrated CIRB intensity between 100 and 1000 μm, where the present estimates show a small scatter, is $\sim (30 \pm 5)\ 10^{-9}\ Watt/m^2/sr$. The addition of the presently un-measurable fraction between 100 and 10 μm using the constraints summarized in Fig. 1 brings the total energy density in the CIRB between 7 and 1000 μm to the value: $\nu I(\nu)|_{FIR} \simeq 40\ 10^{-9}\ Watt/m^2/sr$. This flux is substantially larger than the integrated bolometric emission by distant galaxies between 0.1 and 7 μm (the"optical background"), for which we adopt the value estimated by [14] using the HST integrations in the Hubble Deep Field: $\nu I(\nu)|_{opt} \simeq (17 \pm 3)\ 10^{-9}\ Watt/m^2/sr$. This sets a relevant constraint on the evolution of cosmic sources, if we consider that for local galaxies only 30% on average of the bolometric flux is re-processed by dust into the far-IR. *The CIRB's intensity exceeding the optical background suggests that galaxies in the past should have been much more "active" in the far-IR than in the optical, and very luminous in an absolute sense.*

3   Deep Explorations of the IR Sky

3.1   Deep Surveys by the Infrared Space Observatory (ISO)

A substantial amount of the ISO observing time has been dedicated to deep surveys at mid- and far-IR wavelengths [5,10], providing the first systematic exploration of the IR sky with sensitivities sufficient to detect sources at cosmological redshifts.

In the mid-IR (12-18 μ, $\lambda_{eff} = 15\ \mu$), a set of extragalactic surveys, IGTES, have been performed in the ISOCAM Guaranteed Time, over a total area of 1.5 square degrees with more than one thousand sources detected [5]. The two Hubble Deep Field areas (North and South), including the Flanking Fields for a total of $\sim$ 50 sq. arcmin, have been deeply surveyed by ISOCAM at 15 μm to a sensitivity limit of 100 μJy. The redshift distributions show an excess number of sources between z=0.5 and z=1. At brighter fluxes, the European Large Area ISO Survey (ELAIS) observed a total of 12 square degrees at 15 μm with ISOCAM and at 90 μm with the ISO Spectophotometer (ISOPHOT) [15].

In the far-IR, the most important survey project, FIRBACK, was carried out with ISOPHOT to detect at 170 μm the sources of the CIRB (roughly 200 detected [4]. This survey is limited by extragalactic confusion in the large ISOPHOT beam (90 arcsec) to $S_{170} \geq 100$ mJy.

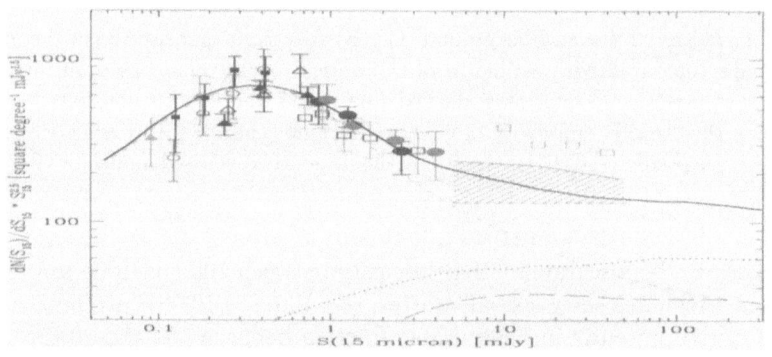

Fig. 2. Normalized differential counts at $\lambda_{eff} = 15\ \mu m$ (see [5]).

The most detailed information on faint IR counts come from the ISO 15 μm surveys. The differential counts normalized to $S^{-2.5}$ (Fig. 2), reveal a roughly euclidean slope from the brightest fluxes down to $S_{15} \sim 10\ mJy$; a sudden upturn at $S_{15} < 3\ mJy$, where the counts increase as $dN \propto S^{-3.1}dS$; and a convergence below $S_{15} \sim 0.3\ mJy$ (where $dN \propto S^{-2}dS$).

3.2   Surveys at Millimetric Wavelengths

Surveys in the sub-millimeter offer the unique advantage to naturally generate volume-limited samples from flux-limited observations, due to the peculiar shape of galaxy spectra [roughly $L(\nu) \propto \nu^{3.5}$]. In addition, local galaxies emit very modestly at these wavelengths. Altogether, a sensitive sub-mm survey preferentially selects sources at very high redshifts, providing a direct picture of the high-redshift universe impossible to obtain at other frequencies.

Important discoveries have come from the implementation of bolometer arrays (SCUBA and MAMBO) on JCMT and IRAM (e.g. [13]). SCUBA has allowed to resolve $\sim 20\%$ of the CIRB background at 850 μm into a population of faint distant, mostly high-z, sources. The extragalactic source counts show a dramatic departure from the Euclidean law [$dN \propto S^{-3} dS$ in the interval from 2 to 10 mJy], a clear signature of the strong evolution and high redshift of mm-selected sources.

4   A Panchromatic View of IR Galaxy Evolution

The deep counts at the various observed wavelengths (e.g. Fig. 2) display very significant departures from the no-evolution expectations. The ISO differential counts at 15 μm shown in Fig. 2, in particular, contain some relevant indications about the properties of the evolving population. The almost Euclidean counts extending from the bright IRAS fluxes down to a few mJy, followed by the sudden upturn below, suggest that it is possibly a locally small fraction of IR galaxies to evolve with high rates back in cosmic time. We have reproduced these data with the contribution of two source populations, one non-evolving with local luminosity function taken from the IRAS 12 μm survey, the other strongly evolving, both in comoving number density and in luminosity, as:

$$\rho(L[z], z) = \rho_0(L_0) \times (1+z)^{5.5}, \quad L(z) = L_0 \times (1+z)^{2.2} \quad z < z_{break}$$

$$\rho(L[z], z) = const, \quad L(z) = const \quad z_{break} < z < z_{max}$$

with $z_{break} = 0.8$ and $z_{max} = 3$, for a Λ-dominated universe ($\Omega_m = 0.3$, $\Omega_\Lambda = 0.7$, $H_0 = 50 \, Km/s/Mpc$). A third source population are type-I AGNs, assumed to evolve in luminosity as $L(z) = L(0) \times (1+z)^3$ up to $z = 1.5$. The local fraction of the evolving starburst population is assumed to be ~ 10 percent of the total, roughly consistent with observations of interacting galaxies.

Deep surveys at various IR/sub-mm wavelengths can be exploited to simultaneously constrain the evolution properties and broad-band spectra of faint IR sources. [10] have fitted the IR multi-wavelength statistics with the above model by adopting suitable SEDs for IR sources. If we assume for the IR evolving sources a more typical starburst spectrum (like the one of M82, by all means similar to those of other luminous starbursts observed by ISO), then most of the observed properties of far-IR galaxy samples and the CIRB spectrum can be appropriately reproduced. Our best-fit to the counts is given in Fig. 2.

The $15\mu m$ counts in Fig. 2 display a remarkable convergence below $S_{15} \sim$ 0.2 mJy. The asymptotic slope flatter than -2 in differential count units implies a modest contribution to the integrated CIRB flux by sources fainter than this limit. A meaningful estimate of the CIRB flux can then be obtained from direct integration of the observed mid-IR counts (the two datapoints at 15 and 7 μm in Fig. 1). This implies a value at 15 μm of 2.6 ± 0.5 $nWatt/m^2/sr$ contributed by LW3 sources brighter than $S_{15} = 40$ μJy ([6]; extrapolating our evolution to the faintest fluxes would bring the value to 3.3 $nJy/m^2/sr$). This value is close to the upper boundary set by the observed TeV cosmic opacity (see Fig. 1): it appears that the ISOCAM surveys have resolved a significant (50-70%) fraction of the CIRB intensity in the mid-IR. Assuming the typical starburst spectrum needed to match the multi-wavelength counts implies that the population detected by ISO is also responsible for a major contribution of the CIRB at any wavelength.

4.1   A Two-Phase Star-Formation: Origin of Galactic Disks and Spheroids

Our best-fit model for IR galaxy evolution implies that *star formation in galaxies has proceeded along two phases: a quiescent one taking place during most of the Hubble time, slowly building stars with standard IMF from the regular flow of gas in rotational supported disks; and a transient actively starbursting phase, recurrently triggered by galaxy mergers and interactions.* During the merger, *violent relaxation* redistributes old stars, producing de Vaucouleur profiles typical of galaxy spheroids, while young stars are generated following a top-heavy IMF.

Because of the geometric (thin disk) configuration of the diffuse ISM and the modest incidence of dusty molecular clouds, the quiescent phase is only moderately affected by dust extinction, and naturally originates most of the optical/NIR background.

The merger-triggered active starburst phase is characterized by a large-scale redistribution of the dusty ISM, with bar-modes and shocks compressing a large fraction of the gas into the inner galactic regions and triggering formation of molecular clouds. As a consequence, this phase is expected to be heavily extinguished and may naturally explain most of the CIRB.

5   Global Constraints

The evolutionary SFR density based on our IR model are reported in Fig. 3. The contribution of IR-selected sources to the high-z luminosity density significantly exceeds those based on optically selected sources, and the excess may be progressive with redshift up to $z \simeq 1$.

The rate of evolution for the IR volume emissivity of galaxies at $z < 1$ (top line) is even higher than that of optical/X-ray AGNs (bottom line). This fast evolution should however level off at higher z, to allow consistency with the observed z-distributions for faint ISOCAM sources and to fit the CIRB spectrum (thick line in Fig. 1, see also [11]). The galaxy activity sampled by

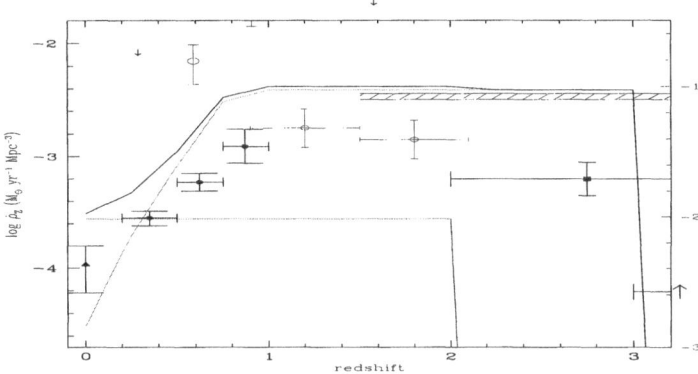

Fig. 3. Evolution of the metal production rates (left axis) and of the star formation rates (in $M_\odot/yr$, right axis) based on our best-fit model. Data points come from optical observations. >From top to bottom, the lines are: total, starburst, quiescent population and AGNs.

long-λ observations is not confined to the very high-redshifts, but shows a peak around $z \sim 1 - 1.5$.

5.1 Energy Constraints from Background Observations

Interesting constraints on the high-redshift far-IR/sub-mm population can be inferred from the observed energy densities in the CIRB and optical backgrounds. Let us assume that a fraction f_* of the universal mass density in baryons Ω_b undergoes at redshift z_* a transformation with radiative efficiency ϵ. For stellar processes, ϵ is determined by the IMF: $\epsilon = 0.001$ for a Salpeter IMF and a low-mass cutoff $M_{low} = 0.1\ M_\odot$, $\epsilon = 0.002$ for $M_{low} = 2$.

Let us schematically assume that the optical/NIR BKG mostly originates by quiescent SF in spiral disks and by intermediate and low-mass stars. As observed in the Solar Neighborhood, a good approximation to the IMF in such moderately active environments is the Salpeter law with $M_{low} = 0.1$ ($\epsilon \sim 0.001$). Then the optical BKG intensity can be obtained by transforming at redshift $z_* \sim 1.5$ a fraction $f_* \simeq 10\%$ of baryons into (mostly low-mass) stars: $\nu I(\nu)|_{opt} \simeq 20\ 10^{-9} h_{50}^2 \left(\frac{\Omega_b}{0.05}\right) \left(\frac{f_*}{0.1}\right) \left(\frac{2.5}{1+z_*}\right) \left(\frac{\epsilon}{0.001}\right)\ Watt/m^2/sr$. A local density in low-mass stars is generated in this way consistent with the observations, $\rho_b(stars) \simeq 7\ 10^{10} f_* \Omega_b \simeq 3.4\ 10^8\ M_\odot/Mpc^3$, with a corresponding density in metals of $\rho_Z(stars) \simeq 7\ 10^{10} f_* \frac{Z}{Z_\odot} \Omega_b\ M_\odot/Mpc^3 \simeq 7\ 10^6 M_\odot/Mpc^3$.

On the other hand, given that luminous starbursts emit a negligible fraction of the energy in the optical-UV and most of it in the far-IR, we coherently assume that the CIRB originates from dusty star-forming galaxies at median $z_* \simeq 1.5$. This process has to explain the huge energy content in the CIRB without exceeding the stellar remnants in local galaxies. A plausible solution would be to assume for the starburst phase a top-heavy IMF (e.g. a Salpeter distribution cutoff below $M_{min} = 2\ M_\odot$, with $\epsilon = 0.002$):

$$\nu I(\nu)|_{FIR} \simeq 40\ 10^{-9} h_{50}^2 \left(\frac{\Omega_b}{0.05}\right) \left(\frac{f_*}{0.1}\right) \left(\frac{2.5}{1+z_*}\right) \left(\frac{\epsilon}{0.002}\right)\ Watt/m^2/sr.$$

This requires that a similar amount of baryons, $f_* \simeq 10\%$, as transformed during the "secular" evolution, are processed with higher efficiency during the starbursting phases, producing a two times larger amount of energy and metals. Most of these metals have to be released by the galaxies into the diffuse cosmic media, as observed for example in the intracluster plasmas.

6   Physical Properties of the IR Source Population

Because of the different K-corrections, faint sources selected by mm telescopes and by ISO display nicely complementary properties as of redshift coverage (typically $z < 1$ for ISO and > 1 for SCUBA sources) and luminosities ($L_{bol} < 10^{12}\ L_\odot$ by ISO, and larger by SCUBA). Having said that, the deep ISO mid-IR surveys provide various advantages when investigating the nature of the IR source populations: they are responsible for a large fraction of the CIRB, their optical counterparts are easy to identify because of the moderate redshift, catalogues of thousands ISO mid-IR sources are available. The situation is different for the high-z objects selected by mm telescopes: of the few tens of sources detected, the optical counterparts are so faint that in only a few cases they have been reliably identified (D. Hughes, these Proceedings), and they contribute in any case a moderate fraction of the CIRB intensity.

The ISO surveys in the Hubble Deep Fields and the CFRS areas have been particularly well studied (e.g. [20,1]). The optical-IR SED of a typical faint 15 μm source at $z = 1.14$ is reported in Fig. 4 (left panel). The dotted line fitting the optical-NIR spectrum and corresponding to the SED of a quiescent spiral (M51) falls short by a factor ~ 10 of explaining the mid-IR emission, whereas SEDs of IR starbursts (e.g. Arp220 and M82) provide better fits. The vast majority of faint ISO sources show similar mid-IR flux excesses. The HST images indicate that a large fraction (30 to 50%) of the sources show evidence of morphological peculiarities and multiple structures [1], in keeping with the local evidence that galaxy interactions trigger luminous IR starbursts. These sources are also typically found in galaxy groups [3].

The baryonic masses in these objects have been estimated by fitting template SEDs of local galaxies to the observed near-IR spectrum (assuming a Salpeter IMF with standard low-mass cutoff $M_{low} = 0.15\ M_\odot$). The values found ($M \sim 10^{11}\ M_\odot$ at $z > 0.4$, with 1 dex typical spread) indicate that massive galaxies preferentially host these sources.

6.1   Evaluating the Star-Formation Activity

While the UV rest-frame spectra of the counterparts of faint ISO sources do not reveal strong emission lines [9], Rigopoulou et al. ([19], see also these Proceedings) have detected strong H_α lines (EW> 50 Å) redshifted into the NIR, demonstrating that *these optically faint but IR luminous sources are indeed powered by ongoing massive dusty starbursts*. Rigopoulou et al. [19] have used the H_α line fluxes to measure the rate of ongoing star-formation: the results of this analysis are reported on the y-axis of Fig. 4 (right panel).

To check the effects of dust extinction on the optical spectra, we have obtained independent evaluations of the SFR by exploiting the mid-IR flux as an alternative to the (heavily extinguished) optical-UV estimators. The capability of the mid-IR flux to measure the bolometric IR emission is discussed by [24] and [6]: the two quantities appear tightly correlated, with the only exceptions of very extinguished peculiar sources (e.g. Arp 220) in which even the mid-IR flux is self absorbed. The bolometric (mostly far-IR) flux provides the most robust measure of the number of massive reddened newly-formed stars.

We report in Fig. 5 our calibrated dependence of the SFR on the 15 μm flux measured by the ISOCAM LW3 filter, as a function of redshift. The application of this relation to observations of 15 μm sources in the HDFS results in the estimates of the SFR plotted on the x-axis of Fig. 4. The rates of SF indicated by the fits to the mid-IR flux for source at $z > 0.5$ range from few tens to few hundreds solar masses/yr, i.e. a substantial factor larger than found for faint optically-selected galaxies.

As shown in Fig. 4, *even after correcting for extinction, the H_α flux turns out to explain only $\sim 20-30\%$ on average of the bolometric far-IR emission: the bulk of SF in luminous IR galaxies is unobservable in the optical*. This is quite consistent with the results obtained by [16] on a local sample of very luminous IR galaxies, indicating a systematic discrepancy between FIR-based estimates of the SFR and those derived from the H_α luminosity.

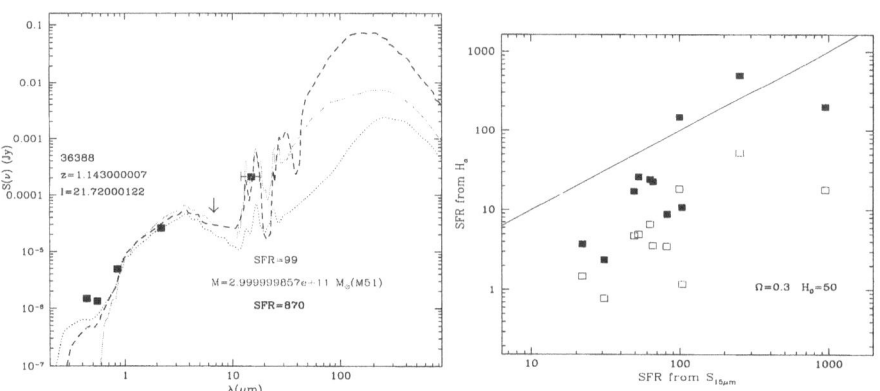

Fig. 4. Left panel: broad-band spectrum of a typical mid-IR source selected by ISO-CAM at 15 μm in the Hubble Deep Field North [1], compared with the SED's of M82, Arp 220 and M51. Estimates of the SF rate and stellar mass are indicated. Right panel: SF rates estimated from the H_α line flux compared with the values based on the mid-IR flux. Open squares: SFR uncorrected for extinction; filled squares: SFR from extinction-corrected H_α flux.

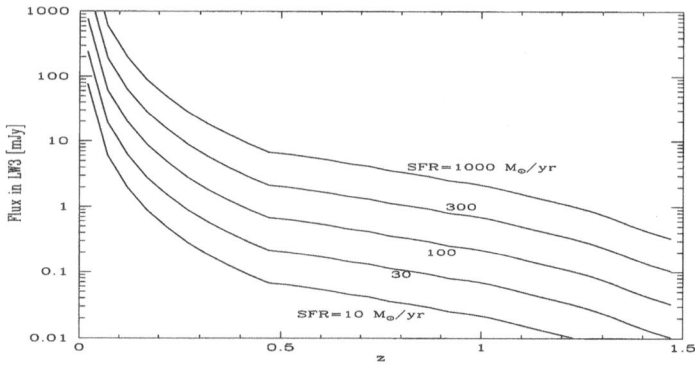

Fig. 5. Relation of the observed flux in the 15 μm LW3 ISO band versus redshift, as a function of the rate of star formation (SFR). This is based on the assumption that the IR flux is dominated by dust-reprocessed emission of young stars. The calibration of the relation has been obtained using the M82 spectral template and the detailed response function of the LW3 filter.

Poggianti, Bressan, Franceschini ([17], see also Poggianti, these Proceedings) interpret the optical spectra of luminous and ultra-luminous IR galaxies as due to selective dust attenuation, where newly-formed stars spend most of their life embedded into optically thick molecular clouds, while older stars, having already disrupted their parent cloud, suffer less extinction.

A variety of facts indicate that UV/optically-selected and IR/mm-selected faint high-redshift sources form almost completely disjoint samples. Chapman et al. [2] observed with SCUBA a subset of z=3 Lyman-break galaxies having the highest predicted rates of SF as inferred from the optical spectrum, but detected only one object out of ten. Van der Werf (these Proceedings) fail to detect with SCUBA two strongly lensed Lyman-break galaxies selected with the same technique. A similar dichotomy is observed in the local universe, where the bolometric flux by luminous IR galaxies is mostly unrelated with the optical emission spectrum [21].

Altogether, the ratio of the integrated IR to optical emissions is very broadly distributed and no clear empirical, nor physical, relationships have yet been established between the two.

7   Conclusions

The galaxy population dominating the faint IR counts and contributing to the bolometric CIRB intensity appears to be composed of luminous ($L_{bol} \sim 10^{11} - 10^{12} \, L_\odot$) starbursts in massive ($M \sim 10^{11} \, M_\odot$ of baryons) galaxies at $z \sim 0.5 - 1.5$, observed during a phase of intense stellar formation ($SFR \sim 100 \, M_\odot/yr$).

This intense stellar activity is possibly related with the assembly of galaxy spheroids in groups and in the field. In particular, the similarity in properties between high-z SCUBA sources and local ultra-luminous IR galaxies argues in favour of the idea that these represent the long-sought "primeval galaxies", originating the local massive elliptical and S0 galaxies. By continuity, the less extreme starbursts ($L \sim 10^{11} - 10^{12} \ L_\odot$) discovered by ISOCAM at lower redshifts may be related to the origin of lower mass spheroids and spheroidal components in late-type galaxies.

Many of the phenomena revealed by the long-wavelength observations were largely unexpected based on UV-optical-NIR observation: among others, the rate of cosmic evolution of IR galaxies, the CIRB energetics, the luminosities and SFRs of LIRGs and ULIRGs. The attempts so far to infer the IR properties based on UV-optical observations are producing modest results: there does not seem to be an alternative to long-wavelength observations if we aim at an exhaustive and reliable description of the history of SF in galaxies.

References

1. H. Aussel, C. Cesarsky, D. Elbaz, J.L. Starck: A&A **342**, 313 (1999)
2. S.C. Chapman et al.: MNRAS **319**, 318 (2000)
3. J. Cohen, et al.: ApJ **538**, 29 (2000)
4. H. Dole, G. Lagache, J.L. Puget, et al.: A&A submitted (2001)
5. D. Elbaz, et al.: A&A **351**, L37 (1999)
6. D. Elbaz, et al.: in preparation (2001)
7. R.S. Ellis: ARAA **35**, 389 (1997)
8. D.J. Fixsen, et al.: ApJ **508**, 123 (1998)
9. H. Flores, F. Hammer, T. Thuan, et al.: ApJ **517**, 148 (1999)
10. A. Franceschini, H. Aussel, C. Cesarsky, D. Elbaz, D. Fadda, J.-L. Puget: A&A, submitted (2001)
11. R. Gispert, G. Lagache, J.-L. Puget: A&A **360**, 1 (2000)
12. M.G. Hauser, R.G. Arendt, T. Kelsall, et al.: ApJ **508**, 25 (1998)
13. S.J. Lilly, et al.: ApJ **518**, 641 (1999)
14. P. Madau, L. Pozzetti: MNRAS **312**, L9 (2000)
15. S. Oliver, et al.: MNRAS **316**, 749 (2000)
16. B.M. Poggianti, H. Wu: ApJ **529**, 157 (2000)
17. B.M. Poggianti, A. Bressan, A. Franceschini: ApJ, in press (astroph/0011160)
18. J.-L. Puget, et al.: A&A **308**, L5 (1996)
19. D. Rigopoulou, A. Franceschini, H. Aussel, et al.: ApJ **537**, L85 (2000)
20. M. Rowan-Robinson, et al.:MNRAS **289**, 482 (1997)
21. D. Sanders, I.F. Mirabel: ARAA **34**, 749 (1996)
22. T. Stanev, A. Franceschini: ApJ **494**, L159 (1998)
23. F. Stecker, O. De Jager, M. Salamon: ApJ **390**, L49 (1992)
24. L. Vigroux, et al.: in "The Universe as seen by ISO", eds. P. Cox & M.F. Kessler, ESA-SP 427, (Noordwijk, ESA 1999), p. 805

Springer Proceedings in Physics